EXTRAS
150.00

Intelligent textiles and clothing

Related titles:

Smart fibres, fabrics and clothing
(ISBN-13: 978-1-85573-546-0; ISBN-10: 1-85573-546-6)
This important book provides a guide to the fundamentals and latest developments in smart technology for textiles and clothing. The contributors represent a distinguished international panel of experts and the book covers many aspects of cutting edge research and development. *Smart fibres, fabrics and clothing* starts with a review of the background to smart technology and goes on to cover a wide range of the material science and fibre science aspects of the technology. It will be essential reading for academics in textile and materials science departments, researchers, designers and engineers in the textiles and clothing product design field. Product managers and senior executives within textile and clothing manufacturing will also find the latest insights into technological developments in the field valuable and fascinating.

Wearable electronics and photonics
(ISBN-13: 978-1-85573-605-4; ISBN-10: 1-85573-605-5)
Building electronics into clothing is a major new concept which opens up a whole array of multi-functional, wearable electro-textiles for sensing/monitoring body functions, delivering communication facilities, data transfer, individual environment control, and many other applications. Fashion articles will carry keypads for mobile phones and connections for personal music systems; specialist clothing will be able to monitor the vital life signs of new-born babies, to record the performance of an athlete's muscles, or to call a rescue team to victims of accidents in adverse weather conditions. A team of distinguished international experts considers the technical materials and processes that will facilitate all these new applications.

Details of these books and a complete list of Woodhead titles can be obtained by:

- visiting our website at www.woodheadpublishing.com
- contacting Customer Services (e-mail: sales@woodhead-publishing.com; fax: +44 (0) 1223 893694; tel.: +44 (0) 1223 891358 ext.30; address: Woodhead Publishing Limited, Abington Hall, Abington, Cambridge CB1 6AH, England)

Intelligent textiles and clothing

Edited by
H. R. Mattila

The Textile Institute

CRC Press
Boca Raton Boston New York Washington, DC

WOODHEAD PUBLISHING LIMITED
Cambridge, England

Published by Woodhead Publishing Limited in association with The Textile Institute
Woodhead Publishing Limited, Abington Hall, Abington
Cambridge CB1 6AH, England
www.woodheadpublishing.com

Published in North America by CRC Press LLC, 6000 Broken Sound Parkway, NW,
Suite 300, Boca Raton FL 33487, USA

First published 2006, Woodhead Publishing Limited and CRC Press LLC
© 2006, Woodhead Publishing Limited
The authors have asserted their moral rights.

This book contains information obtained from authentic and highly regarded sources. Reprinted material is quoted with permission, and sources are indicated. Reasonable efforts have been made to publish reliable data and information, but the authors and the publishers cannot assume responsibility for the validity of all materials. Neither the authors nor the publishers, nor anyone else associated with this publication, shall be liable for any loss, damage or liability directly or indirectly caused or alleged to be caused by this book.

Neither this book nor any part may be reproduced or transmitted in any form or by any means, electronic or mechanical, including photocopying, microfilming and recording, or by any information storage or retrieval system, without permission in writing from Woodhead Publishing Limited.

The consent of Woodhead Publishing Limited does not extend to copying for general distribution, for promotion, for creating new works, or for resale. Specific permission must be obtained in writing from Woodhead Publishing Limited for such copying.

Trademark notice: product or corporate names may be trademarks or registered trademarks, and are used only for identification and explanation, without intent to infringe.

British Library Cataloguing in Publication Data
A catalogue record for this book is available from the British Library.

Library of Congress Cataloging in Publication Data
A catalog record for this book is available from the Library of Congress.

Woodhead Publishing ISBN-13: 978-1-84569-005-2 (book)
Woodhead Publishing ISBN-10: 1-84569-005-2 (book)
Woodhead Publishing ISBN-13: 978-1-84569-162-2 (e-book)
Woodhead Publishing ISBN-10: 1-84569-162-8 (e-book)
CRC Press ISBN-10: 0-8493-9099-0
CRC Press order number: WP9099

The publishers' policy is to use permanent paper from mills that operate a sustainable forestry policy, and which has been manufactured from pulp which is processed using acid-free and elementary chlorine-free practices. Furthermore, the publishers ensure that the text paper and cover board used have met acceptable environmental accreditation standards.

Project managed by Macfarlane Production Services, Dunstable, Bedfordshire
(macfarl@aol.com)
Typeset by Replika Press Pvt Ltd, India
Printed by T J International Limited, Padstow, Cornwall, England

Contents

	Contributor contact details	*xiii*
1	Intelligent textiles and clothing – a part of our intelligent ambience	1
	H Mattila, Tampere University of Technology, Finland	
1.1	Introduction	1
1.2	Intelligent systems	1
1.3	Applications	2
2	Methods and models for intelligent garment design	5
	M Uotila, H Mattila and O Hänninen, Tampere University of Technology, Finland	
2.1	Introduction	5
2.2	Background context	6
2.3	The underpinnings of interdisciplinarity	9
2.4	Scientific practices and research strategies for intelligent garments	12
2.5	Conclusions	15
2.6	References	16
Part I	Phase change materials	19
3	Introduction to phase change materials	21
	M Mäkinen, Tampere University of Technology, Finland	
3.1	Introduction	21
3.2	Heat balance and thermo-physiological comfort	22
3.3	Phase change technology	22
3.4	PCMs in textiles	23
3.5	Future prospects of PCM in textiles and clothing	30
3.6	References	32

4	Intelligent textiles with PCMs	34

W. BENDKOWSKA, Instytut Wlokiennictwa Textile Research Institute, Poland

4.1	Introduction	34
4.2	Basic information on phase change materials	34
4.3	Phase change properties of linear alkyl hydrocarbons	36
4.4	Textiles containing PCM	39
4.5	Measurement of thermoregulating properties of fabrics with microPCMs	55
4.6	Summary	60
4.7	Acknowledgements	60
4.8	References	60

5	The use of phase change materials in outdoor clothing	63

E A MCCULLOUGH and H SHIM, Kansas State University, USA

5.1	Introduction	63
5.2	Methodology	67
5.3	Results	72
5.4	Conclusions	80
5.5	Implications and recommendations	81
5.6	References	81

PART II	Shape memory materials	83

6	Introduction to shape memory materials	85

M HONKALA Tampere University of Technology, Finland

6.1	Overview	85
6.2	Shape memory alloys	86
6.3	Shape memory ceramics	94
6.4	Magnetic shape memory materials	94
6.5	Shape memory polymers and gels	95
6.6	Future prospects of shape memory materials	100
6.7	References	101

7	Temperature sensitive shape memory polymers for smart textile applications	104

J HU and S MONDAL, The Hong Kong Polytechnic University, Hong Kong

7.1	Introduction	104
7.2	A concept of smart materials	105
7.3	Shape memory polymer and smart materials	106

7.4	Some examples of shape memory polymer for textile applications	110
7.5	Potential use of shape memory polymer in smart textiles	115
7.6	General field of application	118
7.7	Challenges and opportunities	120
7.8	Acknowledgement	121
7.9	References	121
8	**Development of shape memory alloy fabrics for composite structures**	**124**
	F Boussu, GEMTEX, France and J-L Petitniot, ONERA, France	
8.1	Introduction	124
8.2	Definition and description of shape memory alloys	125
8.3	Interesting properties of shape memory alloys	126
8.4	Different kinds of alloys	132
8.5	Different kinds of applications of shape memory alloys	134
8.6	Conclusion	138
8.7	Future trends	140
8.8	Internet links	140
8.9	References	141
9	**Study of shape memory polymer films for breathable textiles**	**143**
	J Hu and S Mondal, The Hong Kong Polytechnic University, Hong Kong	
9.1	Introduction	143
9.2	Breathability and clothing comfort	144
9.3	Breathable fabrics	145
9.4	Water vapor permeability (WVP) through shape memory polyurethane	152
9.5	Future trends	162
9.6	Acknowledgement	163
9.7	References	163
10	**Engineering textile and clothing aesthetics using shape changing materials**	**165**
	G K Stylios, Heriot-Watt University, UK	
10.1	Introduction	165
10.2	Innovative design concepts in textiles and clothing	165
10.3	The principles of shape changing materials and their end-uses	166
10.4	Technical requirements for shape changing textiles and clothing	169

viii   Contents

10.5	Engineering textile and clothing aesthetics with shape memory materials	172
10.6	Aesthetic interactive applications of shape changing smart textiles	182
10.7	The concept of mood changing textiles for SMART ambience	184
10.8	Summary	186
10.9	Acknowledgement	187
10.10	References	187

Part III	Chromic and conductive materials	191
11	Introduction to chromic materials	193

P. TALVENMAA, Tampere University of Technology, Finland

11.1	Introduction	193
11.2	Photochromic materials	194
11.3	Thermochromic materials	196
11.4	Colour-changing inks	200
11.5	Electrochromic materials	201
11.6	Conclusion	203
11.7	References	204

12	Solar textiles: production and distribution of electricity coming from solar radiation. Applications	206

R R MATHER and J I B WILSON, Heriot-Watt University, UK

12.1	Introduction	206
12.2	Background	206
12.3	Solar cells	207
12.4	Textiles as substrates	209
12.5	Technological specifications	210
12.6	Challenges to be met	211
12.7	Suitable textile constructions	211
12.8	Conductive layers for PVs	213
12.9	Future trends	214
12.10	Sources of further information	215
12.11	References	216

13	Introduction to conductive materials	217

A HARLIN, Technical Research Centre of Finland, and M FERENETS, Tampere University of Technology, Finland

13.1	Electric conductivity	217
13.2	Metal conductors	220

13.3	Ionic conductors	222
13.4	Inherently conducting polymers	223
13.5	Application technologies for conducting fibre materials	231
13.6	Future trends in conductive materials	236
13.7	References	237

| 14 | Formation of electrical circuits in textile structures | 239 |

T K Ghosh, A Dhawan and J F Muth, North Carolina State University, USA

14.1	Introduction	239
14.2	Development of textile-based circuits	240
14.3	Fabrication processes	240
14.4	Materials used	246
14.5	Characterization	266
14.6	Applications	272
14.7	Potential for the future	276
14.8	Bibliography	277

| 15 | Stability enhancement of polypyrrole coated textiles | 283 |

M Y S Leung, J Tsang, X M Tao, C W M Yuen and Y Li, The Hong Kong Polytechnic University, Hong Kong

15.1	Introduction	283
15.2	Conductivity changes of polypyrrole films on textiles	286
15.3	Stabilisation of the Ppy	290
15.4	Experimental results of stability enhancement	292
15.5	Conclusion	303
15.6	Acknowledgement	304
15.7	References	304

| 16 | Electrical, morphological and electromechanical properties of conductive polymer fibres (yarns) | 308 |

B Kim and V Koncar, ENSAIT-GEMTEX Laboratory, France and C Dufour, Institute IEMN, France

16.1	Introduction	308
16.2	Preparation of conductive fibres – overview	309
16.3	Experimental	311
16.4	Results and discussion	312
16.5	Applications: prototype	320
16.6	Conclusion	320
16.7	Acknowledgements	321
16.8	References	322

Contents

17	Multipurpose textile-based sensors	324

C Cochrane, B Kim and V Koncar, ENSAIT-GEMTEX Laboratory, France and C Dufour, Institute IEMN, France

17.1	Introduction	324
17.2	Conductive polymer textile sensors	326
17.3	Conductive polymer composites (CPCs) textile sensors	331
17.4	Perspective	339
17.5	References	339
18	Textile micro system technology	342

U Möhring, A Neudeck and W Scheibner, TITV Greiz, Textile Research Institut Thuringia-Vogtland, Germany

18.1	Textile micro system technology	342
18.2	Textiles are inherent microstructures	343
18.3	Goal of the application of compliant textile structures	346
18.4	First attempt: textile electronic circuit technology based on copper wires in a lattice structure with interconnections and interruptions	347
18.5	Galvanic modification of yarns	348
18.6	Light effects based on textiles with electrically conductive microstructures	350
18.7	Textile-based compliant mechanisms in microengineering and biomechatronics	351
18.8	References & Sources of further information	354
Part IV	Applications	357
19	WareCare – Usability of intelligent materials in workwear	359

H Mattila, P Talvenmaa and M Mäkinen, Tampere University of Technology, Finland

19.1	Introduction	359
19.2	Objectives	359
19.3	Methodology	360
19.4	Textile materials	361
19.5	Electronics	362
19.6	Usability testing	364
19.7	Conclusions	367
19.8	Bibliography	368

20	Intelligent textiles for medical and monitoring applications	369

J-Solaz, J-M Belda-lois, A-C Garcia, R Barberà, T-V Dorá J-A Gómez, C Soler and J M Prat, A Instituto de Biomecanica de Valencia, Spain

20.1	Introduction	369
20.2	Importance of intelligent textiles for healthcare	370
20.3	Potential applications of intelligent textiles	373
20.4	From medical needs to technological solutions	380
20.5	Summary and future trends	393
20.6	Acknowledgements	394
20.7	References	394

21	Context aware textiles for wearable health assistants	399

T Kirstein, G Tröster, I Locher and C Küng, Wearable Computing Lab, ETH Zürich, Switzerland

21.1	Introduction	399
21.2	Vision of wearable health assistant	399
21.3	Approach	401
21.4	Electronic textile technology	402
21.5	Context recognition technology	414
21.6	Wearable components	414
21.7	Applications	415
21.8	Outlook	418
21.9	Acknowledgement	418
21.10	References	418

22	Intelligent garments in prehospital emergency care	421

N Lintu, M Mattila and O Hänninen, University of Kuopio, Finland

22.1	Introduction	421
22.2	Different cases and situations	422
22.3	Circumstances	422
22.4	Vital functions	422
22.5	Monitoring of vital functions	423
22.6	Selection of monitoring methods	425
22.7	Interpretation of monitored parameters	425
22.8	Telemedicine	425
22.9	Negative effects of transportation on vital parameters	426
22.10	Patient chart	427
22.11	Data security	427
22.12	Day surgery	427
22.13	Protective covering	428

22.14	An integrated monitoring of vital functions	429
22.15	Mobile isolation	429
22.16	Optimal smart solution for prehospital emergency care	430
22.17	Conclusions	431
22.18	References	432
23	**Intelligent textiles for children**	**434**
	C Hertleer and L Van Langenhove, Ghent University, Belgium and R Puers, Katholieke Universiteit Leuven, Belgium	
23.1	Introduction	434
23.2	State of the art	435
23.3	The intellitex suit	436
23.4	Future trends	447
23.5	Acknowledgements	448
23.6	References	448
24	**Wearable biofeedback systems**	**450**
	B J Munro, University of Wollongong and Commonwealth Scientific and Industrial Research Organisation (CSIRO) Textile and Fibre Technology, Australia and J R Steele, T E Campbell and G G Wallace, University of Wollongong, Australia	
24.1	Introduction	450
24.2	Is there a need for biofeedback technology?	450
24.3	Are there problems with current biofeedback devices?	451
24.4	Can we provide biofeedback for joint motion?	452
24.5	The development of a functioning wearable textile sensor	453
24.6	Functional electronics	460
24.7	Interconnections	460
24.8	The Intelligent Knee Sleeve: a wearable biofeedback device in action	462
24.9	Why is the Intelligent Knee Sleeve needed?	463
24.10	Other applications of wearable biofeedback technology	467
24.11	Future directions	467
24.12	References	469
25	**Applications for woven electrical fabrics**	**471**
	S Swallow and A P Thompson, Intelligent Textiles Limited, UK	
25.1	Smart fabric technologies	471
25.2	Active and passive smart fabrics	472
25.3	Electrical smart fabrics	475
25.4	Products and applications	483
25.5	References	487
Index		**489**

Contributor contact details

(* = main contact)

Editor and Chapter 1

Professor Heikki Mattila
Tampere University of Technology
SmartWearLab
Siñitaival 6
33720 Tampere
Finland

E-mail: heikki.r.mattila@tut.fi

Chapter 2

Professor Minna Uotila*, Professor Heikki Mattila and Dr Osmo Hänninen
University of Lapland
PO Box 122 (Siljotie 2)
FIN-96101 Rovaniemi
Finland

Tel: +358 40 556 2893
E-mail: minna.uotila@ulapland.fi

Chapter 3

Mailis Mäkinen
SmartWearLab
Tampere University of Technology
Sinitaival 6
FI-33720 Tampere
Finland

Tel: +358 3 3115 2494
Fax: +358 3 3115 4515
E-mail: mailis.makinen@tut.fi

Chapter 4

Dr Wiesława Bendkowska
Instytut Wlokienictwa
Textile Research Institute
Brzezinska S/15
92–103 Ledz
Poland

E-mail: bendkowska@mail.iw.lodz.pl

Chapter 5

Professor Elizabeth McCullough*
and Dr H. Shim
Kansas State University
Institute for Environmental
Research
64 Seaton Hall
Manhattan, KS 66506
USA

Tel/fax: +1 785-532-2284
E-mail: lizm@ksu.cdu

Chapter 6

Markku Honkala
Tampere University of Technology
Smartwear Lab Siñitaival6
33720 Tampere
Finland
E-mail: markku.honkala@tut.fi

Chapters 7 and 9

Dr Jinlian Hu
Institute of Textiles and Clothing
The Hong Kong Polytechnic
University
Hung Hom
Kowloon
Hong Kong

Tel: 852 27666437
Fax: 852 27731432
E-mail: tchujl@polyu.edu.hk

Chapter 8

F. Boussu* and Dr J-L Petitniot
ENSAIT, GEMTEX Laboratory
9 rue de l'Ermitage
BP 30329
59056 ROUBAIX
Cedex 01
France

Tel: +33 3 20 25 64 76
E-mail: francois.boussu@ensait.fr

Chapter 10

Professor G.K. Stylios
Research Institute for Flexible
Materials
School of Textiles and Design
Heriot-Watt University
Scottish Borders Campus
Galashiels TD1 3HF
UK

E-mail: G.Stylios@hw.ac.uk

Chapter 11

P. Talvenmaa
Tampere University of Technology
SmartWearLab
Sinitaival 6
33720 Tampere
Finland

E-mail: paivi.talvenmaa@tut.fi

Contributor contact details   xv

Chapter 12

Dr Robert Mather* and Professor
John Wilson
School of Engineering and Physical
Sciences
Heriot-Watt University
Riccarton
Edinburgh EH14 4AS
UK

E-mail: r.r.mather@hw.ac.uk

Chapter 13

Professor A. Harlin* and Dr
M. Ferenets
Institute of Fibre Materials Science
Tampere University of Technology
P.O. Box 589
Tampere
33101
Finland

Tel: +358 3 3115 3742
Fax: +358 3 3115 2955
E-mail: ali.harlin@vtt.fi;
marju.ferenets@tut.fi

Chapter 14

Professor Tushar Ghosh, Dr A.
Dhawan* and Dr J.F. Muth
College of Textiles
North Carolina State University
Raleigh, NC 27695-8301
USA

Tel: +1 (919) 515-6568
Fax: +1 (919) 515 - 3733
E-mail: Tushar_Ghosh@ncsu.edu
dhawananuj1976@yahoo.com

Chapter 15

Dr M-Y. S. Leung*, Joanna Tsang,
Professor X-M Tao, Dr C-W. M
Yuen and Yang Li
Institute of Textiles and Clothing
The Hong Kong Polytechnic
University
Hung Hom
Kowloon
Hong Kong

Tel: 852 27666437
Fax: 852 27731432
E-mail: tclens@inet.polyu.edu.hk

Chapter 16

Dr Bohwon Kim*
Laboratory GEMTEX
ENSAIT (Ecole Nationale
Supérieure des Arts et Industries
Textiles)
9 rue de l'Emitage
59056 Roubaix, cedex 1
France

E-mail: bwkim75@yahoo.fr
Tel: +33-(0)3-2025-7587
Fax: +33 (0)3-2027-2597

Professor Vladan Koncar
Laboratory GEMTEX
ENSAIT (Ecole Nationale
Supérieure des Arts et Industries
Textiles)
9 rue de l'Emitage
59056 Roubaix, cedex 1
France

E-mail: vladan.koncar@ensait.fr
Tel: +33 (0)3-2025-8959
Fax: +33 (0)3-2027-2597

Professor Claude Dufour
IEMN/DHS
Avenue Poincaré BP19
59652 Villeneuve d'Ascq Cedex
France

E-mail: claude.dufour@univ-lille1.fr
Tel: +33 (0)3-2019-7908
Fax: +33 (0)3-2019-7878

Chapter 17

Mr Cédric Cochrane*
Laboratory GEMTEX
ENSAIT (Ecole Nationale
Supérieure des Arts et Industries
Textiles)
9 rue de l'Emitage
59056 Roubaix, cedex 1
France

E-mail: cedric.cochrane@ensait.fr
Tel: +33 (0)3-2025-8974
Fax: +33 (0)3-2027-2597

Dr Bohwon Kim
Laboratory GEMTEX
ENSAIT (Ecole Nationale
Supérieure des Arts et Industries
Textiles)
9 rue de l'Emitage
59056 Roubaix, cedex 1
France

E-mail: bwkim75@yahoo.fr
Tel: +33 (0)3-2025-8974
Fax: +33 (0)3-2027-2597

Professor Vladan Koncar
Laboratory GEMTEX
ENSAIT (Ecole Nationale
Supérieure des Arts et Industries
Textiles)
9 rue de l'Emitage
59056 Roubaix, cedex 1
France

E-mail: vladan.koncar@ensait.fr
Tel: +33 (0)3-2025-8959
Fax: +33 (0)3-2027-2597

Professor Claude DUFOUR
IEMN/DHS
Avenue Poincaré BP19
59652 Villeneuve d'Ascq Cedex
France

E-mail: claude.dufour@univ-lille1.fr
Tel: +33 (0)3-2019-7908
Fax: +33 (0)3-2019-7878

Chapter 18

Dr. rer. nat. habil. Andreas
G. Neudeck
TITV Greiz
Textile Research Institute
Thuringia-Vogtland e.V.
Zeulenrodaer Str. 42
D-07973 Greiz
Germany

Tel: (03661) 611 204
Fax: (03661) 611 222
E-mail: a.neudeck@titv-greiz.de

Chapter 19

Professor H Mattila,*
P. Talvenmaa and M. Mäkinen
Tampere University of Technology
SmartWearLab
Sinitaival 6
33720 Tampere
Finland

E-mail: heikki.r.mattila@tut.fi

Chapter 20

Dr Jose S. Solaz*, Mr Juan-Manuel Belda-Lois, Dr/Ana-Cruz Garcia, Mr Ricard Barberà, Dr Juan-Vicente Durá, Mr Juan-Alfonso Gomez, Dr Carlos Soler and Dr Jaime Prat
Instituto de Biomecánica de Valencia (IBV)
Universidad Politécnica de Valencia – Edificio 9C
Camino de Vera s/n
E-46022 – Valencia
Spain

Tel: +34 96 387 91 60
Fax: +34 96 387 91 69
E-mail: jose.solaz@ibv.upv.es

Chapter 21

Dr Tünde Kirstein,* Professor Gerhard Tröster, Ivo Locher Christof Küng
Wearable Computing Lab
ETH Zürich
Gloriastrasse 35
CH-8092 Zürich
Switzerland

E-mail: Troester@ife.ee.ethz.ch
Tel: +41 44 632 5280
Fax: +41 44 6321210

Chapter 22

Niina Lintu,* Dr M. Mattila and Dr O. Hänninen
Department of Physiology
University of Kuopio
P.O. Box 1627
70211 Kuopio,
Finland

E-mail: Niina.Lintu@uku.fi

Chapter 23

Dr Carla Hertleer,* Professor L. Van Langenhove and Professor R. Puers
Ghent University
Technologiepark 907
9052 Zwijnaarde
Belgium

E-mail: Carla.Hertleer@UGent.be

Chapter 24

Dr Bridget J. Munro*
Biomechanics Research Laboratory
University of Wollongong
Wollongong
New South Wales
Australia, 2522

E-mail: bmunro@uow.edu.au

Dr Toni E. Campbell
ARC Centre of Excellence for Electromaterials Science
Intelligent Polymer Research Institute
University of Wollongong
Wollongong
New South Wales
Australia, 2522

E-mail: tonicamp@uow.edu.au

Professor Julie R. Steele
Biomechanics Research Laboratory
University of Wollongong
Wollongong
New South Wales
Australia, 2522

E-mail: jsteele@uow.edu.au

Professor Gordon G. Wallace
ARC Centre of Excellence for
Electromaterials Science
Intelligent Polymer Research
Institute
University of Wollongong
Wollongong
New South Wales
Australia, 2522

E-mail:
gordon_wallace@uow.edu.au;

Chapter 25

Dr Stan S. Swallow* and
Dr A. P. Thompson
Intelligent Textiles Limited
ITL Studio, Brunel Science Park
Runnymede Campus,
Coopers Hill Lane
Egham
Surrey, TW20 0JZ
UK

Tel: ı44 (0)1784 433 262
E-mail:
stan@intelligenttextiles.com

1
Intelligent textiles and clothing – a part of our intelligent ambience

H M A T T I L A, Tampere University of Technology, Finland

1.1 Introduction

Although intelligent textiles and smart clothing have only recently been added to the textile vocabulary, we must admit that the industry has already for several years focused on enhancing the functional properties of textiles. New chemical fibres have been invented. By attaching membranes on textile substrates, fabrics were made breathable and yet waterproof. Three-dimensional weaving technology paved the way for new exciting technical textile developments. These are some examples of a textile-based approach for improving the properties and functionality. Wearable technology, the electronics-based approach, started to add totally new features to clothing by attaching various kinds of electronic devices to garments. The results, however, were often bulky, not very user friendly and often very impractical. The garment was truly wired with cables criss-crossing all over, batteries in pockets and hard electronic devices sticking out from the surface. The piece of clothing had become a platform for supporting electronics and was hardly wearable in a clothing comfort sense. The current objective in intelligent textile development is to embed electronics directly into textile substrates. A piece of clothing remains visibly unchanged and at the end of the day the consumer can still wash it in the washing machine without first removing all the electronics. This of course is very challenging.

1.2 Intelligent systems

Intelligent systems are normally understood to consist of three parts: a sensor, a processor and an actuator. For example, body temperature monitored by the sensor is transferred to the processor, which on the basis of the received information computes a solution and sends a command to the actuator for temperature regulation. To achieve such interactive reactions three separate parts may actually be needed. The sensor may be embroidered on the surface of the T-shirt by using conductive yarns. Signals are transmitted wirelessly between the processor, sensor and the actuators, which could be microscopic flaps that open in order to increase ventilation and temperature transfer. Or the system may work on the basis of physics like phase change materials.

Phase change materials (PCM), shape memory materials (SMM), chromic materials (colour change), conductive materials are examples of intelligent textiles that are already commercially available. This is also reflected in the contents of this book. Part I deals with phase change materials. Part II introduces shape memory materials. Chromic and conductive materials are presented in the next part. The final part deals with applications.

There are numerous research projects on the way around sensors and actuators as can be seen from EU's research records Cordis.[1] Conductive fibres and yarns are equally important. Power supply, perhaps the toughest challenge for intelligent textiles, should also be an integral part of textiles. Flexible solar cells, micro fuel cells and the possibility of transforming body motion into electric power are interesting topics. Infineon Technologies AG has developed a textile embedded power supply based on the temperature difference between the outer and inner surfaces of a garment. Photonics, including textile-based display units, are being developed by many research institutes and companies. Interactive Photonic Textiles, an invention published by Philips in September 2005, contain flexible arrays of inorganic light-emitting diodes, which have been seamlessly integrated into textile structures. The invention turns fabric into intelligent displays to be used for ambient lighting, communication and personal health-care. The textile surface can also be made interactive and Philips has managed to embed orientation and pressure sensors as well as communications devices (Bluetooth, GSM) into the fabric. The jacket display making a man invisible developed at the University of Tokyo is one of most exciting latest inventions.

1.3 Applications

'Where are the commercial applications?' is a frequently asked question. Despite nearly ten years of research and development we have seen only a few smart textile and apparel products on the market. The computerized jogging shoe No. 1 by Adidas is one of them. Interactive Photonic Textiles by Philips may bring a few more around. But countless hours of research and development work is presently allocated to this area by universities, research institutes and companies in different parts of the world. Scientific conferences and commercial events are organized around this theme. One of them was Ambience 05, a scientific conference organized at Tampere, Finland in September 2005. More than 200 participants from 24 different countries

---

1. Community research & development information service (www.cordis.lu)

participated and 42 papers focusing on intelligent textiles, smart garments, intelligent ambience and well-being were presented. In the interactive concluding session regarding future trends in smart textile research the participants were able to express their opinion on key questions through a remote-control on-line voting system. The results of the survey are presented in Table 1.1. It was felt by 79% of the participants that commercially successful

Table 1.1 Future trends in smart textile research according to the participants at scientific conference Ambience 05

Commercially successful smart textile/garment applications will be available in	0–1 years	14.8%
	1–5 years	50.8%
	5–10 years	28.9%
	> than 10 years	3.9%
	Never	1.6%
In which sector do you expect commercially viable smart innovations first to become reality?	Sports and extreme	40.2%
	Occupational clothing	24.8%
	Transportation	6.0%
	Technical textiles	21.4%
	None of these	7.7%
It is possible to miniaturize electronic devices enough to insert them into fibres in	0–1 years	1.5%
	1–5 years	26.7%
	5–10 years	45.0%
	> than 10 years	23.7%
	Never	3.1%
Energy sources can be fully integrated into textile structures in the near future	Agree totally	22.5%
	Agree slightly	43.4%
	Disagree slightly	24.0%
	Disagree totally	10.1%
Phase change materials with real warming/cooling impact will be available in	0–1 years	13.6%
	1–5 years	50.8%
	5–10 years	19.7%
	> than 10 years	8.3%
	Never	7.6%
Shape memory and colour change textiles will generate breakthrough smart garment applications in the near future	Agree totally	18.5%
	Agree slightly	43.1%
	Disagree slightly	28.5%
	Disagree totally	10.0%
Textile embedded sensors and tele-monitoring of patients will be applied in hospitals despite high costs	Agree totally	40.3%
	Agree slightly	38.7%
	Disagree slightly	19.4%
	Disagree totally	1.6%
Breakthrough nano-technology applications in textiles, beside finishing, will be available in	0–1 years	6.4%
	1–5 years	37.6%
	5–10 years	36.7%
	> than 10 years	18.3%
	Never	0.9%

Source: Ambience 05 on-line poll at the interactive concluding session with more than 200 scientific participants.

smart textile and garment applications will be available in the market between five and ten years, most likely in sports and extreme wear, in occupational and professional clothing and in technical textiles. Nano-technology applications and adequate miniaturization of electronic devices for inserting them into fibres were still expected to take a considerable amount of time, while the majority felt that energy sources can be fully integrated into textile structures in the near future. More efficient phase change materials were expected to be available within the next five years, but the majority did not quite believe in breakthrough results with shape memory or colour change materials. Most of the participants expected textile embedded sensors and tele-monitoring of patients to become reality in hospitals despite the high costs.

Intelligent textile and garment research is very cross-scientific. Beside textile knowhow many other skills, such as electronics, telecommunications, biotechnology, medicine, etc., must be brought into the projects. One research institute cannot carry out such projects alone. Networking as well as considerable amounts of financing are required. There are high hopes in the scientific community toward the EU's seventh framework programme for financing and for further networking within the sector. The complexity and broadness of knowledge required for intelligent textile research is also highlighted by this book.

2
Methods and models for intelligent garment design

M UOTILA, H MATTILA and O HÄNNINEN,
Tampere University of Technology, Finland

2.1 Introduction

In recent years, interdisciplinary studies have been the mainstream in research discourses and practices. At the same time, the number of projects with shared expertise has increased enormously. As Klein (1990, 13) states in her book on interdisciplinarity, 'As a result the discourse on interdisciplinarity is widely diffused' and 'the majority of people engaged in interdisciplinary work lack a common identity'. Interdisciplinarity is thus an ambiguous term, applying 'to both the idea of grand unity and a more limited integration of existing disciplinary concepts and theories' (ibid., 27).

Especially in research areas where the research object or the phenomenon explored could be characterised as a complex and hybrid field, the means used in interdisciplinary and multimethodological approaches have been seen as reasonable and useful. According to Klein (1990, 11), educators, researchers, and practitioners have turned to interdisciplinary work, for example, in order to answer complex questions, to address broad issues, to explore disciplinary and professional relations, to solve problems that are beyond the scope of any one discipline, and to achieve unity of knowledge.

When discussing hybrid products, we normally refer to the object in terms of both material and immaterial properties. We speak about intelligent products such as smart houses, vacuum-cleaners, cars, and clothing. Such products could be studied in relation to different contexts, e.g., work, sport and leisure, entertainment, well-being and health, and with regard to fashion design practice (Ullsperger, 2002), to name just some approaches.

Human beings are always in a dynamic state, which can be described as non-linearity, broken symmetry, dissipation of free energy, complexity, orderly disorder and dynamic stability. Even identical twins are phenotypically different. (Yates, 1993) The regulatory functions of homeodynamic responses are pulsatile. From birth to death we are thus in a state of oscillating non-equilibrium. Our reactions are stimulus dependent on and modulated by the central state defined as the total reactive condition, and this state fluctuates.

(Vincent, 1993). Stimuli are collected from the outside world and from within the body. They affect information processing in the brain as well as our behaviour. In cold climates, foresight is evidenced by, among other things, clothing, the construction of shelters, and the discovery and use of fire for heating (Denton, 1993). We can use technology to increase the sensitivity of our sensory systems. Technology can be integrated into garments. Using computing systems, for example, it is possible to develop warning systems that help workers avoid danger. Our sensory and brain mechanisms, as well as motor and vegetative functions, show decline with age. This can be partially compensated for by intelligent garments and integrated computing systems, which can provide warnings or summon expert help if accidents or diseases so require. These computing systems can be in homes, working places, or in fact any location if the information is transmitted in digital form, trends are calculated and smart warning limits have been established. These systems and products are objects of research that clearly go beyond the scope of any single discipline.

2.2 Background context

The aim of the article is to describe the underpinnings of the interdisciplinarity elaborated in the research and design project Methods and Models for Intelligent Garment Design (MeMoGa), funded by the Academy of Finland's Proactive computing program and conducted jointly by University of Lapland, Tampere University of Technology, and University of Kuopio in Finland during the years 2003–2005. The purpose of the project was to analyse the conceptual framework offered by the theoretical bases of the research on clothing and dress and ascertain their applicability to the study of ubiquitous computing and the services, activities and social situations to be found in intelligent environments.

2.2.1 Intelligent garments in the light of clothing theories

An examination of the research on clothing and fashion reveals that in many respects the concept of an intelligent garment has yet to be analysed. The phenomenon whereby the traditional characteristics of a garment are augmented with sophisticated functional features is referred to using terms such as 'wearable computer' (see Suomela *et al.*, 2001) or 'interactive materials' (Nousiainen *et al.*, 2001). The discussion in the field in recent years has revolved around the technological research and knowhow involved, and few if any references can be found in the literature to the conceptual points of departure used in the research on clothing and dress and in fashion design. For example, there has been no research done in the area of clothing theory to clarify how an intelligent garment acquires meanings in social interaction and communication, although new forms and means of communication are more often than not the focus of interest when intelligent garments are mentioned.

Clothing, jewellery and watches have long been part of the everyday life of people in the West. Accordingly, it can be assumed that the theoretical bases and methods developed to study such objects can be profitably applied in solutions for intelligent products and in anticipating the usability and acceptability of such products in everyday life. Research in clothing theory helps us to understand conceptually the social processes which render the wearing of various artefacts natural and familiar. In this way, the approaches and analogies found in research on clothing and dress will prove significant when we proceed from user-centred information technology to proactive applications, i.e., from interaction between people and technology to interaction between people and the environment. Interestingly, an examination of the theoretical approaches of the research on clothing and dress reveals precisely such a transition: whereas in 1930 J.C. Flugel defined the *needs* for clothing and dress in terms of people's desire for protection, modesty and self-decoration (Flugel, 1930), in the 1980s Alison Lurie described clothing as a system akin to *language* (Lurie, 1992). In 1990, Susan B. Kaiser submitted that the meanings of clothing and dressing should be interpreted in *context* (Kaiser, 1990). The underpinnings of the contextual approach lie in symbolic interactionism, which concerns itself with the *communication* between a person's self and the community. Lamb and Kallal (1992) have proposed a Consumer Needs Model that assesses user needs by incorporating the functional, expressive, and aesthetic dimensions of clothing.

One particular focus of the MeMoGa project was to determine how the approaches identified might be used in elaborating models and strategies for research on the usability and acceptability of wearable intelligence and in developing design methods. An additional aim was to develop frameworks that might help initiate R & D projects on new-generation garments and environments for ubiquitous computing in the next few years. The research and design group also created a garment concept in the field of wearable intelligence that can be easily adapted for different users and is acceptable and accessible to as many users as possible.

The researchers in the MeMoGa project focused on the garment needs of workers in heavy industry and supported the concept design of an intelligent garment for them by drawing on Lamb and Kallal's Consumer Needs Model. In Lamb and Kallal's, view functionality (F) encompasses the fit of a garment and the mobility, comfort, protection, and ease of donning and doffing it offers. A garment's aesthetic properties (A) embrace the design principles and artistic elements involved, such as line, pattern, colour and texture. Expressiveness (E) in turn is associated with the communicative and symbolic characteristics of the garment or outfit.

2.2.2 Research and design procedure

The MeMoGa project started with empirical analyses to inform intelligent garment design in which the researchers identified the needs of the selected focus groups, i.e., workers in heavy industry. The data on the target group were collected using semi-structured interviews and a tentative analysis was carried out. The results of the interviews were operationalised into *design criteria* that then guided the concept design process. Here, the term 'concept' refers to the vision stage of the product design process. The purpose of the concept design process was not to produce a detailed description of the product but to facilitate creation of the innovations and visions for forthcoming product design processes (Keinonen and Jääskö, 2004, 21).

In order to test the usability and accessibility of the garment concept, which involved protective clothing for workers in heavy industry, the research team created a multimedia presentation showing the concept in a setting in which the clothing would actually be worn. According to Wright and McCarthy (2005, 19), the use of scenarios and other narrative techniques are different ways 'in which designers can engage with user experience.' The creation of a virtual prototype and 3D modelling has been estimated to bring efficiency to the area of wearable intelligence. Animations in particular have been seen as meriting further research (Uotila *et al.*, 2002). Generally speaking, the research team were much more interested in developing methods that could assist in the early stage of design and could support open communication between users and designers through the concept design and interactive virtual prototypes than they were in producing prototypes or final products (Bannon, 2005, 37). The present findings also indicate that virtual prototypes make it possible to provide initial information about concepts, products and the use of products in different socio-cultural contexts and communities.

The co-operation between the professionals from different fields of research and between the professionals and the end users of the products took place using the virtual working environment Optima. With the partners in the project located at opposite ends of and all over the country, this environment created a good foundation for not only the research and design work but also project management. It provided an appropriate environment for communication by the multidisciplinary research and design team, 'bringing together people with different skills and expertise to discuss together user data, mock-ups, video prototypes, etc.' (Bannon, 2005, 37). Although the environment facilitated real-time communication between the partners, it did not replace face-to-face meetings.

There is no need here to go through the project in depth and present the detailed empirical findings of the research. A number of articles are forthcoming on the topic by the researchers and PhD candidates working in the project

(Mäyrä *et al.*, 2005; Matala *et al.*, 2005; Pursiainen *et al.*, 2005). Instead, we attempt to give a framework for discussion of the concept of interdisciplinarity as regards the research areas mentioned above. The focus of the MeMoGa project is depicted in Fig. 2.1, providing the context for the discussion on interdisciplinarity to follow.

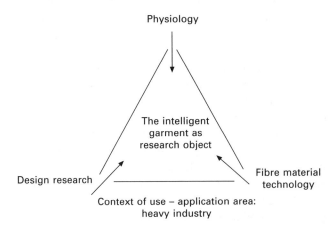

2.1 The intelligent garment as an object of interdisciplinary research from the perspective of three research disciplines: design research, fibre material technology, and physiology.

2.3 The underpinnings of interdisciplinarity

The research on ubiquitous computing and ambient intelligence is very labour intensive and cannot be carried out by just one or two researchers. To gain an orientation to interdisciplinarity it is important to review how researchers view intelligent garments as research objects, and how different disciplines define intelligent products and services integrated into products on the basis of different research paradigms. To this end, this section formulates the concept of the intelligent garment as an object of interdisciplinary research from the perspective of three individual research disciplines: design research, fibre material technology, and physiology. Where scientific models are concerned, the section discusses the scientific practices and models in the various disciplines and the benefits that can be derived from them for smart products, in particular intelligent garments.

2.3.1 Mature sciences vs. pre-sciences

In his publication *The Structure of Scientific Revolutions*, Thomas S. Kuhn (1970) distinguishes mature, normal science from immature, pre-sciences. In any mature science, every scientist will accept the same paradigm, most of the time (Laudan, 1977, 73). In a similar vein, Chalmers (1999, 108) has pointed out, 'A mature science is governed by a single paradigm'. In contrast, in immature sciences there is no consensus on theories, methods, and research objects as there is in more established research areas. Bringing this into the context of the present study, it could be said that where design research is still going through its immature research stage, physiology and fibre material technology are normal sciences in the Kuhnian sense in that they have already reached maturity and are in the process of redefining their concepts and methods.

Then again, it could be said there are no mature sciences. In chemistry and physics one can work with molecules such as DNA and their constituents and calculate their behaviour *in vitro* under experimental conditions. Unfortunately, all life phenomena are non-linear even at the cellular level, and one cannot, for example, extrapolate from DNA what behavioural responses will occur in humans in real life, although most of our genes are similar to those in mice and even closer to those of the primates. Genes are, in a way, prisoners of the successful physiology that carries them, but also of the successful ecological niche in which they find themselves. Integrative physiology is an old and a new science at the same time (Noble and Boyd, 1993).

Another difference is that where the natural sciences are closely connected to empirical research practices and empiricism, design research draws its orientations from design-driven practices, humanism and, quite often, hermeneutics. The natural sciences are usually understood to include physics, chemistry, biology, and their border areas (Hempel, 1966, 1), with the humanities including disciplines such as philosophy, history, and the arts. Although the methods of the natural sciences and the rationale of scientific inquiry have quite often been considered to be more scientific, the questions of knowledge construction, truth and reliability are also essential for human studies. One may ask what the meaning of the hermeneutic dimension is for all knowledge construction, and for modern natural sciences (Gadamer, 2004, 130).

This discussion is significant from the point of view of design research. The positivist doctrine and technical-rational underpinnings of design research have been challenged by reflective practices of design – 'design as discipline, but not design as a science' (Cross, 2001, 54), and with 'its own intellectual culture, acceptable and defensible in the world on its own terms' (ibid., 55). Nowadays design is a broad field of making and planning disciplines (Friedman, 2000, 6). In this vein, Diaz-Kommonen (2002, 27) has asserted, 'There is no single, solid, discursive foundation underlying design, but rather the landscape is one of fluctuating positions, representing discursive formations, in the process of negotiation'. But could there not be some unity of design discipline that sifts multiplicity from the atomistic? (Klein, 1990, 22).

As in any research area, in design research it is possible to distinguish the questions typical of basic research, applied research and development work. Basic research in design includes the underlying theoretical assumptions, i.e., the ontological, epistemological, and methodological underpinnings of the studies. The ontological assumptions about human reality and human values in the foundations of design research also belong on this level. Case studies, which may be understood as design particulars, represent the level of applied design research and development work.

2.3.2 The intelligent garment as an object of interdisciplinary research and design in terms of the Popperian worldview

On the level of basic research it is reasonable to ask, 'What is the object of research?' In his theory of three worlds, Karl Popper distinguishes natural objects as part of world 1, subjective awareness as an aspect of world 2, and cultural products, events and social situations as manifestations of world 3. Crucial to Popper's theory are what he refers to as the emergent features of organisms, that is, the features that produce innovation but which cannot be predicted on the basis of lower-level features or laws. 'Life, or living matter, somehow emerged from nonliving matter; and it does not seem completely impossible that we shall one day know how this happened' (Popper 1987, 150).

Popper's theory of three worlds can also be applied to explain the relationships between designers, artefact (interface) users and the contextual environment of their interactions (Popovic, 2002). The contextual environment of an artefact (interfaces) includes *the artefact's physical environment* (world 1), *social environment* (world 2) and *knowledge environment* (world 3). According to Popovic, *the knowledge environment,* which is analogous to Popper's world 3, consists of a user's and designer's knowledge. In that world, the designer attempts to present her or his knowledge of world 2 (*the social environment*) and world 1 (*the artefact's physical environment*) (ibid. 368).

In general, research in fibre material technology, done by technologists, defines intelligent textiles as interactive material, i.e., material that performs an expected function in response to changes in the environment. Intelligent systems are normally composed of a sensor, a processing unit and an actuator. In intelligent textiles and garments the three may be combined, with the function activated by the physical environment. Phase change materials are one example. Capsules containing paraffin wax are inserted into fibres and when the wax changes its phase from solid to liquid or liquid to solid, heat is either released or absorbed, with the user feeling a warming or cooling effect. In the area of clothing physiology, intelligent garments can help in detecting vital physiological signals in a socially acceptable way, signals which previously could only be recorded with great technical difficulty.

The point of departure in the MeMoGa project was to use Popper's theory of three worlds as the analytical tool for defining the intelligent garment as an object of research. The idea was to view wearable intelligence or intelligent garments more as a social situation and cultural events than as individual objects. Ultimately, the contexts of the user and the identity of the user shape the meanings, uses and impacts of technology.

The research thus adopted a broad conceptual framework by keeping together user-centred and technology-centred approaches. It sought to benefit from the strengths of both approaches while overcoming their limitations by identifying how each provides a crucial part of the context for the other and by recognising the common processes of social change that affect both users and what is known as the smart environment. Accordingly, the notion of intelligent environment and context of use was stressed throughout the R & D project, precisely to avoid the problems of technological determinism that often confront research on technology and new technological innovations.

2.4 Scientific practices and research strategies for intelligent garments

The MeMoGa project provides a framework for discussing not only interdisciplinarity but also collaboration with professionals from different fields of study. Collaboration may be fruitful, but it is always challenging, for each discipline has its own specialists and specialised research procedures and practices.

2.4.1 Crossing scholarly borders

One starting point in defining research through design is to draw a distinction between the semantic, syntactic, and pragmatic dimensions of research. The semantic dimension focuses on the context of end-user, the syntactic dimension on the design process, and the pragmatic dimension on the use of products. The specific methods used in design research for analysing the context of the end user are surveying and interviewing individuals who represent the different target groups. Sometimes studies of the design process can concentrate on the final products to be produced; at other times, they focus on the concept design process, where the aim is to produce visions or ideas of products that have yet to be made rather than to present tangible, final products.

In the field of material technology, the focus is on research methods for testing the properties of different combinations of materials. Testing functional, interactive materials in order to verify their functions and suitability for intelligent garments is particularly important; the test methods have to be specified with particular regard for functionality and safety. In designing future consumer products, aspects of comfort, simplicity, miniaturisation and aftercare should also be taken into consideration.

In the area of ergonomics and clothing physiology, specifying the context (e.g. tasks, repetition, loads, and dynamic-static efforts), environment (e.g. cold or hot), organisation (e.g. individual-group, monotonous-rich work designs) and physiological properties (e.g. gender, age and strength), as well as psychological and social characteristics (e.g. intro- and extroversion, vigour), are essential. Human measures must be the starting point of garment design, especially for female workers (Asikainen and Hänninen, 2001). Assessing the usability of intelligent garments and their prototypes by measuring and analysing their functional efficiency, ease of use, comfort in use, health and safety and their contribution to working life belongs to the area of physiology. The colours and designs of clothing help colleagues recognise each other and thereby increase safety.

Drawing these observations together, it could be said that design research, fibre material technology and physiology have different kinds of relationships to theory and practice. In Kuhnian terms, the testing typical of technology and physiology should be connected to the theory, and this theory must be precise (Kuhn, 1970, 23–24; see also Diaz-Kommonen, 2002, 42). Hence, the testing relies on empirical findings produced by objective research methods. Inductive methods from practice to theory are more typical of design research. Subjective methods are also essential and are used both when design solutions are produced and when the usability of intelligent garments and their prototypes are assessed. In all cases, the current debate in the field suggests that the research questions should be posed by the designer (Diaz-Kommonen, 2002, 43). In this sense, there is justification for design to be understood as a discipline and intelligent garments as research objects of design research, physiology, and fibre material technology.

2.4.2 Dialogue in participatory design

Collaborative frameworks and participatory practice do not apply only to the co-operation between professionals and experts, however. Participatory practice can also be made a part of collaborative design practices and user-centred design. The participatory dimension in design emphasises the proactive role of the end users in the design process and the assessment of product usability. Sanders (2002) has addressed this issue by stating that in participatory experiences the roles of the designer and the researcher become blurred and the user becomes a critical component of the process. Siu (2003, 71) shares this view, stating that 'participation allows users to engage in the design decision making process'. In the area of interaction design, Wright and McCarthy (2005, 20) stress that each user is 'their own expert' in the activity and continue that although the designers may not be their own experts in the user domain, they are 'their own experts' in designing and creating 'possible applications of technology'.

Participatory design provides a number of advantages. Only the end users know all the details of their work and can predict the problems they will face if the protective garments and/or work organisation are changed. Such changes may also change the work of others. For instance, redesigning wheelchair-using clients' standard outerwear decreased the physical work load and strain of their personal helpers. This resulted in savings of effort and energy and increased the service capacity of nurses, with diminished stress levels in the disabled persons as well as the nurses caring for them. (Nevala-Puranen et al., 2003).

In the MeMoGa project, user participation was taken into account by integrating users from heavy industry into the research and design process in the following ways. Firstly, the researchers interviewed the users and gathered information about their clothing needs in the working context. Secondly, the users were allowed to give their comments and opinions during the concept design process. The concept and product design that was carried out in the project had its foundations in the philosophy known as 'design for all', whose point of departure is what people want from technology and what they can or have an opportunity to do with it (Coleman, 1999). The terms 'design for all' or 'universal design' are used to describe design approaches that have the explicit aim of designing products and services which meet the needs and circumstances of the widest possible range of users, including disabled and elderly people. The aim was to design a garment concept in the field of wearable intelligence that can be easily adapted for different users and is accessible to as many users as possible.

Physiological measurements of the old work organisation, the garment designs and the prototypes in collaboration with the workers provided objective data to guide further design work. Wearable sensors and telemetric recordings permitted normal efforts at work without disruption. Textile sensors can even be built into underwear.

Thirdly, the concept designed was evaluated by the users by using a multimedia presentation (Pursiainen et al., 2005) that showed the garment concept in the actual settings of use. The users were able to assess the concept by filling in a questionnaire on the computer that was included as part of multimedia presentation. A screenshot of the multimedia presentation used in the concept assessment is presented in Fig. 2.2. This example of participatory research practice in design may give some ideas on what the basic research activity could be in design-driven research areas that focus on the user experience and phenomenological features.

One should remember, however, that in all working places there are people who have temporary or permanent disabilities. The number of workers near

Methods and models for intelligent garment design 15

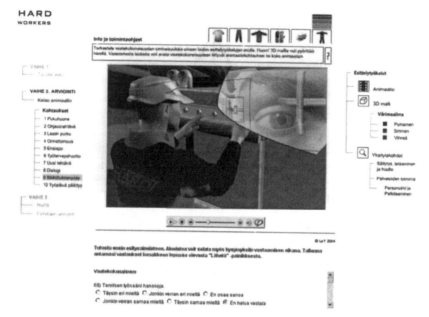

2.2 Screenshot from the multimedia presentation produced by the research and design team (Pursiainen et al. 2005).

– or even past – retirement age will also increase and intelligent garments may in principle help them continue to contribute. These workers and their limitations must be taken into account in planning. Wearable sensors and computer programs are at present reasonably easy to learn to use. The collection of subjective information using visual analogue scales on computers, for example, is also straightforward and such scales are recommended to designers, as they yield numerical information which the designers can then rely upon in their work. A short basic course in clothing physiology adapted to the needs of designers would nevertheless be very helpful. Participatory design programmes would be worthwhile in ergonomics courses.

2.5 Conclusions

At the time of writing, the MeMoGa project is almost completed and knowledge building has started. As the research consortium will not produce any concrete prototypes, it will not be possible to return to the situations in which the products are used and to the everyday world of the user, i.e., working life and work-related situations. Hence, the concrete situations in which the products are used and the different dimensions of usability and the acceptability of the products in industry could not be specified.

What then are the contributions of the MeMoGa project to research in applied technology and intelligent garments? In general, the challenges will be to pursue coherent research thinking, methods, and theory in design research. Where knowledge building is concerned, the challenging question to be answered now and in the future is how technological innovations will become culturally accepted final products, artefacts and natural parts of our everyday life. The second question that could be posed, on a more theoretical level, is how the natural sciences and researchers in those disciplines will meet the humanities and share their knowledge, concepts and working methods in these complex research areas. Accordingly, in further studies a third focus of interest would be to analyse the dialogue between the designer and the user, and study it with reference to the idea of shared expertise through complex products and through interdisciplinary research and design processes.

2.6 References

Asikainen M and Hänninen O (2001), Naisten vaatetuksen mittataulukko N-2001 (The measure table of female clothing N-2001), The Federation of Finnish Textile and Clothing Industries and Laboratory of Clothing Physiology, University of Kuopio, 72.

Bannon J J (2005), A human-centred perspective on interaction design, in Pirhonen A, Isomäki H, Roast C and Saariluoma P (eds), *Future Interaction Design,* London, Springer, 31–52.

Chalmers A F (1999), *What is this thing called Science?* Buckingham, Open University Press.

Coleman R (1999), Inclusive Design – Design for All, in Green W and Jordan P (eds), *Human Factors in Product Design. Current Practice and Future Trends,* London, Taylor & Francis, 159–170.

Cross N (2001), Designerly Ways of Knowing: Design Discipline Versus Design Science, *Design Issues,* vol 17, no 3, 49–55.

Denton D (1993), Control mechanisms, in Boyd C and Noble D (eds), *The Logic of Life, The Challenge of Integrative Physiology,* Oxford, Oxford University Press, 113–146.

Diaz-Komonen L (2002), *Art, Fact, and Artifact Production. Design Research and Multidisciplinary Collaboration,* Publication series of the University of Art and Design, A 37, Helsinki.

Flugel J C (1930), *The Psychology of Clothes.* New York, International Universities Press.

Friedman K (2000), Design knowledge: context, content and continuity, in Durling D and Friedman K (eds), in *Proceedings of the conference Doctoral Education in Design. Foundations for the Future,* La Clusaz France 8–12 July 2000, Stafforsdshire University Press, 5–16.

Gadamer H G (2004), *Hermeneutiikka. Ymmärtäminen tieteissä ja filosofiassa.* (Hermeutics. Understanding in Science and Philosophy), Tampere, Vastapaino.

Hempel C G (1966), *Philosophy of Natural Sciences,* London, Prentice-Hall.

Kaiser S (1990), *The Social Psychology of Clothing. Symbolic Appearances in Context,* New York, Macmillan.

Keinonen T and Jääskö V (eds) (2004), *Tuotekonseptointi,* (Concept design process), Helsinki, Teknologiateollisuus.

Klein J T (1990), *Interdisciplinarity. History, Theory, & Practice,* Detroit, Wayne State University Press.

Kuhn T S (1970), *The Structure of Scientific Revolutions*, Chicago, University of Chicago Press.
Lamb J M and Kallal M J (1992), A Conceptual Framework for Apparel Design, *Clothing and Textiles Research Journal*, 10(2), 42–47.
Laudan L (1977), *Progress and Its Problems. Towards a Theory of Scientific Growth*, London and Henley, Routledge & Kegan Paul.
Lurie A 1992 (1981), *The Language of Clothes. Featuring a New Introduction about Fashion Today*, Frome and London, Butler & Tanner.
Matala R, Mäyrä J, Mäkinen M, Talvenmaa P and Lintu N (2005), Multidisciplinary Intelligent Clothing Design and Applied Fabric Solutions, in *Proceedings of Joining Forces, International Conference on Design Research*, Helsinki, Finland, 22–24 September, 2005.
Mäyrä J, Matala R and Falin P (2005), Utilizing End User Knowledge in the Designing of Intelligent Workwear, in Mazé R (ed.) *Digital proceedings of NORDES, In the Making. First Nordic Design Research Conference*, Copenhagen, Denmark, 29–31 May.
Nevala-Puranen N, Holopainen J, Kinnunen O and Hänninen O (2003), Reducing the physical work load and strain of personal helpers through clothing redesign, *Applied Ergonomics* 34, 557–63.
Noble D and Boyd C (1993), The challenge of integrative physiology, in Boyd C and Noble D (eds) *The Logic of Life, The Challenge of Integrative Physiology*, Oxford University Press, Oxford, 1–13.
Nousiainen P, Mattila H, Mäkinen M, Talvenmaa P and Sahun O (2001), Interactive Textile Materials, in *Proceedings of 90th Anniversary of Academic Textile Research and Education in Finland*, Tampere, Finland, 16–17 August, 2001, 197–208.
Popovic V (2002), Activity and designing pleasurable interaction with everyday artifacts, in Green W S and Jordan P W (eds), *Pleasure With Products: Beyond Usability*, London, Taylor & Francis, 367–376.
Popper K (1987), Natural Selection and the Emergence of Mind, in Radnitzky G and Bartley W W III (eds), *Evolutionary Epistemology, Rationality, and the Sociology of Knowledge*. La Salle, Illinois, Open Court, 139–155.
Pursiainen M and Matala R (2005), Virtual prototype as evaluation tool in design process of intelligent clothing, in *Proceedings of Joining Forces, International Conference on Design Research*, Helsinki, Finland, 22-24 September, 2005.
Pursiainen M, Matala R, Mäyrä J, Latva M, Pohjapelto K, Pyykkönen M, Janhila L, Falin P, Juurikka M and Uotila M (2005), *Hard Workers – Intelligent Clothing Concept for Heavy-industry Workers*, multimedia presentation produced for evaluating usability of intelligent clothing Concepts, University of Lapland, Faculty of Art and Design, Department of Textile and Clothing Design, MeMoGa and CoDes Research Projects. Limited presentation is available from the web site http://www.ulapland.fi/?deptid=17234.
Sanders E B N (2002), From user-centered to participatory design approaches, in Frascara J (ed.) *Design and the Social Sciences. Making connections*, London, New York, Taylor & Francis, 1–8.
Siu K W M (2003), Users' Creative Responses and Designers' Roles. *Design Issues*, vol 19, no 2, 2003, 64–73.
Suomela R, Lehikoinen J and Salminen I (2001), A System for Evaluating Augmented Reality User Interfaces in Wearable Computers, in *The Fifth International Symposium on Wearable Computers*, Zürich, Switzerland, 8–9 October, 2001, 77–84.

Ullsperger I (2002), In search of fundamental innovations for high-tech fashion, in *Digital proceedings of the Avantex International Forum and Symposium for High-tech Apparel Textiles,* May 13–15, 2002. Frankfurt, Germany.

Uotila M, Hildén M, Matala R, Pursiainen M, Ruokanen M, Mäkinen M, and Talvenmaa P (2002), WearCare – The Usability of Intelligent Materials in Workwear. Research and Design Project 2001–2002, in *Digital proceedings of the Avantex International Forum and Symposium for High-tech Apparel Textiles*, May 13–15, 2002, Frankfurt, Germany.

Vincent J D (1993), Endocrinology, in Boyd C and Noble D (eds) *The Logic of Life, The challenge of integrative physiology*, Oxford University Press, Oxford, 147–167.

Wright P and McCarthy J (2005), The value of the novel in designing for experience, in Pirhonen A, Isomäki H, Roast C and Saariluoma P (eds), *Future Interaction Design,* London, Springer, 9–30.

Yates F E (1993), Self-organizing systems, in C Boyd and D Noble (eds), *The Logic of Life, The Challenge of Integrative Physiology*, Oxford University Press, Oxford, 189–218.

Part I

Phase change materials

3
Introduction to phase change materials

M MÄKINEN, Tampere University of Technology, Finland

3.1 Introduction

Generally speaking, phase change materials (PCM) are thermal storage materials that are used to regulate temperature fluctuations. As thermal barriers they use chemical bonds to store and release heat and thus control the heat transfer, e.g., through buildings, appliances and textile products. This chapter focuses on phase change materials used in texiles.

In a cold environment the primary purpose of clothing is to protect the wearer from cold and thus prevent the skin temperature from falling too low. Conventional thermal insulation depends on the air trapped in the clothing layers. When this layer of air gets thinner, e.g., due to windy weather, thermal insulation will be reduced significantly. The situation is the same when the garment becomes wet or perspiration condenses in it. It is possible to increase the thermal comfort by interactive insulation which means use of phase change materials, because compression and water has no effect on the insulation properties of PCM.

Phase change technology in textiles means incorporating microcapsules of PCM into textile structures. Thermal performance of the textile is improved in consequence of the PCM treatment. Phase change materials store energy when they change from solid to liquid and dissipate it when they change back from liquid to solid. It would be most ideal, if the excess heat a person produces could be stored intermediately somewhere in the clothing system and then, according to the requirement, activated again when it starts to get chilly.

The basis of the phase change technology was developed as a consequence of the NASA space research program of the early 1980s. The aim was to protect astronauts and instruments from extreme fluctuations of temperature in space. In 1987 the Triangle Research and Development Corporation (Raleigh, USA) demonstrated the feasibility of incorporating phase change materials within textile fibres and that the fabric's thermal capacity was independent of the amount of still air in the fabric loft. Triangle Research transferred the patent rights of this technology to a company called Gateway Technologies, which is now known as Outlast Technologies (Boulder, Colorado).

3.2 Heat balance and thermo-physiological comfort

It is very important to maintain a relatively even temperature to guarantee human vital functions. The normal human body temperature, 37 °C, fluctuates a little according to the time of the day, being at its lowest early in the morning and its highest in the evening. Also temperatures in different parts of the body fluctuate a little. For example, the temperature of the internal organs in the core of the body is higher than the temperature in other areas. In physical exercise the temperature of the muscles can rise to 39–40 °C. The surface parts of the body and the extremities are from time to time almost hemocryal according to the changes in the ambient air temperature.

Trying to reach thermal equilibrium, man produces and dissipates different amounts of heat depending on the ambient air temperature. Depending of the physical workload, the human being produces a heat quantity of 100 W in the state of rest and up to 600 W by physical effort. Heat production can temporarily be even more, e.g., in skiing up to 1250 W. This resultant quantity of heat has to be dissipated to prevent any marked increase in the rectal temperature and to maintain thermal equilibrium. The human being is in heat balance, when heat production is equal to heat loss. Factors influencing heat balance are the produced heat quantity (physical activity, work), ambient conditions (temperature, wind, humidity), the clothing worn and the individual properties of humans.[1,2]

Clothing is comfortable when humans feel physical, physiological, and mental satisfaction as heat and moisture transfer efficiently from the body to the environment through the clothing.[3] Therefore, development of intelligent fabrics, including thermal storage/release ones, which can adjust and maintain comfort as circumstances change, is very important and necessary.[4]

3.3 Phase change technology

Phase change materials are latent thermal storage materials. They use chemical bonds to store and release heat. The thermal energy transfer occurs when a material changes from a solid to a liquid or from a liquid to a solid. This is called a change in state, or phase.[5] Every material absorbs heat during a heating process while its temperature is rising constantly. The temperature of a PCM rises until it reaches its melting point. During the physical phase change the temperature remains constant until the PCM has totally changed from solid to liquid. Energy is absorbed by the material and is used to break down the bonding responsible for the solid structure. A large amount of heat is absorbed during the phase change (latent heat). If the material is warmed up further its temperature will begin to rise again. The latent heat will be released to the surroundings when the material cools down. The temperature remains constant again until the phase change from a liquid to a solid is complete, when the crystallization temperature of the PCM is reached. PCMs absorb and emit heat while maintaining a nearly constant temperature.

In order to compare the amount of heat absorbed by a PCM during the actual phase change with the amount of heat absorbed in an ordinary heating process, water is used for comparison. If ice melts into water it absorbs approximately a latent heat of 335 J/g. If water is further heated, a sensible heat of only 4 J/g is absorbed while the temperature rises by one degree. Therefore the latent heat absorption during the phase change from ice to water is nearly 100 times higher than the sensible heat absorption during the heating process of water outside the phase range.[6] In addition to water, more than 500 natural and synthetic PCMs are known. These materials differ from one another in their phase change temperature ranges and their heat-storage capacities.[7]

Solid-solid PCMs absorb and release heat in the same manner as solid-liquid PCMs. These materials do not change into a liquid state under normal conditions; they merely soften or harden. Relatively few of the solid-solid PCMs that have been identified are suitable for thermal storage applications. Liquid-gas PCMs are not yet practical for use as thermal storage. Although they have a high heat of transformation, the increase in volume during the phase change from liquid to gas makes their use impractical.[5]

3.4 PCMs in textiles

The most widespread PCMs in textiles are paraffin-waxes with various phase change temperatures (melting and crystallization) depending on their carbon numbers. The characteristics of some of these PCMs are summarized in Table 3.1. These phase change materials are enclosed in microcapsules, which are 1–30 μm in diameter. Compared to our hair the size of the capsule is usually about half the diameter of human hair or it can be 1/20th of it. Phase change materials can be incorporated in textiles only enclosed in these

Table 3.1 Phase change materials[1]

Phase change material	Melting temperature in °C	Crystallization temperature in °C	Heat storage capacity in J/g
Eicosane	36.1	30.6	247
Nonadecane	32.1	26.4	222
Octadecane	28.2	25.4	244
Heptadecane	22.5	21.5	213
Hexadecane	18.5	16.2	237

capsules in order to prevent the paraffin's dissolution while in the liquid state. The shell material of the capsule has to be abrasion and pressure resistant, heatproof and resistant to most types of chemicals.[8,9] Outlast®, Comfortemp® and Thermasorb® are commercially available PCM products based on paraffin-waxes and microcapsule technology.

Hydrated inorganic salts have also been used in clothes for cooling applications. PCM elements containing Glauber's salt (sodium sulphate) have been packed in the pockets of cooling vests.[10]

3.4.1 Textile treatment with PCM microcapsules

Usually PCM microcapsules are coated on the textile surface. Microcapsules are embedded in a coating compound such as acrylic, polyurethane and rubber latex, and applied to a fabric or foam. Capsules can also be mixed into a polyurethane foam matrix, from which moisture is removed, and then the foam is laminated on a fabric.[7] In Fig. 3.1 you can see PCM microcapsules (Outlast) in fabric and in Fig. 3.2 how it works. PCMs-containing microcapsules can be incorporated also into acrylic fibre in a wet spinning process. In this case the PCM is locked permanently within the fibre. The fibre can then normally be processed into yarns and fabrics.[6,7]

3.4.2 Thermal performance

In treating textile structures with PCM microcapsules for garment applications, the following thermal benefits are realized:

3.1 PCM microcapsules in a fabric (Outlast Europe).

Introduction to phase change materials 25

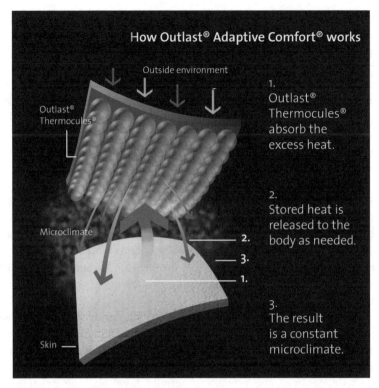

3.2 Functioning principle of PCM in a fabric (Outlast Europe).

- a 'cooling' effect, by absorbing surplus body heat
- an insulation effect, caused by heat emission of the PCM into the textile structure; the PCM heat emission creates a thermal barrier which reduces the heat flux from the body to the environment and avoids undesired body heat loss
- a thermo regulating effect, resulting from either heat absorption or heat emission of the PCM in response to any temperature change in the microclimate; the thermo-regulating effect keeps the microclimate temperature nearly constant.[11]

Body temperature varies considerably: 34–36.5 °C at the head and trunk, 27–30 °C at the thigh and only 25.5–27.5 °C at the hands and feet.[5] The primary purpose of clothing designed to protect the wearer from the effects of a cold environment is to prevent the skin temperature from falling too far. Such clothing normally consists of layers, possibly an outer layer, a lining and a lofted fabric.[6] In this traditional insulation system the lofted fabric provides the vast majority of the insulation (passive insulation). Without phase change materials the thermal insulation capacity of clothing depends on the thickness and the density of the fabric. The thicker the material and the lower the density, the better is the thermal insulation influence of the product. The air within the clothing directly influences the thermal insulation capacity. Normally the thermal insulation capacity of the fabric decreases when the fabric is compressed. It happens in the same way if the material becomes wet, because the thermal conductivity of water is greater than air. Pressure and moisture is not affected by the functionality of PCM. It is possible to improve the passive insulation with an active system of the phase change technology, which reacts immediately to changes in environmental temperature and adapts to the prevailing hot or cold conditions (active insulation).[8]

In a garment system the thermal resistance of textile layers, and the air layers between, limit heat flux from the body to the environment. This passive thermal insulation effect can only be adjusted to the often-changing thermal wearing conditions by adding or removing garment layers but in reality this is not always possible. Thus, strenuous exercise often leads to a thermal stress situation and increased sweat production. In contrast, light exercise combined with low ambient temperatures often results in a decrease in body temperature.[11]

The application of PCM to a garment provides an active thermal insulation effect acting in addition to the passive thermal insulation effect of the garment system. The active thermal insulation of the PCM controls the heat flux through the garment layers and adjusts the heat flux to the thermal circumstances, i.e., the prevailing activity level and the existing ambient temperature. If the heat generation of the body exceeds the possible heat release through the garment layers into the environment, the PCM will absorb and store this excess heat. On the other hand, if the heat release through the garment layers exceeds the body's heat generation during lighter activities, the heat flux through the garment layers is reduced by heat emission of the PCM. The active thermal insulation effect of the PCM results in a substantial improvement of the garment's thermo-physiological wearing comfort.[11]

The steadily-changing intensity of the body's activities leads to a constant charging and recharging of the PCM microcapsules. Furthermore, the change from a cold outdoor environment to a warm indoor environment results in a recharging of the PCM microcapsules. Thus, a necessary PCM quantity can be relatively small.[11]

Thermal performance of active wear

In order to improve the thermal performance of active wear, clothing textiles with thermo-regulating properties are widely used. The thermo-regulating effect provided by these textiles is based on the application of PCM. However, a suitable thermo-regulating effect according to the prevailing wearing conditions can only be realized when specific design principles are applied

Introduction to phase change materials 27

in the development process of such active wear garments. It is necessary, for instance, to match the PCM quantity applied to the active wear garment with the level of activity and the duration of the garment use. Furthermore, the garment construction needs to be designed in a way which assists the desired thermo-regulating effect.[11]

Intensity and duration of the PCM's active thermal insulation effect depend mainly on the heat-storage capacity of the PCM microcapsules and their applied quantity. In addition, performance tests carried out on textiles with incorporated PCM microcapsules have shown that the textile substrate construction also influences the efficiency of the active thermal insulation effect of the PCM. For instance, thinner textiles with higher densities readily support the cooling process. In contrast, the use of thicker and less dense textile structures leads to a delayed and therefore more efficient heat release of the PCM. Furthermore, the phase change temperature range and the application temperature range need to correspond in order to realize the desired thermal benefits.[11]

In order to ensure a suitable and durable active thermal insulation effect of the PCM in active wear garments, it is necessary to apply proper PCM in sufficient quantity. Selecting a suitable substrate requires considering whether the textile structure is able to carry a sufficient PCM quantity and provide an appropriate heat transfer to and from the PCM microcapsules. Further requirements on the textile substrate in a garment application include sufficient breathability, high flexibility, and mechanical stability. The substrate with incorporated PCM microcapsules needs to be integrated into a suitable location of the garment design.[11]

In the first step of the design process, temperature profiles are developed considering possible application temperatures. The temperature profiles are used to determine the temperature ranges in which the PCM is supposed to function because the temperatures in different garment layers vary widely. This must be considered when selecting the PCM for a certain application. For instance, if PCM should be applied to textiles used for underwear, the phase change of selected PCM would have to take place in a temperature range which corresponds with the skin temperature. On the other hand, a PCM applied to the liner material of outer wear needs to exhibit a phase change which takes place within a much lower temperature range.[11]

In order to determine a sufficient PCM quantity the heat generated by the human body has to be taken into account carrying out strenuous activities under which the active wear garments are worn. The heat generated by the body needs to be entirely released through the garment layers into the environment. On one side, this heat release takes place in the form of dry heat flux which is defined by the total thermal resistance of the garment system. Furthermore, heat is also released from the body by means of evaporative heat flux. The amount of heat released by means of evaporative heat flux is determined by the water vapour resistance provided by the garment system. In addition, the garment textiles also absorb surplus body heat with a steady rise in their temperatures. Measurements of the heat and moisture transfer as well as the heat and moisture absorption of the textile layers are used in preliminary tests to determine the total heat flux through the garment system and its thermal buffering. The necessary PCM quantity is determined according to the amount of heat which should be absorbed by the PCM to keep the heat balance equalized.[11]

During strenuous activity heat is mainly generated in the muscles and needs to be released. Because every sport makes use of different muscles, heat generation in distinct parts of the body varies between the strenuous activities. This has to be taken into account in designing active wear garments with PCM. It is mostly not necessary to put PCM in all parts of the garment. Applying PCM microcapsules to the areas that provide problems from a thermal standpoint and thermo-regulating the heat flux through these areas is often enough. It is also advisable to use different PCM microcapsules in different quantities in distinct garment locations. In order to identify the thermal problem areas, heat flux measurements as well as infra-red photographs are taken during strenuous activities.[11]

3.4.3 Test methods

Traditional thermal insulation materials rely upon trapped air for their performance. The non-physiological testing procedures for quantifying the performance of such materials have therefore been designed accordingly. Trapped air insulation is a static system, which relies upon the convection/conduction of heat through air voids and fibre. The TOG and CLO tests therefore are designed to measure this effect. Fabrics containing PCM form a dynamic system that responds to changes in skin temperature and external conditions. It is therefore quite logical that the traditional non-physiological testing procedures for the old trapped air technology do not quantify the benefits of PCM.[12]

The best type of testing for all of these materials is of course human physiological testing, but as always these are very time consuming and costly and could never be used for routine quality control testing. The American Society for Testing and Materials (ASTM) approved a new standard test procedure to measure the amount of latent energy in textile materials in June 2004. Based on years of research and testing textiles containing 'phase change materials' (PCMs) by Outlast Technologies, Inc., and Prof. Dr Douglas Hittle, Director Solar Energy Applications at Colorado State University (http://welcome.colostate.edu), the first 'Test Method for Steady State and Dynamic Thermal Performance in Textile Materials' (ASTM D7024) was established by the ASTM.[13]

Phase-change technology in temperature-regulating textiles with increased latent energy represents an entirely new approach to providing increased comfort and performance. Standard testing procedures used for determining the insulating value of traditional fabrics do not measure the stored energy in these new, innovative 'smart' fabrics. Therefore a new test method and apparatus was required as ASTM D1518 'Standard Test Method for Thermal Transmittance of Textile Materials' determined only the R-value (or CLO value as used in the garment industry) in a steady state. This new test method measures dynamic temperature changes and differentiates and quantifies the temperature-buffering properties of a material in a dynamic environment. It measures the effects of changing temperature and a fabric's ability to absorb, store and release energy. This test provides the measurement to separate PCM technology from unsubstantiated claims of temperature regulation through moisture management, wicking or straight thermal insulation properties of a fabric.

A differential-scanning calorimeter (DSC) is used to measure the heat capacity or enthalpy of the microcapsules and the fibre containing the microcapsules. This is a well established procedure that has been used for many years to quantify the melting and crystallization points, or ranges, of materials as well as the heat absorption and release potential of the same material. The same technique is used to measure the heat capacity of the finished article.[12] Another technique known as thermo-gravimetric analysis (TGA) is used to assess the thermal strength of the micro PCMs. This is important because the process used to manufacture the fibre and the processes through which the fibres containing the microcapsules are subjected to in conversion to yarns and fabrics use heat.[12]

3.4.4 Applications

In textiles phase change materials are used both in winter and summer clothing. On the market can be found clothes and footwear incorporating PCMs mainly for active sports, extreme sports and casual wear. PCM is used not only in high-quality outerwear and footwear, but also in the underwear, socks, gloves, helmets and bedding of world-wide brand leaders. Seat covers in cars and chairs in offices can consist of phase change materials. In the medical textiles field can be found PCMs in acrylic blankets and in bed covers to regulate the micro climate of the patient. Possible applications also include work and protective clothes for both cold and hot environments.

Suitable technical equipment is becoming more and more important for authorities and the military. Not only are electronics, hard- and software playing a large role, there are also increasing demands for apparel. The call for intelligent fabrics is becoming more and more insistent. A new generation of these fabrics feature phase change materials (PCMs) which are able to absorb, store and release excess body heat when the body needs it resulting in less sweating and freezing, while the microclimate of the skin is influenced in a positive way and efficiency and performance are enhanced.[14]

Two companies, Outlast Europe GmbH, Heidenheim, and UCO Sportswear NV, Ghent/Belgium, succeeded in developing denim fabrics with temperature-regulating properties. Around two years of development were necessary for the innovation. One of the greatest challenges was to guarantee the durability of the coating for the usual industrial washes that denim undergoes. Thanks to further development of the coating compound the performance is now guaranteed against the effects of stone, enzyme and other washes. By modifying the coating process of the compound it can be incorporated directly into the fabric structures. This results in the textile touch of the denim fabric being conserved, the coating film being invisible and there is no unpleasant feel on the skin. Comprehensive tests were conducted on coated pure cotton fabrics, cotton/polyester blends and elastic fabrics (weights between 240 g and 360 g) as the summer product has a cooling fresh effect that is very comfortable.[15] Motorcyclists also swear by this intelligent fabric in many situations – on tours through shady woods, sunny roads or when they are exposed to changing influences caused by the wind chill effect while riding or stationary at traffic lights. It was police motorcyclists in the UK who first tested their garments to the limit in all weather conditions providing input from their experiences for the development of innovative motorcycling jackets and trousers.[14]

Cooling vest (TST Sweden Ab) is a comfort garment developed to prevent elevated body temperatures in people who work in hot environments or use extreme physical exertion. The cooling effect is obtained from the vest's 21 PCM elements containing Glauber's salt which start absorbing heat at a particular temperature (28 °C). Heat absorption from the body or from an external source continues until the elements have melted. After use the cooling vest has to be charged at room temperature (24 °C) or lower. When all the PCMs are solidified the cooling vest is ready for further use.[16]

3.5 Future prospects of PCM in textiles and clothing

Retailers and consumers are asking more and more for function, for added values like appearance and improved comfort. The component 'temperature regulation' is continuing to receive increased attention. Outlast Europe, market leader of PCMs has reacted and designed an attractive offer for the season Spring/Summer 2006: the maximum 'lightness' is top priority.[17]

New to the range of thermo-regulating fabrics are two items. For the first time Outlast has created a lightweight coated net lining with a high performance. The hole structure of the fabric is retained due to a special coating process.

The coating on the reverse does not show through and the hand is pleasant and soft. Apparel can be equipped with a net lining, which can absorb excess body heat, store and release it when needed, creating increased freshness in hot summer weather (Fig. 3.3).[17]

3.3 Optimised climate due to Outlast® Adaptive Comfort®: also linings with net optic are now available. In spite of the coating on the back the hole structure is visible. (Outlast Europe)

Another lining has been specially developed for summer-weight clothing. The basis is a polyester nonwoven with significantly reduced weight. In that way the Outlast® fabric can be inserted between the outer layer and the lining to balance temperature swings. The advantage for the manufacturer is freedom to choose the top fabric and lining, without losing added value, the function of Outlast® Adaptive Comfort® (Fig. 3.4).[17]

Textile testing & Innovations, LCC, has developed a cooling undergarment especially for fire fighters, steel mill workers, and workers in nuclear and chemical facilities in order to avoid heat stress and related illnesses while performing their duties. The suit can also been worn by police and army personnel in order to improve the thermo-physiological wearing comfort of ballistic vests and combat suits. A non-combustible salt hydrate was selected for the cooling suit application in order to meet the fire-resistant requirement. Due to the very high latent heat storage capacity of the selected PCM, a durable cooling effect is already obtained with a relatively small amount of PCM which enhances the cooling suit's weight only slightly. After taking off the cooling suit, the PCM is regenerated under room temperature and is seen available to cool again after a short period of time. The PCM is embedded in a polymer matrix from which a film-like structure is made. The polymer film

3.4 Light and soft: the new Outlast® Adaptive Comfort® liner is suited very well for summer jackets regulating the temperature actively (Outlast Europe).

with PCM is applied to fabrics made of fire-resistant fibres. The application of the cooling suit leads to extended wearing times of the protective garment systems resulting in enhanced productivity. The heat stress-related health risks the wearers of protective garments are exposed to is minimized by the application of this cooling suit.[18]

3.6   References

1. Weder, M. and Hering, A., How effective are PCM materials? Experience from laboratory measurements and controlled human subject test. International Man-made Fibres Congress, 13.–15.09, Dornbirn, Austria, 13p.
2. Meinander, H., *Introduction of a new test method for measuring heat and moisture transmission through clothing materials and its application on winter work wear.* Technical research Center of Finland, Publications 24, Espoo, 1985, 63 p.
3. Branson, D. H. and Saltin, B. S., Conceptualization and Measurement of Clothing Comfort: Toward a Metatheory, ITAA Spec. Pub. Int. Textile Apparel Assoc. 4, 1991, pp. 94–105.
4. Choi, Kyeyoun, Thermal Storage/Release and Mechanical Properties of Phase Change Materials on Polyester Fabrics. *Textile Research Journal*, Apr. 2004, 6 p.
5. Energy savers. A consumer guide to energy efficiency and renewable energy. Phase change materials for solar heat storage. Available from: http://www.eere.energy.gov/consumerinfo/factsheets/b103.html Printed out 18.6.2005.
6. Cox, R., Synopsis of the new thermal regulating fiber Outlast. *Chemical Fibers International*, Vol. 48, December 1998, pp. 475–476.
7. Pause, B., New possibilities in medicine: Textiles treated with PCM microcapsules. Lecture No. 627, 10th International Symposium for Technical Textiles, Nonwovens and Textile Reinforced Materials, 7 p.

8. Rupp, J., Interactive textiles regulate body temperature. *International Textile Bulletin* 1/99, pp. 58–59.
9. Nicht zu warm und nicht zu kalt. *Bekleidung und Wäsche* 1/99, pp. 24–26.
10. http://comfortcooling.se/ENG/news.html Printed out 29.6.2005.
11. Pause, B., Tailored to the purpose: Computer-optimized development of thermoregulated active wear. Lecture No. 333. International Avantex-symposium, Frankfurt, Germany, November 27–29.11.2000. 8 p.
12. Cox, R., Outlast – Thermal Regulation where it is needed. 39th International Man-Made Fibres Congress, 13–15 Sept. 2000, Dornbirn, Austria. 7 p.
13. ASTM D 7024-04. *Standard Test Method for Steady State and Dynamic Thermal Performance of Textile Materials.*
14. Outlast technology: New intelligent materials are helping those on duty. Press release. Heidenheim, 28 April, 2005.
15. Outlast/Uco present new developments. Press release. Heidenheim, 8 July, 2005.
16. http://www.tst-sweden.se/Broshyr-brand-eng.pdf Printed out 29.6.2005.
17. Spring/Summer 2006: Outlast focuses on lightness. Press release. Heidenheim, 2 March, 2005.
18. Pause, B., Development of a fire-resistant cooling suit. International Avantex Symposium, Frankfurt, Germany, 6–8 June, 2005.

4
Intelligent textiles with PCMs

W BENDKOWSKA, Instytut Wlokiennictwa,
Textile Research Institute, Poland

4.1 Introduction

In the last decade one of emerging technologies is microencapsulated phase change materials (PCMs), which are being developed to provide significantly enhanced thermal management for fibres, foams and textiles with applications to apparel and technical textiles. Phase change technology originates from the NASA (National Aeronautics and Space Administration) research programme of the 1970s. The aim of this programme was to provide astronauts and instruments with better protection against extreme fluctuations of temperature in space. At present, microencapsulated PCMs have been applied in many fields, including heat management of electronics, telecommunications and microprocessor equipment, solar heat storage systems for buildings, microclimate environmental control for vegetation in agriculture, biomedical and biological carrying systems, and so on.

In this chapter a basic overview of the phase change materials, with particular reference to the linear alkyl hydrocarbons, is presented. The principal functions of the microPCM in textiles are discussed. Special attention is paid to the mode of PCM performance in clothing. The most common methods of incorporating microPCMs into fibrous substrates and the various applications of textiles containing microPCMs are discussed. Additionally, the apparatus for testing thermal properties of the fabrics containing PCMs is presented.

4.2 Basic information on phase change materials

Phase change is a process of going from one physical state to another. The three fundamental phases of matter, solid, liquid and gas, are known but others are considered to exist, including crystalline, colloid, glassy, amorphous and plasma phases. Substances that undergo the process of phase change are known as phase change materials (PCMs). By definition PCMs are materials that can absorb, store and release large amounts of energy, in the form of latent heat, over a narrowly defined temperature range, also known as the phase change range, while that material changes phase or state (from solid to liquid or liquid to solid).

The phase change from the solid to the liquid state occurs when the melting temperature in a heating process is reached. During this melting process the PCM absorbs and stores large amounts of a latent heat. The temperature of the PCM remains nearly constant during the entire process (Fig. 4.1). During the cooling process of the PCM the stored heat is released into the environment within a certain temperature range and a reverse phase change from the liquid to the solid state takes place. During this solidifying process the temperature of the PCM also remains constant. Thus the PCM can be used as an absorber to protect an object from additional heat, as a quantity of thermal energy will be absorbed by the PCM before its temperature can rise. The PCM may also be preheated and used as a barrier to cold, as a larger quantity of heat must be removed from the PCM before its temperature begins to drop. The phase change process results in a density and volume change of the PCM that has to be taken in consideration into its application.

The best-known PCM is water, which at 0 °C becomes ice or evaporates at 100 °C. In addition to water, the number of natural and synthetic phase change materials known today exceeds 500. These materials differ from one another in their phase change temperature range and their heat storage capacities. In order to obtain textiles and clothing with thermal storage and release properties the most frequently used PCMs are solid–liquid change materials. Research on solid–liquid phase change materials has concentrated on the following materials:

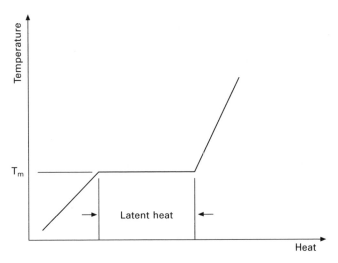

4.1 Heat transfer characteristic of phase change material as it changes from solid to liquid or from liquid to solid.

- linear crystalline alkyl hydrocarbons
- fatty acids and esters
- polyethylene glycols (PEG)
- quaternary ammonium clathrates and semi-clathrates
- hydrated inorganic salts (e.g. lithium nitrite trihydrate, calcium chloride hexahydrate, sodium sulphate decahydrate, zinc nitrate hexahydrate)
- eutectic alloys, containing bismuth, cadmium, indium, lead.

An ideal PCM should meet numerous criteria such as a high heat of fusion, high heat capacity, high thermal conductivity, small volume change at phase transition, be non-corrosive, non-toxic, non-flammable and exhibit little or no decomposition or supercooling. Organic PCMs, such as linear alkyl hydrocarbons, are chemically stable, non-corrosive, and exhibit no supercooling properties. They have a high latent heat per unit weight. Their disadvantages are low thermal conductivity, high changes in volume during phase change and flammability. Inorganic compounds, such as hydrated inorganic salts, have a high latent heat per unit weight and a high thermal conductivity, are inflammable and low-cost. Their utilisation is, however, limited because they suffer decomposition and supercooling which can influence their phase change properties.

In addition to solid–liquid change materials there is another class of substances that are characterised by their high enthalpies or thermal storage and release properties. These substances, commonly called plastic crystals, have extremely high thermal storage or release values that occur prior to or without melting. This thermal effect is believed to be a transition between two solid states (e.g. crystalline or mesocrystalline phase transformation) characterised by large enthalpy changes and hence typically does not become a liquid during a use. Hence these substances are called solid–solid change materials.

Polyhydric alcohols are the solid-solid change materials, which have been recommended for use in textiles.[1–4] Polyhydric alcohols may be selected from the group consisting of pentaerithritol, 2,2-dimethyl-1,3-propanediol, 2-hydroxymethyl-2-methyl-1,3-propanediol and amino alcohols such as 2-amino-2-methyll-1,3-propanediol.

4.3 Phase change properties of linear alkyl hydrocarbons

In the present applications of PCM technology in the textile industry the crystalline alkyl hydrocarbons are used exclusively. The phase change properties of the alkyl hydrocarbons suitable for incorporation into textiles are shown in Table 4.1. These data reveal quite clearly that their melting temperature increases with the number of carbon atoms. Each of the alkyl hydrocarbons

Table 4.1 Thermal characteristics of selected alkyl hydrocarbons

Name	Formula	Temperature of melting, T_m °C	Temperature of crystallisation, T_c °C	Enthalpy, J/g
n-hexadecane	$C_{16}H_{34}$	18.2	16.2	237.05
n-heptadecane	$C_{17}H_{36}$	22.5	21.5	213.81
n-octadecane	$C_{18}H_{38}$	28.2	25.4	244.02
n-nonadecane	$C_{19}H_{40}$	32.1	29.0	222.0
n-eicosane	$C_{20}H_{42}$	36.1	30.6	246.34
n-heneicosane	$C_{21}H_{44}$	40.5		199.86

Data from refs 4 and 5.

is most effective near the melting temperature indicated in Table 4.1. The alkyl hydrocarbons are non-toxic, non-corrosive and non-hygroscopic. In order to realise desired temperature range in which the phase change will take place, the hydrocarbons can be mixed. As a by-product of petroleum refining they are inexpensive. A disadvantage of hydrocarbons is their low resistance to ignition but the addition of fire retardants can solve this problem.

To prevent the liquid hydrocarbons from migrating within a fibrous substrate they need to be microencapsulated. The microencapsulation of the PCMs involves enclosing them in thin and resilient polymer shells so that the PCMs can be changed from solid to liquid and back again within the shells. Figure 4.2 shows a structure of a single-shell microcapsule. A variety of chemical and physical techniques for manufacturing different types of microcapsules exists[6] and can be employed for forming microencapsulated PCMs. Two of the most important chemical methods are coacervation and interfacial polymerisation. In microencapsulation using coacervation, the core particles are uniformly dispersed in an appropriate medium and the coacervate layer is deposited uniformly around the particles. The coating is then hardened by adding a reagent such as formaldehyde resulting in the cross-linking of the

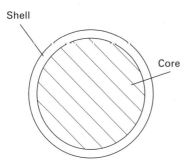

4.2 Structure of microcapsule.

coacervate. In interfacial polymerisation, the capsule wall is formed directly around the core material by polymerisation reactions.

The key parameters of microencapsulated PCM are:

- particle size and its uniformity
- core-to-shell ratio, with PCM content as high as possible
- thermal and chemical stability
- stability to mechanical action.

Diameters of microPCMs can range from 0.5 to 1000 μm. Very small microcapsules ranging from 1 to 10 μm in diameter are used for incorporation within textile fibres.[8] Larger microPCM particle of 10–100 μm can be incorporated into foams or coatings applied on fabrics.[3,8,9] The core of a microcapsule constitutes 60–85% of the particle volume, while the polymer shell is approximately 1 μm thick. Still larger macroencapsulated PCMs ranging from 1 to 3 mm are being developed[11,12] to produce textiles that permit high thermal storage as well as high moisture transport between the capsules.

The microencapsulation process and the size of the microcapsules affect the phase change temperatures. The larger particles, the closer the phase change temperature is to that of the core material. The smaller particles, the greater is the difference between the melting and solidifying temperatures for the PCM. The effective specific heat of encapsulated material undergoing phase change depends on the physical properties of the microcapsule enclosing it. The shell material should conduct heat well and it should be durable enough to withstand frequent changes in the core's volume as the phase change occurs. Experimental results[13] show that microPCMs expand and contract during the phase change process of the core with an order of magnitude of 10%. After solidification of the core on the surface of the microcapsules there are dimples (Fig. 4.3), which are attributed to the lower contract coefficient of the shell than that of the core. Selecting an appropriate shell material to improve the thermal stability of microencapsulated PCMs has been studied intensively. MicroPCMs have been synthesised with urea-formaldehyde,[14] cross-linked nylon,[14] melamine-formaldehyde,[15,17–20] polyurethane,[16,18] urea-melamine-formaldehyde copolymer[17] as shell materials. Thermal stability of microPCMs can be improved by adding the stabilising agent selected from the group consisting of antioxidants and thermal stabilisers.[21]

Microcapsules for textile materials should be stable against mechanical action (e.g. abrasion, shear and pressure) and chemicals. Shin et al.[18] tested the stability of the melamine-formaldehyde microcapsules containing n-eicosane. The results confirmed that microcapsule shells were durable enough to secure capsule stability under stirring in hot water and alkaline solutions. The microcapsules did not show any significant changes in their

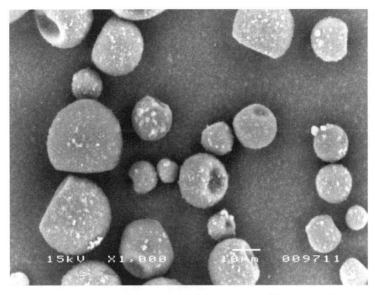

4.3 Microcapsules containing octadecane (magnification: 1000×).

morphology and size, and more than 90% of the heat storage capacity of the microcapsules was retained after testing.

4.4   Textiles containing PCM

Intelligent textiles are able to sense stimuli from the environment, to react to them and adapt to them by integration of functionalities in textile structure. According to this definition textiles containing the PCM are considered as intelligent, because they react immediately to changes in environmental temperature and adapt to the prevailing hot or cold conditions.

4.4.1   Historical background

In the 1980s work on temperature-adaptable fabrics was undertaken at the US Department of Agriculture's Southern Regional Research Centre in New Orleans. The extensive researches were conducted by Vigo and his co-workers. Vigo and Frost[22] filled hollow fibres with inorganic salt hydrates (e.g. lithium nitrite trihydrate, calcium chloride hexahydrate in admixture with strontium chloride hexahydrate). The treated fibres exhibited poor thermal behaviour on repeated thermal cycles, i.e., heating and cooling. Vigo and Frost described also experiments with polyethylene glycols (PEGs).[23,24] They incorporated PEGs having 7 to 56 monomer units with an average molecular weight ranging from 400 to 3350 into the hollow polypropylene and rayon fibres or topically applied PEGs to fabrics consisting of most representative fibre types. The fabrics treated with PEGs were temperature adaptable in both hot and cold environments for 150 heating and cooling cycles, by releasing heat when the temperature drops, and storing heat when the temperature rises. Such treatments however were not durable to laundering or leaching since the PEGs are water-soluble and can thus be removed from the substrate.

Vigo and Bruno[25] reported that low molecular weigth PEGs may be insolubilised on fabrics by their reaction with DMDHEU (dimethyloldihydroxyethyleneurea) under conventional pad-dry-cure treatment conditions to produce textiles thermally adaptable even after launderings. The resultant knitted fabrics, designated Neutratherm, were used for manufacturing underwear.[26] Vigo and Frost have also incorporated into textile fibres some of solid-solid change materials, e.g., 2,2-dimethyl-1,3-propanediol (DMP).[23] However, the application of textiles modified with polyhydric alcohols is limited due to their phase change transition temperature being higher than 40 °C.

Since 1983 considerable research work on including PCMs into textiles has been done at Triangle Research and Development Corporation. Several research programs at TRDC have demonstrated that microcapsules containing selected PCMs (alkyl hydrocarbons and plastic crystals) can be either spun into textile fibres or coated onto them.[2,4] They can also be embedded in a variety of foams.[3] The incorporation of microPCMs within textiles imparted thermal storage and thermoregulating properties. These inventions were awarded several patents that TRDC licensed to Outlast Technologies. TRDC is currently developing other products with microPCMs for enhanced apparel cooling.[11,12]

Since the mid-1990s scientists from Outlast Technologies have developed several processes for incorporating microPCMs into textile products, especially sportswear, protective clothing, home and technical textiles. Recently microPCM technology has drawn the attention of research workers from universities. This has been reflected in an increasing amount of published work on incorporation processes of microPCMs into fabrics[8–20] and properties of textiles containing microPCMs.[33–35]

4.4.2 Function of textile structure with PCM

From comprehensive analysis given by Pause[27,28] it follows that there are several thermal effects that can be obtained by incorporation of PCM into a textile substrate:

- a cooling effect, caused by heat absorption of the PCM
- a heating effect, caused by heat emission from the PCM
- a thermo-regulating effect, resulting from either heat absorption or heat

emission of the PCM which keeps the temperature of the surrounding substrate approximately constant
- an active thermal barrier effect, resulting from either heat absorption or heat emission of the PCM and creating a thermal barrier in surrounding substrate, which regulates the heat flux through the substrate and adapts the heat flux to thermal needs.

The function of the textile substrate with PCM as a barrier or buffer evolves from the characteristic of a phase change process. As the PCM undergoes a phase change, depending upon its initial state, it can either absorb or release a quantity of heat equal to its latent heat of fusion while still maintaining a constant temperature. In such a manner, heat can be lost or gained from either side of this barrier, yet the temperature at this barrier will remain constant. This remains true until all of the PCM has either solidified or melted. After this point, all the latent capacity of the PCM has been consumed and a sensible change in temperature results. The textile substrate with PCM then acts as a conventional textile product. The desired thermal effects have to be identified before selecting the PCM. The efficiency and duration of each of these effects are determined by the total thermal capacity of the PCM incorporated into the textile substrate, the phase change temperature range, the structure of the textile substrate, and the structure of the final product.

The total thermal capacity of the PCM depends on its specific thermal capacity (latent heat of the phase change) and the quantity applied on the textile substrate. The necessary quantity of the PCM can be estimated by considering the application conditions, the desired thermal effect and its duration and the thermal capacity of the specific PCM. In order to obtain the desired thermal performance of the textile product it is necessary:

- to select the appropriate PCM
- to determine the sufficient quantity of the PCM
- to choose the appropriate fibrous substrate
- to design the product.

The choice of which PCM to use in the fibrous structure depends principally on two factors: the latent heat of the phase change and the phase change temperature. In general, the higher latent heat of phase change is, the better PCM because more thermal energy can be stored in it. The choice of the phase change temperature depends largely on the intended application of the textile. The phase change temperature range should correspond with the application temperature range.

To select the appropriate fibrous structure it is necessary to consider whether the textile structure is able to carry a sufficient quantity of the PCM and whether it can provide an appropriate heat transfer to micro PCM.

4.4.3 Mode of PCM performance in clothing

The principal function of clothing assembly is to provide the wearer with protection against undesirable environments. Due to the fact that a human being is homeothermal, the human body regulates temperature in narrow limits around 37 °C. The human body regulates the core temperature by vasomotoric actions, muscle work, general behaviour, sweat production and metabolic heat production.[29] Physical activity and a high temperature of the environment cause an increase in core temperature. Temperature around 38 °C is typical for moderate work. For heavy exercise values up to 39 °C are observed. The amount of heat generated by the human body is determined by metabolic activity. At rest the human body releases the amount of heat needed for the body's basic functions that is equal to about 100 W, and produces little sweat. During strenuous exercise the heat produced by physical muscular activity adds quickly to the basic metabolic heat production and at strenuous work the values over 1000 W are observed.[30] The increased heat must be released into the ambient environment. It can be done either by higher blood flow through extremities or through evaporation of liquid sweat on the skin. Due to the body's ability to sweat, large amounts of heat can be dissipated from the body: 1 dm^3 of sweat that evaporates on the skin corresponds to approximately 670 W heat which is withdrawn from the body.[29]

Physiological regulation of the body temperature is partially or completely inhibited by clothing. Clothing, an intermediate medium between the skin and the ambient environment, provides resistance to heat and to evaporation of perspiration escaping from the human body. The presence of sweat on the skin is a significant discomfort factor. Typically, a skin temperature between 33 and 35 °C and no deposition of liquid sweat on the skin surface are associated with thermal comfort.[31,32]

Fabrics containing PCMs appear to be effective in contributing to apparel comfort by buffering and reducing overheating, the cause of perspiration. The fabric with PCM reacts immediately to changes in environmental temperature and adapts to the prevailing hot or cold conditions. When temperature rise occurs as a result of body activity or a higher environmental temperature, PCM reacts by absorbing the heat. Storing this surplus energy the PCM liquefies. This phase change produces a temporary cooling effect in the clothing layers. Once the PCM has completely melted, the storage of heat stops. The PCM releases the stored heat with a drop of environmental temperature or when a body is at rest, and a temporary warming effect occurs in the clothing layers. This heat exchange produces a buffering effect in clothing layers, minimising changes in skin temperature. The PCM incorporated into clothing layers should be able to operate at or near the temperature of human skin. This temperature is different for different parts of the body.[29] The average head temperature is 35 °C, the average abdominal skin temperature is 34–35 °C, whilst the average temperature for hands and feet is 31–32 °C. It is therefore important to choose the PCM with the relevant melting point.

The garments made up from fabrics with incorporated microPCM are intended to moderate skin temperature when wearers experience varying levels of activity. The clothing layer containing microPCM has to go through the transition temperature range before the PCM will change phase and either generate or absorb heat. The PCMs have no effect under steady state conditions. In the last decade the fabrics containing microPCM have appeared in garments sold by well-known companies. However, there is little published work on thermal performance of the garments made up from PCM fabrics.

Shim and McCullough[33] investigated the effects of PCM on heat and moisture transfer in clothing during sensible temperature transients. In their study they used an open-cell, hydrophilic, expanded PU foam containing an optimum level of 60% microPCM. They measured the effect of one and two layers of PCM clothing materials on reducing the heat loss or gain from a thermal manikin as it moved from a warm chamber to a cold chamber and back again. The results indicated that the heating and cooling effects lasted approximately 15 minutes, and that the heat release by microPCM in a cold environment depended on the number of PCM layers, their orientation to the body and the amount of body surface area covered by PCM garments. They concluded that microPCM can produce small, temporary heating and cooling effects in clothing layers when the temperature of layers reaches the PCM transition temperature. They supposed that the effect of phase change materials will probably be maximised when the wearer is repeatedly going through temperature transients (i.e. going back and forth between a warm and cold environment) or intermittently touching or handling cold objects.

Ghali et al.[34] analysed the effect of microPCM on the thermal performance of fabric during periodic ventilation. The results obtained indicated that the presence of microPCMs in fabric causes a temporary heating effect when subjected to a sudden change from a warm environment to a cold environment. This effect is revealed in a decrease in the sensible heat loss during the phase change process compared to a fabric without microPCM. The duration interval of the phase change process decreases with increased frequency of ventilation, while duration increases with an increased percentage of PCM in fabric and with an increased environmental temperature.

Li and Zhu[35] presented the mathematical model of heat and moisture transfer in porous textiles containing microPCM. On the basis of a finite difference scheme, the thermal buffering effect of PCM is simulated. With specification of initial and boundary conditions, the distributions of temperature, moisture concentration and water content in the fibres can be numerically computed for different amounts of PCM in textiles. They compared the predictions of temperature changes during combined moisture and temperature transients with experimental measurements and they found reasonable agreement. In their opinion this model can be used in the design of new fabrics containing microPCM and intelligent clothing products.

4.4.4 Manufacture of textiles containing microPCMs

In applications of PCM technology to textile industry, microcapsules containing selected PCM can be applied to fibres in a wet-spinning process, incorporated into foam or embedded into a binder and applied to fabric topically, or contained in a cell structure made of a textile reinforced synthetic material.

Incorporation of microPCMs into fibres

In 1988 Bryant and Colvin[2] presented the conceptual feasibility of fibres with microencapsulated PCMs such as alkyl hydrocarbons (e.g. eicosane or octadecane) integrally incorporated into the matrix of fibres during manufacturing. Plastic crystalline materials such as DMP (2,2-dimethyl-1,3-propanediol) and HMP (2-hydroxymethyl-2-methyl-1,3-propanediol and the like may also be used. In manufacturing the fibre, the selected PCM microcapsules are added to the liquid polymer or polymer solution, and the fibre is then expanded according to the conventional methods such as dry or wet spinning of polymer solutions and extrusion of polymer melts.

Fabrics can be formed from the fibres containing PCM by conventional weaving, knitting or nonwoven methods, and these fabrics can be applied to numerous clothing applications. The advantages of incorporating microPCM into fibres are as follows:

- the micro PCMs are permanently locked within the fibres
- the fibre is processed with no need for variations in yarn spinning, fabric knitting or dyeing
- properties of fabrics (drape, softness, tenacity, etc.) are not altered in comparison with fabrics made from conventional acrylic fibres.

At present, microencapsulated PCMs have been incorporated only into acrylic fibres (trade name Outlast).[8,36] Figure 4.4 shows microphotographs of Outlast fibre. The microcapsules incorporated into the spinning dope of acrylic fibres have an upper loading limit of 5–10% because the physical properties of the fibres begin to suffer above that limit, and the finest fibre available is about 2.2 dtex. Due to the small content of microcapsules within the fibres their thermal capacity is rather modest, about 8–12 J/g.

The incorporation of microPCMs into acrylic fibres is readily accomplished due to the wet solution process of forming acrylic fibres. However, it is more difficult to incorporate PCMs into melt-spun fibres, because during a melt-

Intelligent textiles with PCMs   45

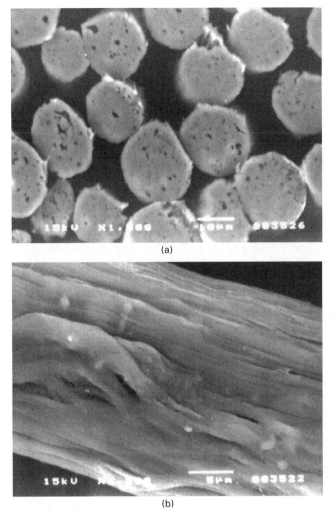

4.4 Microphotographs of Outlast fibre. (a) cross-section, (b) longitudinal view.

spinning process temperatures involved are typically in the range 200–380 °C and pressures may be as high as $2 \cdot 10^4$ kN/m^2. Such processing conditions may damage the microcapsules shell and induce degradation of PCM. Degradation of PCM may lead to inadequate thermal properties. Nevertheless, scientists continue trying to develop melt-spun fibres with microPCMs. The process of forming the polypropylene fibres containing alkyl hydrocarbons on a laboratory spin draw device was studied by Leskovsek *et al.*[37] Microcapsules 2 μm in diameter were chosen as the most appropriate for spinning. The shell of the microcapsules was melamine-formaldehyde resin, while the alkyl hydrocarbon with a melting point of 50 °C (tetracosane) constituted the core. The key problem was achieving the homogeneous distribution of microcapsules into fibres. Microcapsules showed the potential to form clusters that became bigger at the fibre spinning and consequently caused discontinuity of the process, filament breaks and enlarged the thickness of fibres.

Lamination of PU foam containing microPCMs onto a fabric

Colvin and Bryant[3] described a method of fabrication of the foam insulation materials containing microPCMs. The selected microPCMs are added to the liquid polymer or elastomer and mixed therein to ensure wetting and equal dispersion throughout the mixture. After mixing the microcapsules will be wetted and substantially spaced apart from each other. Typical concentrations of microPCMs range from 20% to 60% by weight. Next, the base polymeric material is foamed. Common methods of foaming include adding a hardening agent which causes a chemical reaction, thermally setting the base material with heat, or bubbling a gas through the liquid polymer (or elastomer) while hardening. Microcapsules should be added to the liquid polymer or elastomer prior to hardening.

After foaming microcapsules will be embedded within the base material matrix so that they are individually encapsulated and embedded within the base material and the space between neighbouring adjacent microcapsules will be base material, not the foaming gas. The foam pad containing PCMs may be fabricated from neoprene or polyurethane. The application of the foam pad is particularly recommended because:

- a greater amount of microcapsules can be introduced
- different PCMs can be used, giving a broader range of regulation temperatures
- microcapsules may be anisotropically distributed in the layer of foam. The possibility of anisotropic insertion of the microcapsules in the foam can reinforce the thermoregulation effect by concentrating PCMs towards the interior of the clothing.

The foam pad with microPCMs may be used as a lining in a variety of clothing: gloves, shoes, hats, outerwear. Before incorporation into clothing or footwear the foam pad is usually attached to the fabric, knitted or woven, by any conventional means such as glue, fusion or lamination. Figure 4.5 shows the cross-section of the knitted fabric laminated with polyurethane foam containing PCM microcapsules.

Incorporation of microPCMs to fibrous structure by different processes

In the last decade several methods of incorporating microPCMs into fibrous structures were developed to produce fabrics having enhanced thermal

Intelligent textiles with PCMs 47

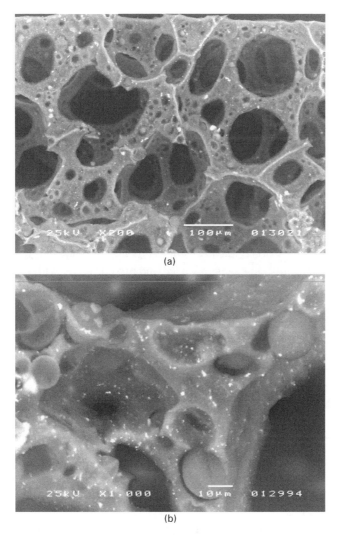

4.5 Cross-section of a PU foam containing microPCM.
(a) magnification 200×, (b) magnification 1000×.

properties. A number of experimental conditions should be optimised before these methods are developed to produce fabrics with the desirable structure and properties to meet practical applications. Most work in this field can be found in patent literature and only a few papers in published literature report the formulation of PCM microcapsules, finishing of fabrics and the evaluation of their characteristics, including their thermal properties and durability.

The PCM microcapsules are applied to a fibrous substrate using a binder (e.g. acrylic resin). All common coating processes such as knife over roll, knife over air, screen-printing, gravure printing, dip coating may be adapted to apply the PCM microcapsules dispersed throughout a polymer binder to fabric. The conventional pad–mangle systems are also suitable for applying PCM microcapsules to fabrics. The formulation containing microPCMs can be applied to the fabric by the direct nozzle spray technique.

The method for manufacturing a coating composition was described by Zuckerman et al.[10] A coating composition for fabrics includes wetted microcapsules containing PCMs dispersed throughout a polymer binder, a surfactant, a dispersant, an antifoam agent and a thickener. As a result of experiments Zuckerman has found that wetting PCM microcapsules with water and maintaining a uniform dispersion of the microcapsules in a wet coating minimises the tendency of such microcapsules to destabilise the binder polymer. A coating composition is prepared by mixing dry microcapsules with water to induce the microcapsules to swell. A surfactant and a dispersant are added to the water prior to mixing with the microcapsules. The surfactant decreases surface tension of the layers of microcapsules and thereby promotes their wetting. An antifoam agent is added and mixed slowly with the mixture to remove air trapped as dispersed bubbles in the mixture. A thickener is added to adjust the viscosity of the mixture and to prevent the microcapsules from floating or sinking in the mixture. Adjusting the pH of the mixture to 8.5 or greater promotes swelling of microcapsules. Swelling is typically completed within 6 to 24 hours. Thereafter, the microcapsules dispersion is added to a mixture of a polymer dispersion, surfactant and dispersant having a pH approximately the same as the pH of the microcapsules dispersion. The polymeric binder may be in the form of a solution, dispersion or emulsion. The most frequently used are acrylic resins or acrylic/butadiene copolymer.

The most preferred ratios of components of the coating composition are:

- 70 to 300 parts by dry weight of microcapsules for each 100 parts by dry weight of acrylic resin
- 0.1 to 1% dry weight of each surfactant and dispersant to dry weight of microcapsules
- water totalling 40% to 60% of the final wet coating composition
- antifoam agent 0.1% to 0.5% of the final wet coating composition.

The use of polymeric binders has some drawbacks. The amount used should be enough for good fixation of the microcapsules, but the properties of fabrics such as drape, air permeability, breathability, thermal resistance, softness and tensile strength can be affected adversely as the percentage of binder add-on increases. Pushaw[9] reported that it was difficult to maintain durability, moisture vapour permeability, elasticity and softness of coated fabrics when the coating was loaded with a sufficiently high content of microPCM.

Choi et al.[19] synthesised, by the interfacial polymerisation method, the melamine-formaldehyde shell on microspheres of octadecane. The mean diameter of the microcapsules ranged from 1 to 1.5 µm and their shapes were

Intelligent textiles with PCMs   49

almost spherical. The microcapsules were mixed with acrylic binder and urethane binder and the coating mixture was applied on polyester fabrics by the knife-over-roll and screen printing methods. They found that thermal properties of treated fabrics increased as microcapsules concentration increased, but there was no great difference between these two adhesive methods. The thermal properties of coated fabrics, analysed by DSC, were rather poor, because latent heat of fabric samples hardly reached 7.5 J/g. It is obvious that shear stiffness and bending rigidity of coated fabrics were greater compared with printed fabrics.

Chung and Cho[20] investigated the possibilities of producing thermally adaptable, vapour-permeable and water-repellent fabrics. Microcapsules containing octadecane were added to polyurethane and the coating mixture was dissolved in DMF (dimethyl formamide). This was coated onto the 100% nylon fabrics. The average diameter of microcapsules ranged from 3 to 5 µm. Due to the greater volume of microcapsules their latent heat was higher and the latent heat of the coated fabric was about 14 J/g. This amount of latent heat was maintained after 30 launderings. The water repellence of a fabric coated with microPCMs was the same as for a coated fabric without PCM, but the water resistance decreased dramatically.

Shin et al.[18] prepared melamine–formaldehyde microcapsules containing eicosane by interfacial polymerisation. The mean diameter of the microcapsules was 1.89 µm and most of the microcapsules had a particle size of 0.1–10 µm. The core to shell ratio was about 53% and the heat storage capacity of the microcapsules was 143 J/g. The prepared microcapsules were added to polyester knitted fabrics by the conventional pad–dry–cure process. The treated fabrics had heat storage capacities of 0.91–4.41 J/g, depending on the addition of the microcapsules. After five launderings the treated fabrics retained 40% of their heat storage capacities. The results suggest that microcapsules with a higher core-to-shell ratio should be used to improve the thermoregulating properties of fabrics.

Lotenbach and Sutter[38] described a novel method of screen-printing on textile surfaces during which the print is provided in the form of naps containing a high concentration of microPCM. In their opinion, the accumulation of the PCM in the form of a number of naps largely preserves the elasticity as well as the substrate's ability to exchange vapour and moisture. Figure 4.6 shows the SEM picture of a nonwoven sample containing microPCM applied by the screen-printing technique. The PCM microcapsules occur on one side only, forming a very thin layer of 0.10–0.14 mm. In the sample of nonwovens prepared by the spraying method (Fig. 4.7) microPCMs occur on both sides in thin layers ca. 1.4 mm thick. In the samples of nonwoven prepared by pad-mangle method (Fig. 4.8) microPCMs occur in the entire cross-section of tested nonwovens.

50 Intelligent textiles and clothing

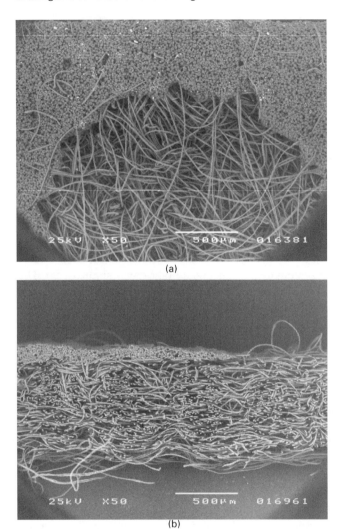

4.6 Microphotographs of nonwoven sample with microPCM incorporated by screen printing. (a) surface of the nonwovens; (b) cross-section of the nonwovens.

4.4.5 Applications of textiles containing PCMs

Fabrics containing microPCMs have been used in a variety of applications including apparel, home textiles and technical textiles. Some exemplary applications are presented below.

Intelligent textiles with PCMs 51

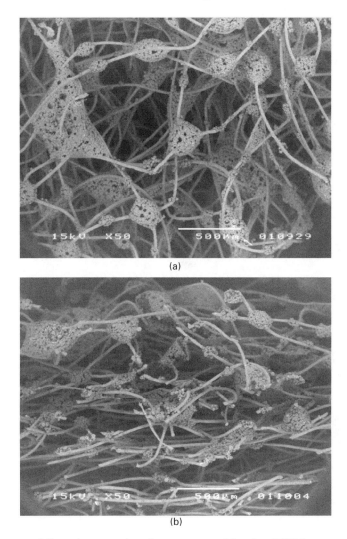

(a)

(b)

4.7 Microphotographs of nonwoven with microPCM incorporated by spraying. (a) surface of the nonwoven, (b) cross-section of the nonwoven.

Apparel

Major end-use areas include:

- life style apparel – smart jackets, vests, men's and women's hats, gloves and rainwear
- outdoor activewear apparel – jackets and jacket lining, boots, golf shoes, trekking shoes, socks, ski and snowboard gloves
- protective garments.

52 Intelligent textiles and clothing

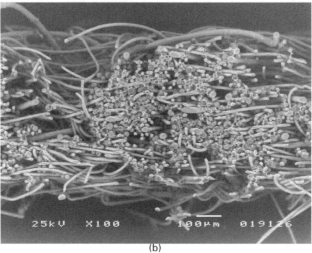

4.8 Microphotographs of nonwoven with microPCM incorporated by padding (a) surface of the nonwoven, (b) cross-section of the nonwoven.

In protective garments PCMs functions are as specified:

- absorption of body heat surplus
- insulation effect caused by heat emission of the PCM into the fibrous structure
- thermo-regulating effect, which maintains the microclimate temperature nearly constant.

Pause[39] described the application of PCMs in nonwoven protective garments used in pest control or the treatment of hazardous waste. In order to improve the wearing comfort of these protective garments, PCM was incorporated into a thin polymer film and applied to the inner side of the fabric system via lamination. The test results indicated that the cooling effect of the PCM can delay the temperature rise and hence limit the moisture rise in the microclimate. As a result, the wearing time of the garment can be extended without the occurrence of heat stress as a serious health risk.

Safety helmets have a thermal resistance of approximately 1.0 m^2 k/W and due to their structure the heat generated by the wearer can be dissipated only by means of convection. The results of tests carried out by Weder and Herring[40] indicated that an incorporation of microPCM in the helmet liner leads to substantial reduction of the microclimate temperature in the head area.

In the case of chemical or biological protective clothing a conflict between the protective function of clothing and the physiological regulation of body temperature may occur. The conflict led to discomfort and physical strain and in extreme cases can put the person at risk from heat stress. Colvin and Bryant[12] developed and patented microclimate cooling apparel for the military and civilians that can be used beneath protective garments to provide significant microclimate cooling for 1–3 hours under unusually high heat conditions. This cooling apparel uses macroenapsulated PCMs uniformly distributed within lightweight vests, helmet liners, cowls and neck collars. The diameter of macrocapsules ranges from 2 to 4 mm. The macrocapsules containing octadecan change phase at 26–28 °C. Other temperatures can also be selected by changing macroPCMs or by using mixtures of them to optimise their performance with different environmental conditions. The macrocapsules can be recharged without refrigeration.

Recently, the US Navy has investigated microPCMs inside divers' dry suits to thermal protection in extremely cold water applications. Nuckols[41] developed an analytical model of a dry suit system to predict its thermal performance in simulated ocean environments. The results of this study indicated that the foam with embedded microPCMs can reduce diver heat loss during the initial phase of dive by releasing the latent heat in microcapsules during the cold exposure. The result of this release of latent energy is lower temperature gradients on the inside of the suit. The reduction of diver heat loss will continue until the PCM in microcapsules solidifies, at this point the foam will behave as conventional suit insulation.

Domestic textiles

Employment of domestic textiles containing microPCM leads to an improvement of the thermal comfort of dwellings. Blinds and curtains with microPCMs can be used for reduction of the heat flux through windows. In the summer months large amounts of heat penetrate the buildings through windows during the day. At night in the winter months the windows are the main source of thermal loss. Results of the test carried out by Pause[42] on curtains containing microPCM have indicated a 30% reduction of the heat flux in comparison to curtains without PCM.

In rooms there is usually a temperature difference between floor and ceiling. During the heating period, this temperature difference can exceed 5 °C. The comfort sensation of human beings is especially dependent on the temperature gradient between floor and ceiling. The higher the temperature gradient the greater the feeling of discomfort. PCM can be used to reduce the temperature gradient between floor and ceiling and keep it constant over a specific period. By using floor coverings laminated with PU foam containing microPCM it is possible to improve the room climate.[42]

PCM is also used to improve the thermal comfort of upholstery products. This idea has been used in the development of the new office chair. While sitting in a chair, the heat flux from the body through the seat into the environment is essentially reduced leading to a rapid temperature increase in microclimate. The moisture transfer from the body through the seat into the environment is also reduced resulting in a substantial moisture increase in the microclimate. The cooling effect provided by the heat absorption of the PCM incorporated in the chair cushion leads to the lower temperature increase in the microclimate above the chair cushion and the reduction of the microclimate moisture content.[42]

The microencapsulated PCM can also be used to improve the thermal comfort of bedding, e.g., mattresses, mattress pads, pillows, blankets. Rock and Sharma[43] developed heating/warming textile articles with phase change components. A fibrous electric blanket with electrical resistance heating elements also contains a phase change component which releases and absorbs latent heat in cycles corresponding to on/off operation of a power source thus saving energy. The heating element is formed of a conductive yarn. Upon application of electrical power to the heating elements, heat is generated to increase the temperature within the blanket. During this 'on' period of heat generation the PCM incorporated into the blanket is also caused to change phase from solid to liquid. When a desired temperature is achieved, application of electrical power is discontinued. During this 'off' period heat is released from the blanket. The rate of heat loss – cooling action – is delayed by the release of latent heat by the PCM as it changes phase with cooling, from liquid back to solid.

Medical products

Buckley described[5] a therapeutic blanket made of a flexible PCM composite. If such a blanket contains a microPCM having a transition temperature below normal skin temperature, it can be used for cooling febrile patients in a careful and controlled manner. A careful selection of the phase change temperature makes it possible to avoid the danger of overcooling the patient that is inherent with ice packs. Alternatively, a blanket with PCM can be useful for gently and controllably reheating hypothermia patients. Another therapeutic medical use of the PCM is to incorporate it into elastic wraps or orthopaedic joint supports. Thus hot or cold therapy can be combined with supporting bandage for joints or muscles.

Automotive textiles

Barbara Pause demonstrated that there are substantial benefits in exploiting PCMs to moderate the temperature and comfort of the interior of an automobile.[44] Studies were carried out in order to select the appropriate PCMs, the necessary amounts, and a suitable location in the passenger compartment.[45] Results of these studies show that by using microPCMs containing hydrocarbons integrated into textiles, particularly in headliner and seating materials, interior temperature reductions of 2–4 °C can be obtained, along with less moisture build-up, reducing the load on the air-conditioning unit and thereby saving energy.

Air-conditioning buildings with PCM

Recently PCMs have been studied for application to solar thermal storage and air conditioning in domestic buildings. By PCM applications in coatings for textiles used in roof covering, the thermal insulation value may be greatly enhanced. After the PCM has absorbed the surplus heat during the day it can be recharged by the overnight cooling effect. Pause described[42] a special panel system with PCM, which can be used for increasing the thermal resistance of lightweight construction walls. The main element of this panel is the cell structure, which is made of a textile reinforced material and filled with PCM. The quantity of the PCM contained in the panel is equivalent to a thermal storage of 700 kJ. Computer simulations have indicated that application of this panel can provide energy savings of about 20%.

4.5 Measurement of thermoregulating properties of fabrics with microPCMs

Fabrics containing PCM microcapsules present a unique challenge to the standard test procedures used for determining the thermal properties of fabrics. In the case of traditional fabrics, the thermal properties are investigated by standard steady state procedures involving the use of guarded hot plate apparatus. Steady state procedure is inadequate in assessing the dynamic performance of the fabrics containing PCMs, because PCM is a highly productive thermal storage medium. Hittle and Andre[46] formulated a model of heat transfer through a textile containing PCM, and based on this model the new test instrument and testing procedure has been developed. In 2004 ASTM standardised this test method.[47]

4.5.1 Heat transfer through textiles containing PCM

Heat conduction through a one-dimensional homogeneous material is governed by the following second-order partial differential equation:

$$\frac{\partial^2 T(x, t)}{\partial x^2} = \frac{1}{\alpha}\frac{\partial T(x, t)}{\partial t} \qquad 4.1$$

where:
T = temperature at position x
t = time
α = thermal diffusitivity $\alpha = \frac{k}{\rho C_p}$
k = thermal conductivity (W/m K)
ρ = density (kg/m^3)
C_p = specific heat (J/kg K).

The heat flux at position x and time t is given by eqn 4.2:

$$q(x, t) = -k\frac{\partial T(x, t)}{\partial x} \qquad 4.2$$

Hittle and Andre assumed that in both equations k, ρ i C_p are constant. Additionally, they assumed that a constant and comparably large C_p in the temperature region of the phase change is a reasonable approximation for the energy storage of the PCM in the fabric. This assumption leads to a convenient metric called temperature regulating factor (TRF).

Hittle and Andre gave a solution to the above equations for sinusoidal boundary conditions. In order to characterise the thermoregulation effect they have proposed the use of the quotient of the amplitude of the temperature variation and the amplitude of the heat flux variation present in this solution. The smaller the quotient the better the regulation effect.

Dividing this quotient by the value of the steady state thermal resistance of the fabric (R) they obtained TRF value:

$$TRF = \frac{(T_{max} - T_{min})}{(q_{max} - q_{min})}\frac{1}{R} \qquad 4.3$$

TRF is a dimensionless number varying in range (0,1). TRF shows how well a fabric containing microPCM moderates the hot plate temperature. A TRF value of 1, means the fabric has no capacitance and poor temperature regulation.

If TRF equals zero, it means that the fabric has infinite capacitance and that a body being in contact with it will remain at a constant temperature. It is obvious that all fabrics fall somewhere between these extreme values. TRF is a function of the frequency of the sinusoidal variation of the heat flux into the hot plate. The temperature regulation increases with increasing frequency. Hence TRF increases with increasing cycle time of sinusoidal variation, going exponentially from 0 to 1.

4.5.2 Principle of measuring

Determination of the temperature regulating factor (TRF) of apparel fabrics is done by means of the instrument which uses a dynamic heat source (Fig. 4.9). This instrument simulates an arrangement: skin–apparel–environment. The fabric sample is sandwiched between a hot plate and two cold plates, one on either side of the hot plate. These cold plates at constant temperature simulate the environment outside the apparel. Sinusoidally varying heat input to the hot plate simulates human activity. To measure the steady state thermal resistance of the fabric (R), the controlled heat flux is constant and the test

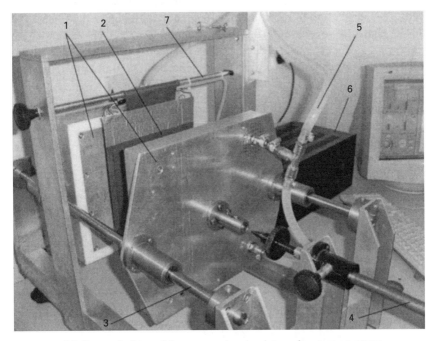

4.9 General view of the apparatus to determine temperature regulating factor (TRF). 1, cold plates; 2, hot plate; 3, guide bars; 4, cold plates pressure adjustment; 5, cooling water supply to cold plates; 6, thermostat with Peltier cells; 7, sample holder.

proceeds until steady state is reached. To assess temperature regulating ability, the heat flux is varied sinusoidally with time and the temperature regulating factor (TRF) is determined. This is a function of the frequency of the sinusoidal variation of the heat flux into the hot plate. The temperature regulation increases with increasing frequency. Hence TRF increases with increasing cycle time of sinusoidal variation, going exponentially from 0 to 1.

Examples

Since 2002 a systematic study involving nonwoven fabrics containing PCMs has been conducted at Instytut Włókiennictwa (Textile Research Institute, Poland). In this study microencapsulated octadecane and eicosane was used. MicroPCMs dispersed in acrylic-butadiene copolymer were applied on polyester hydroentangled nonwovens by the screen-printing method or the pad-mangle method.

Figure 4.10 shows TRFs as a function of cycle time for nonwoven samples treated by the pad-mangle method. Tested nonwoven samples differ in microPCM mass contained in one square metre of the nonwoven; this means they differ in thermal capacitance. The diagram also shows, for comparison needs, the curve of reference nonwoven, i.e., not containing microPCM. We can see that nonwovens with microPCM exhibit lower TRF values in the whole range of cycle times of heat flux changes when compared to the reference nonwoven. This is due to the increase of nonwoven thermal

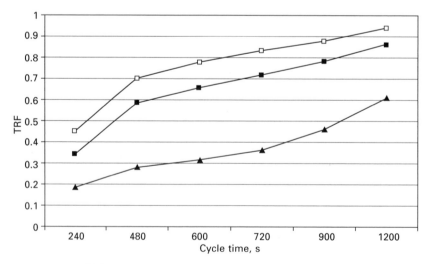

4.10 TRF as a function of cycle time (padded nonwoven samples). ☐ NP41/0 nonwoven sample without microPCM; ▲ NP41/95 padded nonwoven sample with microPCM (24.3% microPCM); ■ NP15/95 padded nonwoven sample with microPCM (36.1% microPCM).

Intelligent textiles with PCMs 59

Table 4.2 Characteristics of tested nonwovens

Sample	Area weight g/m^2	Thickness mm	Percentage of microPCM (on weight) %	Latent heat/m^2 of nonwovens kJ/m^2
NP41/0	97	0.78	0	–
NP41/95	133	0.78	24.3	6.0
NP15/95	182	0.83	36.1	12.0

capacitance resulting from the incorporation of microPCM. It can be observed that with the increase in microPCM mass contained in one square metre of nonwoven there occurs an improvement of thermoregulating properties defined by the lower TRF value.

Figure 4.11 shows the results of determination of the temperature regulating factor for the assemblies of nonwovens. The whole assembly was composed of two nonwovens:

1. hydroentangled nonwoven with microPCM incorporated by screen-printing, denoted DR1/95
2. needled nonwoven without microPCM, denoted NAdg.

Measurements of TRF performed on the assembly: needled nonwoven without microPCM + printed nonwoven indicated that the location of the surface

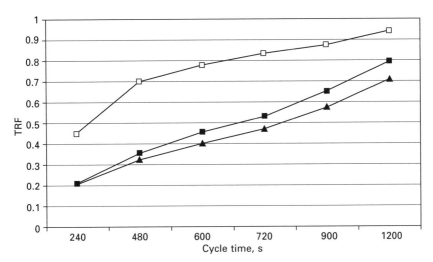

4.11 TRF as a function of cycle time for nonwoven assemblies. □ nonwoven assembly without PCM (DR1/0 + Nadg); ■ nonwoven assembly DR1/95 + Nadg, printed side in contact with hot plate; ▲ nonwoven assembly DR1/95 + Nadg, nonprinted side in contact with hot plate (Nadg – needled nonwoven without PCM).

covered with a microPCM layer towards the hot plate has a substantial influence on the TRF value. When a printed surface contacts the hot plate TRF values are higher than in the case where a non-printed surface contacts the hot plate. The examples above show that the new testing method can provide information for fabric and garment designers and can be useful in quality control during manufacture of fabrics with microPCMs.

4.6 Summary

It is evident from the literature reviewed that studies on the application of microPCM technology in the textile industry and properties of textiles containing microPCMs are at the beginning. Fabrics containing PCMs are particularly attractive to the apparel industry because of their ability to improve clothing comfort. Preliminary studies clearly demonstrate the suitability of fabrics with PCMs for protective clothing and home textiles. To assess the impact of PCMs for specific applications a basic study on heat and moisture transfer in textiles with PCMs has to be continued. Probably over the next decade many new products based on microPCMs technology will be produced and commercialised.

4.7 Acknowledgements

The author would like to thank Mr Henryk Wrzosek (Technical University, Łódź) for preparing SEM microphotographs, as well as the research team, which has contributed to the experimental results described. The author also appreciates the contribution of Ms Katarzyna Grzywacz in creating the manuscript for this chapter. For financial support, the author is grateful to the Polish Ministry of Scientific Research and Information (research project No. 4 T08E 063 23).

4.8 References

1. Vigo T.L., "Intelligent fibrous materials", *J. Text. Inst.*, 1999, **90**, (3), 1–13.
2. Bryant Y.G., Colvin D.P., "Fibre with reversible enhanced thermal storage properties and fabrics made therefrom", US Patent 4756 958, 1988.
3. Colvin D.P., Bryant Y.G., "Thermally enhanced foam insulation", US Patent 5 637 389, 1996.
4. Bryant Y.G., Colvin D.P., "Fabric with reversible enhanced thermal properties", US Patent 5 366 807, 1994.
5. Buckley T., "Phase change thermal control materials, method and apparatus", US Patent 6 319 599, 2001.
6. Clark T., McKervey M.A., "Saturated hydrocarbons" in *Comprehensive organic chemistry* vol. 1, 38-111, ed. by Stoddart J.F., Pergamon Press Ltd., Oxford 1979.
7. Herbig J.A., "Microencapsulation" in *Encyclopaedia of Polymer Science and Technology*, vol. 8, 719-736, New York 1968.

8. Cox R., "Synopsis of the new thermal regulating fibre Outlast", *Chemical Fibers Int.*, 1998, **48**, 475–479.
9. Pushaw R.J., "Coated skived foam and fabric article containing energy absorbing phase change material", US Patent 5 677 048, 1997.
10. Zuckerman J.L., Pushaw R.J., Perry B.T., Wyner Daniel M., "Fabric coating containing energy absorbing phase change material and method of manufacturing same", US Patent 6 514 362, 2003.
11. Colvin V.S., Colvin D.P., "Microclimate temperature regulating pad and products made therefrom", US Patent 6 298 907, 2001.
12. Colvin D.P., Bryant Y.G., "Microclimate cooling garment", US Patent 5 415 222, 1995.
13. Mulligan J.C., Colvin D.P., Bryant Y.G. "Microencapsulated phase-change material suspensions for heat transfer in spacecraft thermal systems", *J. of Spacecraft and Rocket*, 1996, **332**, (2), 278–284.
14. Brown R.C., Rasberry J.D., Overmann S.P., "Microenapsulated phase-change materials as heat media in gas fluidised beds", *Powder Technology*, 1998, **98**, 217–222.
15. Yamagishi Y., Takeuchi H., Pyatenko A.T., Kayukawa N., "Characteristics of microencapsulated PCM slurry as a heat-transfer fluid", *AIChE Journal*, 1999, **45** (4), 696–707.
16. Cho J., Kwon A., Cho Ch., "Microencapsulation of octadecane as phase change material by interfacial polymerisation in an emulsion system", *Colloid Polym. Sci.*, 2002, **280**, 260–266.
17. Zhang X.X., Tao X.M., Yick K.L., Wang X., "Structure and thermal stability of microencapsulated phase change materials", *Colloid Polym. Sci.* 2004, **282**: 330–336.
18. Kim J.H., Cho G.S., "Thermal storage/release, durability, and temperature sensing properties of thermostatic fabrics treated with octadecane-containing microcapsules", *Textile Res. J.* **72**(12), 1093–1098, 2002.
19. Choi K., Cho G., Kim P., Cho CH., "Thermal storage and mechanical properties of phase change materials on polyester fabrics", *Textile Res. J.* 2004, **74**(7), 292–296.
20. Chung H., Cho G., "Thermal properties and physiological responses of vapour permeable water repellent fabrics treated with microcapsule containing PCMs", *Textile Res. J.* 2004, **74** (7), 571–575.
21. Hartmann M.H., "Stable phase change materials for use in temperature regulating synthetic fibres, fabrics and textiles", US Patent 6 689 466, 2004.
22. Vigo T.L., Frost C.M., "Temperature-sensitive hollow fibres containing phase change salts", *Textile Res. J.*, 1982, **52** (10), 633–637.
23. Vigo T.L., Frost C.M., "Temperature-adaptable textile fibres and method of preparing same", US Patent 4 871 615, 1989.
24. Vigo T.L., Frost C.M., "Temperature-adaptable fabrics", *Textile Res. J*, 1985, **55** (12), 737–743.
25. Vigo T.L., Bruno J.S., "Temperature-adaptable textiles containing durably bound polyethylene glycols", *Textile Res. J.*, 1987, **57** (7), 427–429.
26. Harlan S.L., "A new concept in temperature-adaptable fabrics containing polyethylene glycols for skiing and skiing-like activities", Chapter 15 in *High-Tech Fibrous Materials* eds Vigo T.L., Turbak A., ACS, Washington DC, 1991.
27. Pause B., "Thermal insulation effect of textiles with phase change materials", *Technical Tex. Int.*, 1999, (9), 21–26.
28. Pause B., "Computer-optimized design of thermo-regulated active wear", *Technical Tex. Int.*, 2001, (11), 23–26.

29. Havenith G., "An individual model of human thermoregulation for the simulation of heat stress response", *Journal of Applied Physiology*, 2001, **90**: 1943–1954.
30. Holmes D.A., "Performance characteristics of waterproof breathable fabrics", *J. of Ind. Tex.*, 2000, **29** (4), 306–316.
31. DeMartino R.N., "Improved comfort polyester. Part III. Wearer trials", *Text. Res. J.*, 1984, **44** (5), 447–458.
32. Slater K., "Comfort properties of textiles", *Textile Prog.*, 1977, **9**, (4).
33. Shim H., McCullough E.A., Jones B.W., "Using Phase Change Materials in Clothing", *Textile Res. J.*, 2001, **71** (6), 495–502.
34. Ghali K., Ghaddar N., Harathani J., Jones B., "Experimental and numerical investigation of the effect of phase change materials on clothing during periodic ventilation", *Textile Res.J.*, 2004, **74** (3), 205–214.
35. Li Y., Zhu Q., "A model of heat and moisture transfer in porous textiles with Phase Change Materials", *Textile Res. J.*, 2004, **74** (5), 447–457.
36. Cox R., "Repositioning acrylic fibres for the new millennium", *Chemical Fib. Int.*, 2001, **51** (2), 118–120.
37. Leskovsek M., Jedrinovic G., Stankovic-Elesin U., "Spinning polypropylene fibres with microcapsules", Proceedings of 2nd ITC & DC, ed. Faculty of Textile Technology, University of Zagreb, 2004, 84–88.
38. Lottenbach R., Sutter S., "Method for producing temperature-regulating surfaces with phase change material", Patent WO 02095314, 2002.
39. Pause B., "Nonwoven Protective Garments with Thermo-regulating Properties", *J. of Industrial Tex.*, 2003, **33** (2), 93–99.
40. Weder M., Hering A., "How effective are PCM materials? Experience from laboratory measurements and controlled human subject tests", 39th International Man-Made Fibres Congress, 13–15.09.2000, Dornbirn/Austria.
41. Nuckols, M.L., "Analytical modeling of a diver dry suit enhanced with micro-encapsulated phase change material", *Ocean Engineering*, 1999, **26**, 547–564.
42. Pause B., "Possibilities for air-conditioning buildings with Phase Change Material", *Technical Tex. Int.*, 2001, **44** (1), 38–40.
43. Rock M., Sharma V., "Heating/warming textile articles with phase change components", US Patent 6 723 967, 2004.
44. Pause B., "Thermal control of automotive interiors with phase change material", Patent WO 02083440, 2002.
45. Pause B., "Driving more comfortably with phase change materials", *Technical Tex. Int.*, 2002, **45** (3), 24–28.
46. Hittle D.C., Andre T.L., "A new test instrument and procedure for evaluation of fabrics containing phase change materials", *ASHRAE Trans.* 2002, **107**, (1), 175–182.
47. ASTM D 7024-04 "Standard Test Method for Steady State and Dynamic Thermal Performance of Textile Materials".

5
The use of phase change materials in outdoor clothing

E A MCCULLOUGH and H SHIM,
Kansas State University, USA

5.1 Introduction

Phase change materials (PCMs) store and release thermal energy as they go through solid-liquid transitions. PCMs used in textiles are combinations of different types of paraffins – each with different melting and crystallization (i.e., freezing) points. By changing the proportionate amounts of each type of paraffin in the phase change material (e.g., hexadecane, octadecane), desired melting and freezing points can be obtained. The PCMs are enclosed in a protective wrapping, or microcapsule, a few microns in diameter. The microcapsule prevents leakage of the material during its liquid phase (Bryant and Colvin, 1992). Microcapsules of phase change materials can be incorporated into the spinning dope of manufactured fibers (e.g., acrylic), incorporated into the structure of foams, and coated onto fabrics.

When the encapsulated PCM is heated to the melting point, it absorbs heat energy as it goes from a solid state to a liquid state. This phase change produces a temporary cooling effect in the clothing layers. The heat energy may come from the body (e.g., when the wearer first dons the garment) or from a warm environment. Once the PCM has completely melted, the storage of heat stops. If the PCM garment is worn in a cold environment where the temperature is below the PCM's freezing point, and the fabric temperature drops below this transition temperature, the micro-encapsulated liquid PCM will change back to a solid state, generating heat energy and a temporary warming effect. The developers claim that this heat exchange produces a buffering effect in clothing, minimizing changes in skin temperature and prolonging the thermal comfort of the wearer.

The clothing layer(s) containing PCMs must go through the transition temperature range before the PCMs will change phase and either generate or absorb heat. Consequently, the wearer has to do something to cause the temperature of the PCM fabric to change. PCMs are a transient phenomenon; they have no effect under steady-state thermal conditions.

5.1.1 Change in environmental temperature

Some people have jobs where they go to and from a cold storage facility or transport vehicle and a warm building or outside environment on an intermittent basis. PCM protective garments should improve the comfort of workers as they go through these environmental step changes (e.g., warm to cold to warm, etc.). For these applications, the PCM transition temperature should be set so that the PCMs are in the liquid phase when worn in the warm environment and in the solid phase in the cold environment.

The developers and producers of phase change materials in textiles claim that garments made with PCMs or 'dynamic insulation' will keep a person warm longer than conventional insulations when worn in cold environments. They also claim that the use of PCMs in outdoor clothing will decrease the thickness and weight of the clothing required (Pause, 1998). There is no question that a phase change material will liberate heat when it changes from a liquid to a solid. However, in order for PCMs to improve the thermal comfort characteristics of a clothing ensemble in a cold environment, they must produce enough heat in the garment layers to reduce heat loss from the body to the environment. Even if all of the garments in a clothing ensemble were treated with PCMs, not all of the PCM microcapsules would go through phase changes when the wearer went from an indoor environment to a cold environment. (See Fig. 5.1.) Heat flows from the warm body to the cooler environment, and there is a temperature gradient from the skin surface through the clothing layers to the environment. The PCMs closest to the body (e.g., in long underwear or socks) will probably remain close to skin temperature and stay in the liquid state even when the wearer moves to the colder environment. PCMs in the outermost layers of clothing will probably get

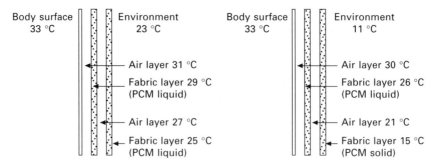

5.1 In a warm indoor environment (left), the two PCM fabric layers are above the 20°C transition temperature, and the PCMs are in the liquid state. When the wearer enters a cold environment, the temperature gradient from the warm body to the cold air changes and the temperature of the outermost fabric layer drops below 20 °C; the PCMs solidify, producing heat.

cold and solidify, thus producing some heat. However, in the outer layers of clothing, a portion of the heat would be lost to the cold environment. Once the PCMs have changed from a liquid to a solid, the liberation of heat energy is over. Consequently, a good insulation system with adequate thickness is still necessary for thermal comfort and safety during extended exposure to cold environments.

When workers wear flame resistant clothing and are exposed to high environmental temperatures for short periods of time (e.g., race car drivers), PCMs may improve their comfort and/or safety by providing a short-term cooling effect. In these circumstances, changing the PCM back to a solid and liberating heat would not be necessary or desirable. However, the addition of paraffin-based PCMs to fabrics may increase their flammability. If this becomes a problem, the PCMs could be used in garment layers under a flame resistant outer shell.

Some jobs require workers to handle hot or cold objects. In these instances, the PCM transition temperature should be set between the ambient environment and the temperature of either hot or cold objects. If the workers touch the objects repeatedly for short periods of time, (so that the PCMs have time to change back to their initial state in the ambient environment), the resulting buffering effect would continue, and thinner materials could probably be used in the gloves.

5.1.2 Change in skin temperature

In some jobs and sports (e.g., skiing), people are very active and inactive on an intermittent basis. As the activity level of a person increases, so does the amount of heat produced by the body. The companies that produce PCM garments claim that PCMs will improve the comfort of people as they go through changes in activity and metabolic heat production. They are assuming that PCM garments will absorb some of the extra body heat produced during exercise, change to the liquid state, and produce a temporary cooling effect (Cox, 1998). When the person becomes inactive, the PCMs will solidify again and liberate heat (assuming the air temperature stays below skin temperature). This has been called the 'thermal regulating effect' of PCMs.

For PCMs to work in this scenario, the transition temperature would have to be set between the highest and lowest skin temperatures resulting from changes in metabolic heat production. A person's skin temperature changes only a few degrees during exercise, even when a large amount of heat is being generated by the body. For example, when a person is downhill skiing, he/she may produce about five times more body heat than he/she would produce while sitting. However, the skin temperature does not increase by a proportionate amount. First, the blood vessels near the skin surface vasodilate, raising the skin temperature slightly, and increasing the temperature gradient

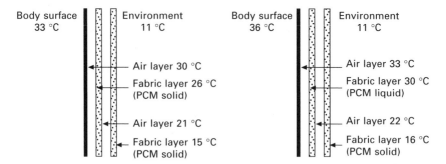

5.2 When a person is sitting in a cold environment, his skin temperature is 33 °C and the two PCM layers are below the 30 °C transition temperature. When he/she becomes active, the skin temperature rises to 36 °C and the temperature of the inner layer increases to 30 °C; the PCMs liquefy, producing a cooling effect.

for heat loss to the environment. (See Fig. 5.2.) However, this physiological response is not adequate for getting rid of the excess body heat produced during vigorous exercise. Consequently, the body begins producing sweat at the skin surface. The process of changing liquid sweat to vapor is a phase change in itself which takes heat from the body surface, providing a natural cooling effect. The sweat is the same temperature as the skin and has a minimal effect on the temperature of the clothing layers as it diffuses through them. Even if the PCMs in the innermost clothing layer liquefy and produce a cooling effect, the PCMs in the outermost layer may not.

Skin temperatures vary on different parts of the body. For example, the average temperature of the head is higher than that of the feet. In addition, the variability of skin temperatures from person to person is often greater than the change in skin temperature of one person due to exercise. Therefore, producing PCM thermal regulating effects that result from small changes in skin temperature is a difficult challenge for designers and manufacturers of protective clothing.

5.1.3 The need for research

Most of the published research that is available in the scientific literature was conducted on small pieces of fabric – not on garments – where it is relatively easy to demonstrate and quantify the heating and cooling effect (Pause, 1994, 1995, 1998; Cox, 1998). The magnitude and duration of the phase change heating and cooling effects on a clothed body have not been documented in the scientific literature. Factors such as the amount of PCM in a garment layer, the melting/freezing points of the PCMs incorporated in each garment layer, the number of PCM garment layers in a clothing ensemble, the amount of body surface area covered by garments with PCMs, the looseness or tightness of fit, and the effect of mixing garment layers with and without PCMs will influence the amount of heat in the clothing that actually affects body heat loss and the thermal comfort of the wearer.

The addition of the phase change materials to fibers, fabrics, or foams may change certain textile properties, so these effects need to be measured. Studies have shown that PCMs increase the weight of textile structures (Pause, 1994, 1995), and decrease the strength and elongation of fabrics (Bryant and Colvin, 1992). Changes in physical properties will vary depending upon what percentage of PCM by weight is used in the textile, and they should be measured prior to use in garments.

5.1.4 Purpose

Currently, phase change materials are being used in a variety of outdoor apparel items (e.g., gloves, boots, jackets, earmuffs, etc.) under the trade names Outlast™ and ComforTemp^R. The addition of PCMs to fibers, foams, and fabrics substantially increases the price of the textile. The price increase varies based on the volume being produced and the percent by weight of PCM that is added. Considering product safety, performance, and cost issues, the effect of PCMs in types of garments worn in cold environments on thermal comfort should be investigated prior to their use. Therefore, the purpose of this project was to quantify the effect of PCMs in fabric-backed foams on selected fabric characteristics (Phase I), on heat loss from a thermal manikin's surface to the environment during environmental temperature transients (Phase II), and on human subjects' physiological responses and comfort perceptions during environmental temperature transients and changes in activity (Phase III). Identical fabrics and garments – with and without the PCMs – were compared.

5.2 Methodology

5.2.1 Materials and garments

An open-cell, hydrophilic polyurethane foam was produced directly on a fabric substrate of polyester knitted fleece. The experimental foam contained 60% PCM microcapsules and 40% foam (by weight), whereas, the control foam contained 100% polyurethane foam. Approximately 40% of the PCM microcapsules contained 75% octadecane and 25% of other chemicals that change phase at 28.3 °C (83 °F), and 60% of them contained 75% hexadecane and 25% of other chemicals that change phase at 18.3 °C (65 °F).

The experimental suits for the manikin tests consisted of a long-sleeved, fitted top that was about one inch longer than waist level and a pair of fitted long pants. Two sets of each garment were produced with the experimental and control materials (i.e., four suits total). The second suit was slightly larger than the first to minimize compression when the two garments were worn together. The PCM treated foam side of the fabric was on the outside of each garment (i.e., away from the body). The garments in the ensemble were designed to minimize overlap so that the layering effects of treated and untreated garments could be controlled. The manikin's head was covered with a knitted wool hat. His hands were covered with knitted polyester fleece gloves. His feet were covered with ankle length knitted acrylic socks and athletic shoes. Manikin tests were conducted on one layer and two layer suits, and the orientation of the PCM layer to the body and the amount of PCM coverage were evaluated using two layer combinations.

Ski jackets and ski pants – with and without the PCMs – were made for the manikin and human tests. The ski garments were made of the same fabric-backed foam as the experimental suits with a woven nylon shell fabric used as the outside layer. The ski garments were worn with a 50% cotton/ 50% polyester, long-sleeve, turtleneck knitted shirt, men's briefs, and the auxiliary garments listed above.

5.2.2 Phase I: measurement of textile properties

The textile samples were conditioned at the standard temperature of 21 °C (69.8 °F) and relative humidity (65%) at least 24 hours prior to testing, according to ASTM D 1776, Standard Practice for Conditioning Textiles (ASTM, 2000). Fabric stiffness was measured according to ASTM D 1388, Standard Test Method for Stiffness of Fabrics. Four specimens measuring 2.54 cm × 20 cm (1 in. × 8 in.) were cut from each fabric with the long direction parallel to the wales and four with the long direction parallel to the courses. To conduct a test, a specimen was moved parallel to its long dimension on the platform of the cantilever and allowed to bend 41.5° under its own weight. Then the overall flexural rigidity (i.e., stiffness) was calculated. Fabric thickness was measured according to ASTM D 1777, Standard Test Method for Measuring Thickness of Textile Materials, using a pressure foot 7.6 cm (3 in.) in diameter under 0.117 kPa (0.017 psi) of pressure. The average thickness was determined from ten readings on each fabric. Fabric weight was measured according to ASTM D 3776, Standard Test Methods for Mass Per Unit Area (Weight) of Woven Fabric, option C on small swatch of fabric. Two specimens measuring 13 cm × 13 cm (5 in. × 5 in.) were weighed on a Mettler balance. The mean weight per unit area for each fabric was reported. Fabric flammability was measured according to ASTM D 2863, Standard Test Method for Measuring Oxygen Concentration to Support Candle-like Combustion of Plastics (Oxygen Index). Oxygen Index is defined as the minimum concentration of oxygen, expressed as volume percent, in a mixture of oxygen and nitrogen that will support flaming combustion of a material initially at room temperature under equilibrium conditions of candle-like burning. Ten specimens measuring 5.1 cm × 12.7 cm (2 in. × 5 in.) were tested, and the average oxygen index values for the fabrics were reported. The lower the OI value, the higher the flammability of a fabric.

The fabric insulation value (i.e., resistance to dry heat transfer) was measured using the constant temperature method specified in Part A of ASTM F 1868, Standard Test Method for Thermal and Evaporative Resistance of Clothing Materials Using a Sweating Hot Plate (ASTM, 2000). The hot plate measurement section was a 25.4 cm × 25.4 cm (10 in. × 10 in.) square surrounded by a 5 in. (12.7 cm) guard section. The three specimens measuring 50.8 cm × 50.8 cm (20 in. × 20 in.) were tested at a 20 °C (68 °F) air temperature, 50% relative humidity, and 0.7 m/s (140 ft/m) average air velocity under the hood. The plate temperature was maintained at 35 °C (95 °F). Each fabric was placed on the horizontal, flat plate so that the fleece side was next to the plate. When the system had reached steady-state, data including the plate surface temperature, air temperature, dew point temperature, and current and voltage to the test section were collected by computer for 30 minutes. The insulation value for the fabric alone was determined by subtracting the mean dry resistance value measured for the air layer (i.e., bare plate test) from the mean insulation value for the total fabric system.

The evaporative resistance of the fabrics was measured using the method specified in Part B of ASTM F 1868. A liquid barrier (PTFE film laminated to a tricot knit fabric) was placed on the plate to prevent the water (which is supplied through the porous metal in the plate surface) from wetting the fabric. In this way only water vapor had contact with the fabric sample, not liquid water. The procedures were basically the same as those for the dry tests. However, the air temperature was the same as the plate temperature so that there was no temperature gradient for dry heat loss, and the relative humidity was 40%. This is called an isothermal test.

5.2.3 Phase II: environmental step change tests with a manikin

A life-size, computerized, thermal manikin was used to simulate the heat loss from a human being to a cooler environment and to measure the insulation (clo) value of the clothing systems. Fred is a full-size male manikin with 18 electrically separate segments which provide independent temperature control and measurement. The manikin's skin temperature distribution corresponds to the temperatures on different parts of the body when a person is sedentary and comfortable.

In order to make the phase change materials go from liquid to solid and vice versa, they must go through a temperature change, and a transient test must be conducted. The environmental temperature transient was achieved by using two adjacent environmental chambers. One chamber simulated a warm indoor environment: 25 °C (77 °F) air temperature, 50% relative humidity, and 0.2 m/s (39.4 ft/min) air velocity. For the warm condition, we selected the highest 'indoor' temperature that would still permit heat loss from all of Fred's body segments when he was dressed in the most insulative ensemble (i.e., the ski ensemble). The 25 °C (77 °F) temperature maximized the amount of PCM that actually liquefied when the garments were worn on the manikin while maintaining a temperature gradient between the manikin's skin temperature and the environment that allowed body heat loss to occur. Another chamber simulated a cold outdoor environment: 10 °C (50 °F) air temperature, 75% relative humidity, and 0.2 m/s air velocity (39.4 ft/min). For the cold condition, we selected the lowest 'outdoor' temperature that would allow the manikin's body segments to reach their set points when he was dressed in the least insulative ensemble (i.e., the one layer experimental suit).

An overhead track was mounted between the two chambers, and the manikin hung from this track on a roller hooked to his head. This arrangement allowed the manikin to be moved from one chamber to the other very quickly (i.e., 1–2 minutes) and with minimal disruption. The air temperature and dew point temperature were measured continuously in each chamber. The garments were conditioned in the warm chamber. To conduct a test, the manikin was placed in the warm chamber and dressed in a specific set of garments. For the tests, the manikin was heated to an average skin temperature of 33.2 °C (92 °F).

Equilibrium was maintained for at least one hour prior to testing. First, the insulation (clo) value of the clothing and the average level of body heat loss were measured according to ASTM F 1291, Standard Test Method for Measuring the Thermal Insulation of Clothing using a Heated Manikin (ASTM, 2000). Data were collected by computer every 30 seconds for a 30 minute test. The total thermal insulation value of the clothing plus the surrounding air layer was calculated using eqn 5.1 and multiplying by 6.45 to convert SI units to clo units.

$$R_t = \frac{(T_s - T_a) A_s}{H} \qquad 5.1$$

where
R_t = resistance to dry heat transfer provided by the clothing and air layer, m² °C/W
A = surface area of the manikin, m²
T_s = surface temperature of the manikin, °C
T_a = air temperature, °C
H = heat flow, W

The amount of power that it takes to keep the manikin heated to the proper skin temperature is equal to the amount of heat loss from his body surface.

After the baseline insulation and heat loss data were collected under steady-state conditions in the indoor environment, the manikin was moved quickly to the adjacent cold chamber. The power level to the manikin was recorded every minute during the transient. After the transient was over, the insulation value and the average power level at steady-state in the cold environment were measured again. The process was repeated as the manikin was moved back to the warm chamber. The entire transient test took about six hours. Two replications of each test were conducted.

5.2.4 Phase III: exercise/rest tests with human subjects

A one-way treatment structure in a randomized complete block design (where subjects served as blocks) was used to determine the effect of phase change materials (independent variable) on the skin temperatures, thermal sensations, and clothing comfort sensations perceived by 16 male subjects (dependent variables) during an exercise/rest protocol that simulated skiing. In addition, the amount of unevaporated sweat in the ski ensemble was determined by weighing the garments before and after the experiment. The amount of sweat per unit area was determined by weighing a cotton pad before and after placing it on a subject's back. Each subject wore two ski ensembles – one with the PCM fabric-backed foam and one without the PCM – in a random order.

The experiment began in a warm chamber with an air temperature of 25 °C (77 °F), 50% relative humidity, and an air velocity of 0.2 m/s (39.4 ft/min). Then the subjects moved to an adjacent cold chamber with an air temperature of –4 °C (24.8 °F), 65% relative humidity, and an air velocity of 0.2 m/s (39.4 ft/min). When the subjects arrived for a test session, they went inside the warm environmental chamber and took off their clothes (except briefs). Then the experimenter attached thermocouples on the subjects' skin with transpore tape on the pectoral region of the chest, the radial surface of the forearm, the fibular surface of the calf, and the tip of the middle finger. Cotton cosmetic pads measuring 5.1 cm × 5.7 cm (2 in. × 2.25 in.) were oven-dried at 93 °C (200 °F) for one hour, placed in an air-tight container, and weighed. The cotton pads were taped to the back of each subject at the same time, ten minutes before the experiment began. Plastic vinyl measuring 6.35 cm × 7.0 cm (2.5 in. × 2.75 in.) was used to cover the cotton pad, and plastic tape was used to seal it tightly to the skin. Then the subjects put on the ski ensemble, except for the gloves and hat.

The activities of the subjects were designed to simulate the metabolic heat production of a male who is alternately sitting (as on a ski lift) and skiing. Sitting produces about 1 MET of heat and skiing produces about 5.5 MET (McArdle *et al.*, 1981). The speed and incline of the treadmill that would generate 5.5 MET of heat production was determined using procedures given

72   Intelligent textiles and clothing

in *Guidelines for Exercise Testing and Prescription* (American College of Sports Medicine, 1995). The exercise/rest protocol is listed below:

 0–15 minutes, 1 MET: subjects sat in the warm, ambient environment.
 16–30 minutes, 1 MET: subjects put on their hat and gloves and sat in the cold chamber.
 31–45 minutes, 5.5 MET: subjects walked on a treadmill at 4.0 mph and a 3% incline in the cold.
 46–60 minutes, 1 MET: subjects sat in the cold.
 61–75 minutes, 5.5 MET: subjects walked on the treadmill again in the cold.
 76–90 minutes, 1 MET: subjects sat in the cold.

After going through the exercise/rest protocol, the subjects returned to the warm environment and removed their clothes and thermocouples. The experimenter put each subject's turtleneck knitted shirt, jacket, and pants in a plastic garbage bag, sealed it, and weighed it. The experimenter removed the cotton pad from each subject's back and put it in an air-tight container and weighed it. The turtleneck knitted shirts and socks were laundered and conditioned in preparation for the next test. The jackets and pants were dried in a tumble drier on low heat to remove the moisture prior to conditioning for the next test.

The subjects' four skin temperatures were measured every minute during the exercise/rest protocol using a computerized data acquisition system. Thermal sensation and clothing comfort responses were measured at the end of each 15 minute period using ballots. Thermal sensation was rated from #1 very cold to #5 neutral to #9 very hot. The clothing comfort of the ski ensemble was rated from #1 uncomfortable to #5 comfortable and from #1 clammy to #5 dry using a semantic differential scale.

5.3   Results

5.3.1   Phase I: measurement of textile properties

Separate one-way analyses of variance were used to determine the effect of PCMs on various textile properties. (See Table 5.1.) The PCM foam had a significantly higher insulation value and evaporative resistance value than the control foam. The insulation value may have been higher in the PCM foam because it was significantly thicker. The evaporative resistance may have been higher in the PCM foam because it was thicker and because the PCM microcapsules displaced air in the foam structure and inhibited diffusion. This might be a problem when PCMs are coated on fabrics also, but it would not be a problem when PCMs are put inside manufactured fibers. The PCM foam had a significantly higher weight and stiffness than the control foam.

Table 5.1 Physical properties of fabric-backed foam – with and without phase change materials

Physical property	Foam with PCM	Foam without PCM
Fabric thickness	4.13**	3.44**
Fabric weight (g/m^2)	455.2**	267.4**
Fabric stiffness (mg $\cong$ cm)	2924.9**	2089.4**
Oxygen index (flammability)[a]	18.5**	19.2**
Fabric insulation (m^2 · °C/W)	0.120*	0.111*
(clo units)	(0.774)	(0.714)
Fabric evaporative resistance (m^2 · kPa/W)	0.0120**	0.0096**
Insulation per unit thickness (clo/mm)	0.187	0.208
Insulation per unit weight (clo/g/m^2)	0.002	0.003

[a]The lower the index number, the higher the flammability.
*Means were significantly different at the 0.05 level.
**Means were significantly different at the 0.01 level.

It was also significantly more flammable. The PCM foam had an oxygen index value of 18.5 which was similar to that of a cotton fabric. Therefore, the addition of the PCM microcapsules to the fabric-backed foam significantly altered its properties as compared to those of the fabric-backed foam without PCMs. The increases in stiffness, flammability, evaporative resistance, and weight are not desirable for cold-weather clothing, but a higher insulation value is desirable.

5.3.2 Phase II: environmental step change tests with a manikin

Steady-state manikin tests

Insulation values for the PCM suits were slightly higher than those measured for the suits without PCM, probably because the PCM fabric-backed foam had a higher insulation value, as measured in the hot plate test. The one layer suit with PCM was 1.57 clo and without PCM was 1.48 clo. The two layer suit with PCM was 2.07 clo and without PCM was 1.95 clo. In addition, the heat loss values were lower for the PCM foam. The insulation values for the ski ensembles with and without PCM were the same (2.19 clo). The ski ensemble fit more loosely and had more air layers between the garments compared to the tight-fitting experimental suits. Air layers have a major effect on the thermal insulation of a clothing system, and they were about the same in both ensembles since the same pattern was used to make them. Also the auxiliary garments (e.g., hat, gloves) were the same in both ensembles, so small differences in the fabric insulation did not have much effect on the overall ensemble insulation.

Treatment of transient manikin data

Heat loss data from two replications of the environmental transient tests with the manikin were averaged for each minute of the data collection period. Data were collected for one hour in each chamber, but only the first 30 minutes were graphed because the effect of the PCM was usually over in 15 minutes. The experimental suits – with and without the PCM – were not exactly identical with respect to their heat loss and insulation values under steady-state conditions. Therefore, the heat loss curves for the PCM suits were slightly lower than they would have been if they had been identical to the control suits. (See Fig. 5.3.)

To correct this problem, the average heat loss measured for each suit at steady-state was used to correct the transient data (Shim *et al.*, 2001). Then the difference between the PCM curve and the no-PCM curve (i.e., the buffering effect of the phase change material) was plotted for each suit type. The difference curves indicated that the effect of the phase change materials on body heat loss lasted approximately 15 minutes. Therefore, the magnitude of this buffering effect was calculated and reported as the integrated total difference in heat loss for the 15 minute period (J) and as the average difference in the rate of heat loss during the 15 minute period (W). (See Table 5.2.) The PCM heating effect (during a warm to cold environmental transient) and the PCM cooling effect (during a cold to warm transient) are the resulting differences in body heat loss to the environment. This is not the amount of

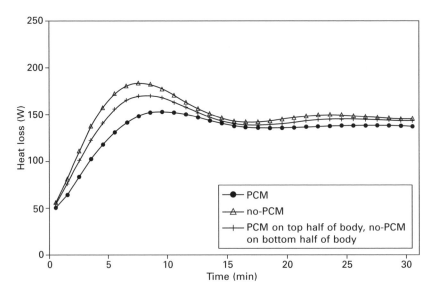

5.3 Warm to cold: average heat loss from the manikin wearing the two-layer suits – with PCM, with PCM on top half of body, and no-PCM control.

Table 5.2 Heating/cooling effects of clothing during the first 15 minutes of environmental temperature transients

Garments with PCM compared to garments without PCM	Average heating effect during warm to cold transient (W)	Total energy gain for 15 minutes during warm to cold transient (J)	Average cooling effect during cold to warm transient (W)	Total energy loss for 15 minutes during cold to warm transient (J)
One-layer suit with PCM vs. no-PCM control	6.5	5,850	7.6	6,840
Two-layer suit with PCM vs. no-PCM control	13.2	11,880	11.0	9,900
Two-layer suit (inner layer with PCM, outer layer without PCM) vs. two-layer suit without PCM	4.5	4,050	2.0	1,800
Two-layer suit (outer layer with PCM, inner layer without PCM) vs. two-layer suit without PCM	9.9	8,910	7.1	6,390
Two-layer suit with PCM or top 39% of body and without PCM on bottom 42% of body vs. two-layer suit without PCM	6.3	5,670	3.4	3,060
PCM ski ensemble vs. ski ensemble without PCM	5.9	5,310	7.4	6,660

76   Intelligent textiles and clothing

heat actually produced or stored by the PCM in the material. Some of this heat is lost to or gained from the environment. Therefore, the effect on body heat loss measured in this study is the actual heating or cooling 'buffering' effect of the PCM that the wearer would experience.

Results of transient manikin tests

The difference in heat loss between the one-layer PCM suit and the control suit without PCM is graphed in Fig. 5.4. The PCM one-layer suit produced an average heating effect of 6.5 W in the first 15 minutes after the manikin was moved from a warm environment to a cold environment. In other words, the manikin lost 6.5 W less heat during the first 15 minutes of wearing the PCM suit as compared to the control suit. When the manikin was moved from the cold chamber to the warm one, the PCM suit produced an average cooling effect of 7.6 W. (See Table 5.2.)

The difference in heat loss between the two-layer PCM suit and control suit is graphed in Fig. 5.5. The PCM two-layer suit produced an average heating effect of 13.2 W in the first 15 minutes after the manikin was moved from a warm environment to a cold environment. When the manikin was moved from the cold chamber to the warm one, the PCM suit produced an average cooling effect of 11.0 W. The effect of the PCMs on body heat loss was greater in the two-layer suits than in the one-layer suits. This result was

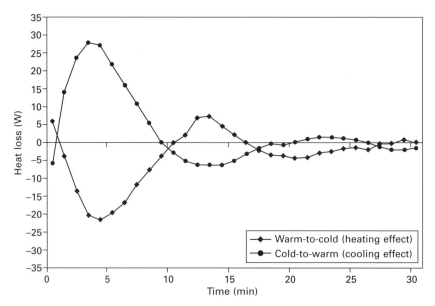

5.4 Difference in heat loss from the manikin wearing the one layer-suits – with PCM and no-PCM control.

The use of phase change materials in outdoor clothing   77

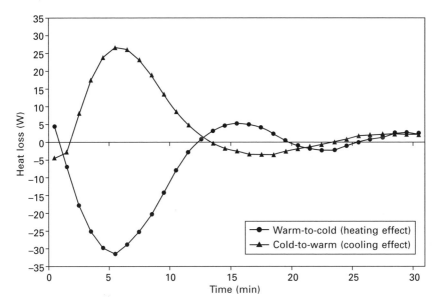

5.5 Difference in heat loss from the manikin wearing the two-layer suits – with PCM and no-PCM control.

expected because there was more PCM present in the two-layer ensemble. In addition, more of the PCM microcapsules were away from the body in the two-layer suit so that they could respond to changes in the environment. (See Table 5.2.)

As expected, the heating and cooling effects of the two-layer suits containing only one layer of PCM were less than the effects produced by the two-layer PCM suit. There are less PCM microcapsules in one layer of clothing (as compared to two layers) to go through the phase change. In addition, there was a larger heating and cooling effect when the PCM layer was on the outside than when it was on the inside, next to the body. The suit with a PCM inner layer produced an average heating effect of 4.5 W in the first 15 minutes of the warm to cold transient, whereas the suit with the PCM outer layer produced an average heating effect of 9.9 W. These layering differences may not be as great in clothing systems that fit more loosely on the body and contain thicker air layers. In addition, the heating and cooling effects of a two-layer suit containing PCM on the top half of the body only were less than the effects produced by the two-layer PCM suit with full coverage.

The PCM ski ensemble produced an average heating effect of 5.9 W in the first 15 minutes after the manikin was moved from a warm environment to a cold environment. The ski jacket and pants contained one layer of fabric-backed foam and an outer shell fabric. A turtleneck knitted shirt was worn under the jacket. The heating and cooling effects associated with phase change materials in the ski garments were similar to the effects produced by the tight-fitting, one-layer suit.

In conclusion, the magnitude of the heating and cooling effects associated with phase change materials in clothing increases as the number of PCM garment layers increases and as the amount of body coverage with PCM fabrics increases. Consequently, if a person wears only a jacket containing PCM, the effect on body heat loss will be smaller than if he/she wears a sweat shirt, jacket, and long pants containing PCM. In addition, it is apparently more effective to have one layer of PCM on the outside of a tight-fitting, two-layer ensemble than it is to have it as the inside layer. This may be because the PCMs closest to the body do not change phase.

5.3.3 Results of phase III: exercise/rest tests with human subjects

Skin temperature data and subjective data

The average skin temperatures of the subjects were plotted for the 90 minute exercise/rest experiment. Then an analysis of variance was conducted on the skin temperature data collected during the 12th minute of each 15 minute period. The 12th minute was selected because the effect of the PCM on body heat loss was probably complete, and the subjects were changing conditions during the 14th and 15th minute of each period. The skin temperatures varied considerably from subject to subject. In addition, an analysis of variance was conducted on the subjects' thermal sensation responses and clothing comfort responses collected near the end of each 15 minute period in the 90 minute experiment.

Environmental effect

When the subjects moved from a seated position in a warm environment to a seated position in a cold environment, their average leg skin temperature was significantly higher when wearing the PCM ski ensemble than the ensemble without PCMs (F value = 6.53, $p = 0.022$). The average leg temperature was 31.5 °C (88.7 °F) under the PCM ensemble and 30.3 °C (86.5 °F) under the control. The PCMs probably solidified in the cold environment, producing a heating effect. The other skin temperatures were not statistically different, probably because a turtleneck knitted shirt was worn between the jacket and the body and because subject variability was high and the average heating effect of the PCM ski ensemble was low (about 6 W). No statistically significant differences in the subjects' thermal sensations or comfort levels were found when wearing the PCM ensemble as compared to the control ensemble during the environmental step change at low activity.

The use of phase change materials in outdoor clothing 79

Exercise/rest effect

The results of this study indicated that there were no statistically significant differences in the skin temperatures of males wearing PCM ski ensembles vs. control ensembles without PCMs during exercise/rest periods in the cold. When the subjects were walking at 5.5 MET in the cold environment, they were producing more than five times the amount of heat than they did while sitting (1 MET). This extra body heat did not raise their skin temperatures to proportionately higher levels and heat the PCMs so that they would change back to liquid. Instead, the body dissipated the extra heat energy by producing sweat at the skin surface. The process of changing liquid sweat to vapor is a phase change in itself which takes heat from the body surface, providing a cooling effect. The sweat vapor is generated at the same temperature as the skin and has a minimal effect on the temperature of the clothing layers as it diffuses through them. The temperature of a PCM garment layer has to increase to its melting temperature in order to turn to liquid and store heat. Consequently, the PCMs probably did not change from solid to liquid when the subjects were exercising, and the phase change activity in the cold environment was over.

There were no statistically significant differences in the thermal sensation and clothing comfort responses of the subjects wearing the PCM ensemble as compared to the control ensemble in the cold. However, the subjects did feel significantly more clammy (as opposed to dry) during the last two test periods in the cold chamber when they were exercising in the PCM ensemble (F value = 12.10, p = 0.003) and sitting in the PCM ensemble (F value = 5.79, p = 0.030). The PCM microcapsules increased the evaporative resistance of the PCM fabric-backed foam which probably contributed to these results.

Sweat accumulation in garments

A cotton pad taped to each subject's back was used to determine sweat rate during the 90 minute experiment. There was no statistically significant difference in the amount of sweat absorbed in the pad under the PCM ensemble (0.428 g, 0.015 oz.) and the amount absorbed under the control ensemble (0.355 g, 0.013 oz.). However, there was significantly more sweat in the PCM jacket and pants (5.14 g, 0.18 oz.) as compared to the control garments (2.75 g, 0.10 oz.) (F value = 16.61, p = 0.002). There also was significantly more sweat in the turtleneck knitted shirt (worn under the jacket) when the subjects wore the PCM ski ensemble (2.38 g, 0.08 oz.) than the control (1.55 g, 0.05 oz.) (F value = 6.74, p = 0.023). More moisture may have been trapped in the PCM garments and the shirt worn under the PCM jacket because the evaporative resistance of the PCM fabric-backed foam was higher than the control. The higher level of sweat in the PCM ensemble probably contributed to the subjects feeling clammier at the end of the exercise/rest protocol.

5.4 Conclusions

The addition of phase change materials (PCMs) to a fabric-backed foam significantly increased the weight, thickness, stiffness, flammability, insulation value, and evaporative resistance value of the material. According to the transient tests with a thermal manikin, PCMs produced a small, temporary heating or cooling effect when garments made of the material went through a step change in temperature. The PCM heating and cooling effects changed body heat loss by an average 2–13 W for the first 15 minutes of the environmental transient, and then the effects were over. The magnitude of the effect increased as the number of garment layers and the amount of body coverage with PCM fabrics increased. In addition, it was more effective to have one layer of PCM on the outside of a tight-fitting, two layer ensemble than to have it as the inside layer. This may be because the PCMs closest to the body did not change phase.

When 16 male subjects moved from a seated position in a warm environment to a seated position in a cold environment, their average leg skin temperature was significantly higher when wearing the PCM ski ensemble as compared to the control ensemble. The PCMs probably solidified in the cold environment, producing a heating effect. However, the subjects' chest, forearm, and finger skin temperatures and their thermal sensations and comfort perceptions were not significantly different during this environmental step change at low activity. There were no statistically significant differences in the skin temperatures and thermal sensations of males wearing the PCM ski ensemble compared to the control during the four remaining exercise/rest periods in the cold. When the 90 minute experiment was over, there was significantly more unevaporated sweat in the PCM ensemble than the control, and the subjects felt significantly clammier when wearing it during the last exercise period and rest period of the test.

The increase in a person's metabolic heat production during exercise in the cold may not increase the skin temperature enough to warm the PCM garment layers to the melting point and keep the cooling and heating cycle going. The effect of phase change materials in clothing on the physiological and subjective thermal responses of people would probably be maximized if the wearer was repeatedly going through temperature transients (i.e., going back and forth between a warm and cold environment) or intermittently touching hot or cold objects with PCM gloves.

5.5 Implications and recommendations

The magnitude and duration of PCM heating and cooling effects in clothing systems are dependent upon several textile and design-related factors. Each of these variables needs to be considered in relation to the type of temperature transient anticipated during use (i.e., environmental temperature and/or skin temperature changes). Product variables include

- the transition temperatures (i.e., melting/freezing points) of the PCMs incorporated into each garment layer
- the effect of mixing PCMs with different transition temperatures in one garment layer
- the amount and purity of the PCMs in a garment layer (percent add-on)
- the number of PCM garment layers in a clothing ensemble
- the placement order of PCM and non-PCM garment layers from the body surface to the environment
- the amount of body surface area covered by garments with PCMs.

Considering product safety, performance, and cost issues, the effect of PCMs in protective clothing on the thermal comfort of the wearer should be investigated carefully prior to their use. In addition, changes in the performance characteristics of fabrics with PCMs (e.g., weight, stiffness, flammability, evaporative resistance, strength) and their durability during use and maintenance should be determined.

5.6 References

American College of Sports Medicine (1995), *ASCM's guidelines for exercise testing and prescription*, Baltimore, Williams & Wilkins.

American Society for Testing and Materials (2000), *Annual Book of ASTM Standards, Parts 7.01, 7.02, 8.02, 11.03*, Conshohocken, PA, ASTM.

Bryant Y G, and Colvin D P (1992), 'Fibers with enhanced, reversible thermal energy storage properties', *Techtextil-Symposium*, 1–8.

Cox R (1998), 'Synopsis of the new thermal regulating fiber outlast', *Chemical Fibers International*, 48, 475–479.

McArdle W D, Katch F I, and Katch V L (1981), *Exercise physiology: energy, nutrition, and human performance*, Philadelphia, Lea & Febiger.

Pause B H (1994), 'Investigations of the heat insulation of protective textiles with microencapsulated PCM', *Techtextil-Symposium*, 245, 1–9.

Pause B H (1995), 'Development of heat and cold insulating membrane structures with phase change material', *Journal of Coated Fabrics*, 25 (7), 59–68.

Pause B H (1998), 'Development of new cold protective clothing with phase change material', *International Conference on Safety & Protective Fabrics '98*, Baltimore, 78–84.

Shim H, McCullough E A, and Jones B W (2001), 'The use of phase change materials in clothing', *Textile Research Journal*, 71, 495–502.

Part II

Shape memory materials

6
Introduction to shape memory materials

M HONKALA, Tampere University of Technology, Finland

6.1 Overview

Shape memory materials (SMMs) are a set of materials that, due to external stimulus, can change their shape from some temporary deformed shape to a previously 'programmed' shape. The shape change is activated most often by changing the surrounding temperature, but with certain materials also stress, magnetic field, electric field, pH-value, UV light and even water can be the triggering stimulus [1–5]. When sensing this material specific stimulus, SMMs can exhibit dramatic deformations in a stress free recovery. On the other hand, if the SMM is prevented from recovering this initial strain, a recovery stress (tensile stress) is induced, and the SMM actuator can perform work. This situation where SMA deforms under load is called restrained recovery [6].

Because of the wide variety of different activation stimuli and the ability to exhibit actuation or some other pre-determined response, SMMs can be utilised to control or tune many technical parameters in smart material systems in response to environmental changes – such as shape, position, strain, stiffness, natural frequency, damping, friction and water vapour penetration [1, 2]. Today, a large variety of alloys, ceramics, polymers and gels have been found to exhibit shape memory behaviour. Both the fundamental theories and engineering aspects of SMMs have been investigated extensively and a rather wide variety of different SMMs are presently commercial materials.

Commercialised shape memory products have been based mainly on metallic shape memory alloys (SMAs), either taking advantage of the shape change due the *shape memory effect* or the *superelasticty* of the material, the two main phenomena of SMAs. Shape memory polymers (SMPs) and shape memory gels are developed at a quick rate, and within the last few years also some products based on magnetic shape memory alloys have been commercialised. Shape memory ceramic (SMC) materials, which can be activated not only by temperature but also by elastic energy, electric or magnetic field, are mainly at the research stage.

Although the largest commercial successes have no doubt been seen in the fields of bioengineering and biomedical applications, shape memory materials are becoming increasingly important in many other technological fields including high-performance aircraft and automotive components, space applications, vibration and seismic applications, micro-electromechanical systems (MEMS), telecommunications, polymer matrix composites and many others, including textiles and the clothing sector.

6.2   Shape memory alloys

6.2.1   History

The first milestone in the history of shape memory materials was the discovery of the pseudoelastic behaviour of the Au-Cd alloy in 1932 by Swedish physicist Arne Olander [7]. Later in 1938 Greninger and Mooradian observed the formation and disappearance of a martensitic phase by decreasing and increasing the temperature of a Cu-Zn alloy. In 1949 Kurdjumov and Khandros reported more widely the phenomenon of the memory effect and the thermoelastic behaviour of the martensite phase, soon followed by Chang and Read in 1951 [8].

However, it was not until the early 1960s that the 'revolution' of shape memory materials really began. William J. Buehler and his co-workers at the US Naval Ordnance Laboratory accidentally discovered the shape memory effect in an equiatomic nickel-titanium (NiTi) alloy [9]. Later this alloy was commercialised under the trade name Nitinol (an acronym for NIckel TItanium Naval Ordnance Laboratories). Since the birth of Nitinol, extremely intensive research has been done worldwide to clarify the characteristics of the basic behaviour of the shape memory effect. Today NiTi alloys are the most studied and best characterised of all the alloys ever found with shape memory behaviour.

6.2.2   General principles of shape memory alloys

Shape-memory alloys are metal compounds, which can memorise a predetermined shape, and after being bent, stretched or otherwise mechanically deformed they can return to this shape under certain temperature conditions. This shape-memory effect is due to a phenomenon known as a *thermoelastic martensitic transformation*, which is a reversible, diffusionless transformation between two different crystal microstructures that occurs when a shape-memory alloy is heated or cooled beyond alloy specific *transition temperatures*. These temperature dependent crystal structures or phases are called *martensite* (low temperature) and *austenite* (high temperature) [9].

Shape memory alloy is quite strong and hard in its austenite (parent) form, but in the martensite form it is soft and ductile and can easily be deformed. SMAs also exhibit *superelasticy* (or *pseudoelasticity*) giving the material a rubber-like behaviour. The application specific overall austenitic shape *(parent shape)* of SMA is formed and locked (programmed) through a specific high-temperature tempering process.

During the early years of SMAs, a wide variety of alloys have been developed to exhibit the shape memory effect, but only those that can recover substantial amounts of strain or that generate significant force when changing shape are of commercial interest. Today, alloys fulfilling this criterion are mainly the nickel-titanium alloys (NiTi) and copper-base alloys such as CuZnAl and CuAlNi. The characteristics of SMAs that are given in the following pages are mainly based on NiTi alloys.

Hysteresis

When martensite NiTi is heated, it begins to change into austenite (Fig. 6.1). The temperature at which this phenomenon starts is called austenite start temperature (A_s). The temperature at which this phenomenon is complete is called austenite finish temperature (A_f). When austenite NiTi is cooled, it begins to change into martensite. The temperature at which this phenomenon starts is called martensite start temperature (M_s). The temperature at which martensite is again completely reverted is called martensite finish temperature (M_f) [9].

The temperature at which the martensite-to-austenite transformation takes place upon heating is somewhat higher than that for the reverse transformation when cooling. The difference between the transition temperatures when heating and cooling is called hysteresis. Hysteresis is generally defined as the difference between the temperatures at which the material is 50% transformed to austenite upon heating and 50% transformed to martensite upon cooling (see Fig. 6.1). This difference can be up to 20–30 °C [9]. When thinking of medical

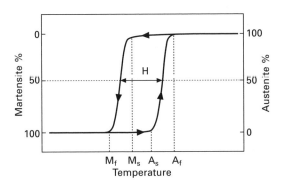

6.1 Martensitic transformation and hysteresis (= H) upon a change of temperature.

applications, this practically means that an alloy designed to be completely transformed by body temperature upon heating (Af < 37 °C) would require cooling to about +5 °C to fully retransform into martensite (M_f) [8].

Thermoelastic martensitic transformation

The unique behaviour of NiTi is based on the austenite-to-martensite phase transformations on an atomic scale. The SMA behaviour during the martensitic transformations is very complicated and still not completely understood. The austenite-to-martensite phase transformation can be either temperature induced, known as *thermoelastic martensitic transformation,* or between certain temperature limits also stress induced transformation (pseudoelasticity or superelasticity).

The phenomena causing the shape recovery in thermoelastic transformation is a result of the need of the crystal lattice structure to accommodate to the minimum energy state for a given temperature [7]. If we could look into a NiTi alloy in the atomic scale we could see nickel and titanium atoms arranged alternately in a crystal lattice structure. In the austenite high-temperature form the lattice structure is a simple cubic structure (Fig. 6.2(a)) having a strong symmetry, which prevents it from shifting or changing orientation to accommodate an applied stress. This is the reason why austenitic alloys are much more rigid than martensitic ones.

When a high-temperature austenite is cooled to M_s, the austenite-to-martensite phase transformation begins. This is a diffusionless transformation in which the atoms move only short distances in order to join the new phase (on the order of the interatomic spacing). When the alloy is cooled below M_f, it is entirely in its martensitic phase being now soft and malleable, just waiting to become mechanically deformed out of its original programmed shape to a new 'temporary shape'. During the cooling process the overall macroscopic shape of the alloy does not change significantly. However, in order to maintain the overall shape during cooling and to accommodate to the minimum energy state, the alloy deforms (self-accommodates) the structure of its crystal lattice to a diamond shape structure (Fig. 6.2(b)). The mechanism by which single martensite variants deform is called *twinning*, and it can be described as a mirror symmetry displacement of atoms across a particular atom plane, the twinning plane [9, 10].

Now, if we mechanically deform the soft martensite to a desired temporary shape, the alloy system minimises its energy by continuing twin boundary movement (Fig. 6.2(c) and 6.2(d)). When most metals deform by slip or dislocation, NiTi responds to stress by simply changing the orientation of its crystal structure through the movement of twin boundaries. Next, whenever this deformed martensite is heated above the temperature A_s, the martensite-

Introduction to shape memory materials   89

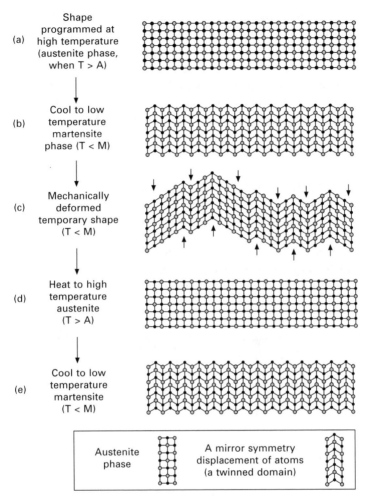

6.2 Schematic illustrations of the phase transitions and crystal lattice structures during the shape-memory effect.

to-austenite phase change starts and the alloy returns to its original programmed austenite crystal structure and to the original macroscopic and rigid shape (Fig. 6.2(e)).

Shape memory effect

The above described macroscale phenomenon, where materials exhibit shape memory upon heating, is called a *one-way shape memory effect* (SME). When the shape change follows a process of free recovery, no reverse shape change in subsequent recooling normally occurs, so the specimen needs to be strained again to repeat the SME. However, a *two-way shape memory effect*, where the alloy remembers its low-temperature and high-temperature configurations, can be imparted to an SMA with appropriate training. Special alternative training processes can be used to build up two-way SME, first a specific heat treatment, and then several complicated and time-consuming thermomechanical cycles. Two-way SMAs have much lower recoverable strain and the low-temperature recovery forces are very small compared to the values in the one-way shape memory effect [11].

Superelasticity

Another important phenomenon affecting the shape change of SMA is *Superelasticity*. As can be seen from Fig. 6.3(a), an austenite SMA normally stays in austenite form when the temperature is above M_s. However, when the temperature is only slightly above M_s, a martensitic phase transformation from austenite-to-martensite can be initiated by applying outer mechanical stress to the alloy. An increase in the applied stress produces an effect analogous to a decrease in temperature. However, only one martensite variant (stress

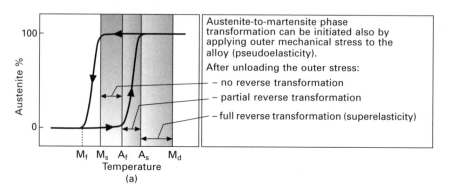

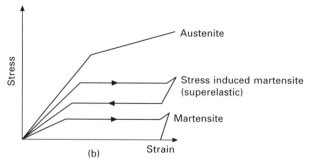

6.3 Austenite-to-martensite phase transformation due to outer mechanical stress: (a) temperature ranges for stress-induced phase transformations; (b) stress-strain behaviour of different phases of SMA at constant temperature.

induced martensite, SIM) grows in the direction most favourable for the applied stress. When the temperature is in the range of $M_s < T < A_s$, SIM will be formed, but after unloading the material will remain as martensite. Partial reversion to austenite will take place upon stress release when temperature is in the range of $A_s < T < A_f$. The complete reverse transformation takes place immediately upon stress release at temperatures approaching or greater than A_f. This full reverse transformation upon stress release is called superelasticity. The alloy provides the same force of the deformation stress and has a rubber-like property (Fig. 6.3b). Superelasticity can be observed only over a specific temperature area. The highest temperature where martensite can be stress induced is called M_d. Above this temperature the austenite alloy can be deformed like ordinary materials by slipping [6, 8, 12].

Mechanical properties of shape memory alloys

Since the birth of NiTi several other alloy types have been discovered. Yet only two main types of SMAs, copper-based alloys and nickel-titanium (NiTi) alloys, are considered to have potential for commercial engineering applications. The copper based alloys, such as Cu-Zn-Al, Cu-Al-Ni, Cu-Zn-Si, Cu-Zn-Sn, Cu-Sn, Cu-Zn, Mn-Cu, have a more extensive range of potential transformation temperatures and they are made of relatively inexpensive materials. They have lower processing costs than NiTi SMAs, partially because they can be melted and extruded in air instead of being processed in a vacuum. However, in spite of higher costs and more complicated fabrication, NiTi alloys have many unique characteristics, compared to the other alloys, that make them maybe the only practical shape memory alloys. The essential benefits of Niti alloys are the following [13, 6]:

- high shape memory strain, up to 8% (copper-base alloys up to 5%)
- very good shape memory/superelastic characteristics (only after proper thermomechanical treatments)
- low elastic anisotropy
- high ductility (60% cold working possible)
- excellent corrosion and abrasion resistance
- better thermal stability and ageing properties
- biomedical compatibility
- potentiality to amorphisation
- high damping capacity
- relatively high electrical resistance.

The mechanical properties of shape-memory alloys are principally determined by their composition, which allows them to be tailored to suit many different applications. Table 6.1 lists some of the typical properties of these alloys, manufactured by medical technologies n.v. (formerly Advanced Materials and Technologies, AMT) [www.amtbe.com].

Table 6.1 Properties of different shape memory alloys (by medical technologies n.v.)

	Ni–Ti	Cu–Zn–Al	Cu–Al–Ni
Melting temperature (°C)	1,300	950–1,020	1,000–1,050
Density (g cm^{-3})	6.45	7.64	7.12
Resistivity (μΩ cm)	70–100	8.5–9.7	11–13
Thermal conductivity (W cm^{-1} per °C)	18	120	30–43
Young's modulus (GPa)	83 (austenite) 26–48 (martensite)	72 (beta phase) 70 (martensite)	85 (beta phase) 80 (martensite)
Yield strength (MPa)	195–690 (austenite) 70–140 (martensite)	350 (beta phase) 80 (martensite)	400 (beta phase) 130 (martensite)
Ultimate tensile strength (MPa)	895	600	500–800
Shape-memory strain (% maximum)	8.5	4	4
Transformation range (°C)	−200–110	<120	<200
Transformation hysteresis (°C)	30–50	15–25	15–20

Fabrication of shape memory alloys

SMA alloy materials can be fabricated into bars, strips, fibres or wires, tubing, foils, thin films, particles and even porous bulks. In the next few paragraphs the aspects of SMA fabrication (melting, basic fabrication, secondary processing methods, finishing technologies) are based mainly on the fabrication of NiTi.

Molten Nitinol is highly reactive and must be processed in a vacuum. Both vacuum induction melting (VIM) and vacuum consumable arc melting (VAR) processes are commonly used for production [14]. The most important properties of SMA are those two transformation temperatures from martensite to austenite and vice versa. They can be altered by changing the composition of the alloy. At certain concentration levels as tiny as one weight percent deviation, variations in Ni or Ti content can result in about a 100 °C shift in transformation temperatures. The transformation temperatures can be measured by monitoring practically any property of the alloy, such as electrical resistance, or stress-strain in a standard tensile test at temperatures across the transformation, but the most accurate means is differential scanning calorimetry (DSC) [6, 14].

After melting, the NiTi ingot is usually forged and rolled into a bar or a slab at elevated temperatures around 800 °C where the alloy is easily workable. The final dimensions with the desired physical and mechanical properties are obtained by cold working and appropriate heat treatments. Cold working of NiTi is quite challenging due to multiple reductions and frequent interpass annealing at 600–800 °C [14]. NiTi can be machined using conventional techniques such as milling, turning, drilling, grinding, sawing and water jet cutting. Laser and electro-discharge machining and photochemical etching processes are used to fabricate NiTi components such as tubular stents, baskets and filters. Various powder metallurgy processes have also been developed for NiTi. Porous Nitinol has attracted recent attention to its potential as an implant material [14].

Corrosion resistance is significantly affected by methods of surface preparation. After different heat treatments the surface of NiTi contains oxide layers, which can be removed by mechanical means, such as grit blasting and polishing, or by chemical etching. Mechanically polished surfaces, although they can be highly smooth, appear to be most susceptible to corrosion attack while the chemically etched surfaces appear to be the most resistant. Electropolishing alone does not sufficiently enhance corrosion resistance [14].

NiTi can be quite easily soldered using halogen-based fluxes. It can also be welded to itself by using CO_2 laser, but joining NiTi to dissimilar metals is significantly more challenging. Welding NiTi to stainless steel is especially difficult due to the formation of brittle intermetallic compounds [14]. Thermal spray is a conventional thick film coating technology. NiTi coating can be applied by plasma spraying or physical vapour deposition [15] The process has also been used to fabricate NiTi foils and thin wall mill products.

More recently, thin film SMA has become a promising material in the field of micro-electro-mechanical system (MEMS) applications (such as microgrippers, micro-pumps, micro-mirror, sensors and actuators). Thin film can be patterned with standard lithography techniques and fabricated in a batch process. Thin film SMA has only a small amount of thermal mass to heat or cool, thus the cycle (response) time can be reduced substantially and the speed of operation may be increased significantly. In thin film applications the fabrication of NiTi actuators have been made by first depositing NiTi on silicon, glass or polymeric substrates by sputter deposition. Commercial devices consisting of Nitinol film on silicon substrate are then fabricated by photolithographic techniques [14, 31].

SMA applications

The first successful industrial application of a shape memory alloy took place in 1970, when *Cryofit* tube coupling was demonstrated as a part of high-pressure hydraulic system on a US Navy F-14 fighter aircraft. In the following years, these demonstrations led to the production of over a million couplings [16, 17]. Since then the range of applications for SMAs has been increasing and one major area of expansion has been medicine. However, despite thousands of patents issued for every conceivable application for shape memory alloys, the list of the truly successful commercial devices is quite short. In SMA applications, superelastic devices have been the most significant so far in both material consumption and commercial value [16, 17].

The bulk shape-memory alloys exhibit large strokes and forces but suffer from a slow response. The bandwidth is usually near 1 Hz, which limits the material's applicability in many situations. The bandwidth limitation is due to the relatively slow cooling processes related to surface area-to-volume ratios. SMA thin films provide a small amount of thermal mass to be cooled, and hence the cycle lifetime can be decreased substantially.

6.3 Shape memory ceramics

Shape memory effects can also be found in ceramic materials. Shape memory ceramics (SMC) can be categorised as viscoelastic, martensitic, ferroelectric or ferromagnetic depending on their activation mechanism [1]. Shape memory ceramics (SMC) can tolerate much higher operating temperatures than other shape-memory materials, but their recoverable strain is quite small. On the other hand, the actuation by electric field, for instance, can be much faster than by heat.

6.4 Magnetic shape memory materials

The principle of magnetic shape memory (MSM) alloys was presented in 1996 by Ullakko *et al.* [18]. MSM alloys belong to a novel group of shape memory materials that can change their shape in less than a millisecond and generate up to 10% strain when exposed to a magnetic field. Today there are several different alloy types, such as Ni-Mn-Al, Ni-Mn-Al, Fe-Pd, Co-Ni-Ga, Ni-Co-Al, but the most promising material seems to be Ni-Mn-Ga [19–20].

The MSM mechanism is based on the martensite twin boundary motion driven by the external magnetic field, when the material is in complete martensitic state [21]. Magnetic materials, even without an external magnetic field, are characterised by local magnetic moments, which have a certain preferable direction (the so-called easy direction) in respect to the crystal lattice. If the easy direction and the twin direction in a twinned microstructure are parallel, the lattice orientations of the twin variants are different and therefore the magnetisation directions also differ, as shown in Fig. 6.4(a) [22].

When an external magnetic field is applied, the magnetic moments try to align in the field direction. However, if the energy required to rotate the magnetisation out of the easy direction (the *magnetic anisotropy energy* MAE) is higher than the energy required to move a twin, it is energetically more favourable to move the twin boundaries instead of rotating the magnetisation. The fraction of twins, where the easy axis is in the direction of the field, will grow at the expense of the other twin variants. This process

Introduction to shape memory materials 95

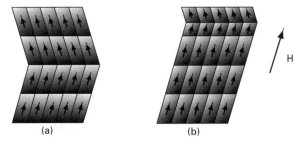

6.4 (a) Magnetic moments without the external field;
(b) redistribution of the variants in an applied field [22].

results in large shape changes as shown schematically in Fig. 6.4(b) [22].

There are two basic requirements for the appearance of the MSM effect. First, the material should be (ferro)magnetic and exhibit martensitic transformation, and second, the magnetic anisotropy energy should be higher than the energy needed to move the twin boundaries. Also from the practical applications point of view the material should be in its martensitic phase at room temperature [22].

MSM materials can be made to change their shape in different ways, to elongate axially, bend or twist. So far, actuators developing linear axial motion are the most common. The characteristics of present MSM actuators are [20]:

- large strokes and large forces
- high frequency operation (rise time is normally less than 0.2 ms; it is limited by flux generation and the inertia of the moving mass rather than MSM mechanism of the material)
- electromechanical hysteresis produce losses in MSM material, it also complicates the control of some positioning system applications
- hysteresis increases vibration-damping capacity of the MSM material
- wide operating temperature range (between –70 °C and 100 °C)
- MSM actuators can operate long times at high frequencies without significant fatigue of the actuating element (> 200 million cycles).

One of the pioneering companies in the field of magnetic shape memory research and applications is Adaptamat Ltd in Finland. It produces commercial MSM materials and different types of actuators.

6.5 Shape memory polymers and gels

6.5.1 General properties of shape memory polymers

Shape memory polymers (SMPs) were first introduced in 1984 in Japan. Shape memory behaviour can be observed for several polymers that may differ significantly in their chemical composition. In SMPs, the shape memory effect is not related to a specific material property of single polymers; it rather results from a combination of the polymer structure and the polymer morphology together with the applied processing and programming technology [2]. Just like SMAs, the most common stimulus in SMP applications is heat. However, there is much ongoing research on systems, which may respond also to other stimuli, such as UV light, water, pH, electric or magnetic field. Some success has been reported on light and water induced SMPs [3, 5].

A thermally induced polymer undergoes a shape change from its actual, deformed temporary shape to its programmed permanent shape after being heated above a certain activation temperature T_{trans} [2]. SMPs are characterised by two main features, triggering segments having the thermal transition T_{trans} within desired temperature range, and cross-links determining the permanent shape. Depending on the kind of cross-links, SMPs can be thermoplastic elastomers or thermosets [29].

Segmented polyurethane thermoplastic SMPs have two separated molecular phases, a hard segment and a soft segment, with different glass transition temperatures, $T_{g,hard}$ being higher than $T_{g,soft}$. The polymer can be processed using conventional techniques (injection, extrusion, blow moulding) to desired shapes. During the processing stage, the material is at or above the melting temperature, T_{melt}, and all of the polymer chains have high degrees of mobility. Once the material cools down to $T_{g,hard}$, the configuration of the hard segments is 'stored' by physical cross-links. However, at temperatures between $T_{g,soft}$ and $T_{g,hard}$, the soft segments still allow the material to deform to a temporary shape while the physical cross-links of the hard segments store strain energy. Below $T_{g,soft}$, the material is completely glassy, and will hold a deformed shape without external constraint. When the material is heated back above $T_{g,soft}$, the soft segments are too mobile to resist the strain energy stored in the bonds of the hard segments, and an unconstrained recovery from the temporary deformed shape to the original 'stored' shape occurs. At temperatures higher than $T_{g,hard}$, the physical cross-links of the hard segments are released, thus erasing the 'memory' of the polymer. As the polymer is a three-dimensional network, a SMP can fully recover near 100% strain in all three dimensions [23].

The typical representation of the thermomechanical cycle of an SMP is shown in Fig. 6.5 [24]. Before starting the cycle the SMP is first heated to $T_{g,soft}$. The first step of the cycle describes the high-strain deformation of the SMP to the desired temporary shape. During step 2 the material is cooled under constraint to hold the deformation. The stress required to hold this earlier deformed shape diminishes gradually to zero as temperature decreases. The temporary shape is now 'locked' and the constraint can be removed. In the final step of the cycle, the SMP is subjected to a prescribed constraint level and then heated again towards $T_{g,soft}$. In Fig. 6.5, the two limiting cases

Introduction to shape memory materials   97

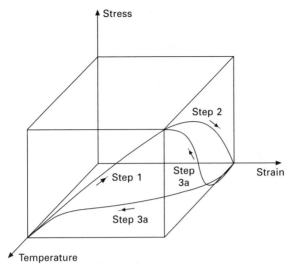

6.5 Stress-strain behaviour of different phases of NiTi at constant temperature.

of constraint are shown, namely a constrained recovery (step 3a) and an unconstrained recovery. Constrained recovery implies the fixing of the pre-deformation strain and the generation of a gradually increasing recovery stress. Unconstrained recovery implies the absence of external stresses and the free recovery of the induced strain. With the increase of temperature, the strain is gradually recovered. After the shape recovery step, the remaining strain is called residual strain. Recovered strain is defined as the pre-deformation strain minus the residual strain [24]. Figure 6.6 further illustrates the material's behaviour during an unconstrained shape recovery [25].

The benefits of SMPs over SMAs [1–3, 14, 23–27]:

- much lower density
- very high shape recoverability (maximum strain recovery more than 400%)

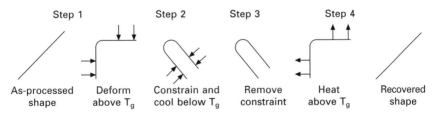

6.6 Schematic picture of the idealised thermo mechanical cycle leading to unconstrained strain recovery for a shape memory polymer [25].

- the shape recovery temperature can be engineered to occur over a wide range
- the recovery temperature can be customised by adjusting the fraction of the hard and soft phases
- less complicated (and more economical) processing using conventional technologies
- fast programming process
- some polymer networks are also biocompatible and biodegradable.

Drawbacks of SMPs:

- low recovery time
- low recovery force; SMP's ability to generate a 'recovery' stress under strain constraint is limited by their relatively lower stiffness, the shape-recovery property is lost when rather a small amount of stress (<4 MPa) is applied to the polyurethane components [1]. However, the stiffness and recovery stress of shape memory polymers can be substantially increased, at the expense of recoverable strain, by the inclusion of hard ceramic reinforcements [27].
- Polyurethane SMP may lose its shape fixing capability after being exposed to air at room temperature (about 20 °C) for several days. It can, however, fully regain its original properties after being heated up to its melting temperature [28].

6.5.2 Applications of shape memory polymers

In literature and other media a large number of potential applications of shape memory polymers have been presented to suit various products in almost every aspect of daily life: industrial components like automotive parts, building and construction products, intelligent packing, implantable medical devices, sensors and actuators, etc. Yet, only a few shape memory polymers have so far been brought to commercial markets, and the number of implemented applications is still very small. SMPs are used in toys, handgrips of spoons, toothbrushes, razors and kitchen knives, also as an automatic choking device in small-size engines [30]. The main reason for the lack of 'killer applications' is most likely the fact that so far SMPs are suitable only for applications where free recovery or very low recovery force meets the requirements.

One of the most well known examples of SMP is a clothing application, a membrane called Diaplex. The membrane is based on polyurethane based shape memory polymers developed by Mitsubishi Heavy Industries. A part of the membrane is an ultra-thin nonporous polymer. Diaplex takes advantage of Micro-Brownian motion (thermal vibration) occurring within the membrane when the temperature rises above a predetermined activation point. As a result of this motion, the molecules form free spaces (micropores) in the membrane, which allows water vapour and body heat to escape (see Fig. 6.7). Because permeability increases as the temperature rises, the membrane is able to respond intelligently to changes in the wearer's environment and body temperature. Water vapour inside the garment is absorbed before it has a chance to condense. The absorbed water vapour is conducted into and diffused throughout the membrane and then emitted from the surface of the membrane [31].

Among the few suppliers making shape memory polymers are companies such as:

- *Mitsubishi Heavy Industries* (polyurethane based shape memory polymers – e.g. Diaplex)

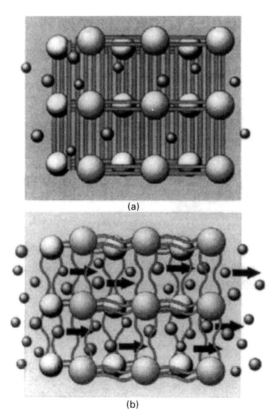

6.7 (a) At low temperatures the polymer molecular chains form a continuous surface that restricts the loss of body warmth by stopping the transfer of vapour and heat. (b) At increasing temperatures the molecular configuration changes resulting in the formation of free space. This allows the transfer of heat and vapour from perspiration and helps to prevent discomfort and clamminess within the garment.

- *Composite Technology Development* (CTD) (Elastic Memory Composite (EMC) materials, shape memory thermosets)
- *CRG Industries LLC*, Veriflex™ shape memory thermosets, Veritex™ dynamic composites (Veriflex™ resin is used as the matrix)
- *The Polymer Technology Group Inc.* Calo•MER™ shape-memory thermoplastics
- *Bayer MaterialScience*, shape-memory thermoplastics
- *MnemoScience GmbH*, biodegradable thermoplastic shape memory polymers, etc.

6.5.3   Shape memory gels

Smart polymeric gels have an ability to react to infinitesimal changes in their environmental conditions by considerable volume changes, swelling or shrinkage. Volume changes can be triggered besides temperature also by a variation in the pH value, the ionic strength, biochemical element or the quality of the solvent. For certain gels the triggering stimulus can also be light or electric field or stress, depending upon the precise structure of the gel. The main negative aspect of gels is their poor mechanical stability [1, 2].

Gels are capable of conversion between chemical energy and mechanical work. By undergoing phase transitions, which are accompanied by reversible, continuous or discrete volume changes by three orders of magnitude, the gels can provide actuating power capacities comparable to that of human muscles [1].

6.6   Future prospects of shape memory materials

The potential future benefits of shape memory materials, their smart structures and systems, seem to be remarkable in their scope. The research on smart materials is growing all over the globe. However, shape memory materials do have their shortcomings to be overcome before their engineering significance is more widely recognised in the industrial world.

Within shape memory alloys there are problems mainly in the engineering and modelling aspects; fabrication and processing of high-quality and lower-cost materials. We need to understand the still unclear issues like hysteresis and ageing effects better. In addition, scientific research is required to get answers to a few 'mysteries', for example, if there is a common microscopic origin for all the martensitic transformations, and what the relationships between electronic structure and the competing martensite structures are [13].

Advancements in process development, especially in NiTi processing, allow SMAs to break into cost-sensitive commercial volume markets. The medical device market can tolerate higher cost but has a strict demand for superb quality and consistency. Thin film technology is about to become widely commercialised; process stability and quality control on thin film properties will probably attract far more attention and awareness. Thin film alloy bandwidths are reaching from 100 Hz to the theoretical predictions of the 1 kHz limit. These large bandwidths, in conjunction with large stress and strain output, provide films producing enormous values of power per unit mass. More and more compliant mechanisms are being developed using MEMS process technology. The relatively large power provides unique opportunities for small-scale industrial and biomedical applications (e.g. miniature motors, heart valves and drug delivery systems) [32].

The future applications of magnetic shape memory alloys will be broadened from actuators to new application areas, such as fluidics, vibrators, couplers, positioning devices, sensors and generators. Complicated mechanical structures can be simplified by using MSM materials. One future vision is a sewing machine: a traditional sewing machine has a rotating electric motor and a large number of parts in a rather complicated mechanical transmission system, although the desired motion of the needle is up and down. A sewing machine based on MSM technology could consist of an electromagnet (coil) and a needle made of MSM material.

Much future effort with SMPs will probably be applied to increasing the recovery force of SMPs. One possible way of doing this is reinforcing the polymer matrix with carbon or glass fibres or ceramic filler particles. [24, 25] The two-way shape-memory effect and stimuli other than temperature, such as light or electromagnetic fields, will play a major role in future research. Degradable shape-memory polymers will provide interesting advantages over metal implants, because much of today's follow-on surgery in removing the implants can thus be avoided. SMP-based biological micro-electromechanical systems (Bio-MEMS) are also expected to be a potential medical application. On the other hand, SMP micro-grippers have potential use also in industrial applications where objects must be manipulated in unreachable locations in micro-system assembly.

In the history of man, textiles were the driving force to industrialisation – now it seems that smart textiles and intelligent clothing can play an important role again when developing smart materials for all-round applications.

6.7 References

1. Z. G. Wei, R. Sandström, S. Miyazaki (1998), Shape-memory materials and hybrid composites for smart systems, Part I, *Journal of Material Science* 33, 3743–3762.
2. A. Lenlein, S. Kelch (2002), Shape-Memory Polymers, *Angew. Chem. Int. Ed.* 41, 2034–2057.
3. A. Lendlein, H. Jiang, O. Jünger, R. Langer (2005), Light-induced shape memory polymers, *Nature*, Vol. 434, 14 April 2005.

4. K. Ullakko, J. K. Huang, C. Kantner, R. C. O'Handley, V. V. Kokorin (1996), Large Magnetic-field-induced strains in Ni_2MnGa single crystals, *Appl Phys. Lett.* 69 1996–8.
5. W. M. Huang, B. Yang, L. An, C. Li, Y. S. Chan (2005), Water-driven programmable polyurethane shape memory polymer: Demonstration and mechanism, *Appl. Phys. Lett.* 86, 114105 (2005).
6. T. L. Turner (Langley Research Center) (2001), *Thermomechanical Response of Shape Memory Alloy Hybrid Composites*, NASA/TM-2001-210656.
7. K. Otsuka, C. M. Wayman, (1998), *Mechanism of Shape Memory Effect and Superelasticity, Shape Memory Materials*, Cambridge University Press, Cambridge, (1998) pp. 27–48.
8. J. Ryhänen, (1999), *Biocompatibility Evaluation of Nickel-titanium Shape Memory Metal Alloy* (Ph.D. thesis), University of Oulu, Oulu University Library, 1999.
9. W. J. Buehler, F. E. Wang (1967), A Summary of Recent Research on the NITINOL Alloys and their Potential Application in Ocean Engineering, *Ocean Engineering*, 1, (1967) pp. 105–120, Pergamon Press.
10. G. F. Andreasen, J. L. Fahl (1987), Alloys, Shape Memory. In: Webster J. G. (ed.) *Encyclopedia of medical devices and instrumentation*, Volume 2. Wiley, New York, p. 15–20.
11. C.-Y. Chang, D. Vokoun, C.-T. Hu, (2001), Two-Way Shape Memory Effect of NiTi Alloy Induced by Constraint Aging Treatment at Room Temperature, *Metallurgical and materials transactions A*, Volume 32A, July 2001.
12. T. W. Duerig, A. R. Pelton, D. Stockel (1996), The utility of superelasticity in medicine, *Biomed. Mater. Eng.* 6: 255–266.
13. K. Otsuka, X. Ren (2005), Physical metallurgy of Ti-Ni-based shape memory alloys, *Progress in Materials Science* 50 (2005) p. 511–678.
14. Ming H. Wu, (2001), Fabrication of Nitinol materials and components *Proceedings of the International Conference on Shape Memory and Superelastic Technologies*, Kunming, China, p. 285–292 (2001).
15. G. Julien, Post Processing For Nitinol Coated Articles, *United State Patent 6,254,458* (2001).
16. Ming H. Wu, L. McD. Schetky (2000), Industrial Applications for Shape Memory Alloys, *Proceedings of the International Conference on Shape Memory and SuperelasticTechnolgies,* Pacific Grove, California, p. 171–182 (2000).
17. K. Otsuka, T. Kakeshita (guest editors) (2002), Science and Technology of Shape-Memory Alloys: New Developments, *MRS Bulletin*, February 2002 p. 91–100.
18. K. Ullakko, J. K. Huang, C. Kantner, R. C. O'Handley, V. V. Kokorin, *Appl. Phys. Lett.* 69, 1966 (1996).
19. H. X. Zheng, W. Z. Ma, Y. Y. Lu, M. X. Xia, J. G. Li (2004), New approach to preparing unidirectional NiMnFeGa magnetic shape memory alloy, *Journal of Materials Science* 39 (2004) p. 2557–2559.
20. O. Söderberg, Y. Ge, A. Sozinov, S.-P. Hannula, V. K. Lindroos (2004), Recent breakthrough development of the magnetic shape memory effect in Ni-Mn-Ga alloys, *Smart Mater. Struct.* 14 (2005) p. S223-S235.
21. I. Suorsa, J. Tellinen, K. Ullakko, E. Pagounisa (2004), Voltage generation induced by mechanical straining in magnetic shape memory materials, *Journal of Applied Physics*, Volume 95, Number 12, 15. June 2004.
22. J. Enkovaara (2003), Atomistic Simulations Of Magnetic Shape Memory Alloys (Dissertation for the degree of Doctor of Science in Technology, Department of

Engineering Physics and Mathematics, Helsinki University of Technology), Dissertation 119 (2003), Otamedia Oy Espoo

23. B. C. Heaton, A Shape Memory Polymer for Intracranial Aneurysm Coils (2004), *Thesis for the Degree of Master of Science*, Georgia Institute of Technology, July 2004.
24. Y. Liu, K. Gall, M. L. Dunn, P. McCluskey (2003), Thermomechanics of shape memory polymer nanocomposites, *Mechanics of Materials* 36 (2004) p. 929–940.
25. K. Gall, M. Dunn, Y. Liu, D. Finch, M. Lake, N. Munshi (2002), Shape memory polymer nanocomposites. *Acta Materialia* 50: 5115–5126.
26. A. Lendlein, R. Langer (2002), Biodegradable, Elastic Shape-Memory Polymers for Potential Biomedical Applications, *Science* Vol. 296 31 May 2002.
27. K. Gall, M. L. Dunn, Y. Liu, G. Stefanic, D. Balzar (2004), Internal Stress storage in Shape Memory Polymer Nanocomposites, *Applied Physics Letters* Volume 85, Number 2, 12 July 2004.
28. B. Yang, W. M. Huang, C. Li, C. M. Lee, L. Li (2003), On the effects of moisture in a polyurethane shape memory polymer, *Smart Mater. Struct.* 13 (2004) 191–195.
29. A. Lendlein, A. M. Schmidt, R. Langer (2001), AB-polymer networks based on oligo(ε-caprolactone) segments showing shape-memory properties, *PNAS*, January 30, 2001 Vol. 98 no. 3, p. 842–847.
30. S. Hayashi, Y. Tasaka, N. Hayashi, Y. Akita (2004), Development of Smart Polymer Materials and its Various Applications, *Mitsubishi Heavy Industries, Ltd. Technical Review* Vol. 41 No.1 (Feb. 2004).
31. Y. Fu, H. Du, W. Huang, S. Zhang, M. Hu (2004), TiNi-based thin films in MEMS applications: a review, *Sensors and Actuators*, A 112 (2004) 395–408.
32. G. P. Carman (2004), Thin Film Active Materials, *Tenth annual symposium on US frontiers of engineering*, September 9–11, 2004, Irvine, California.

7
Temperature sensitive shape memory polymers for smart textile applications

J HU and S MONDAL, The Hong Kong Polytechnic University, Hong Kong

7.1 Introduction

Material scientists predict a prominent role in the future for intelligent materials [1]. Environmentally responsive materials that respond to the external conditions and exhibit various functions are called intelligent [2–3] or smart materials. Research on such materials is actively beginning in various fields. Shape memory materials are one kind of such smart materials, which have the ability to remember their original shapes. When the material deforms into a temporary shape it returns to its original shape by external stimuli. This feature, known as the shape memory effect, was first observed in samples of gold-cadmium in 1932 and 1951, and in brass (copper-zinc) in 1938.

It was not until 1962, however, that William J. Buehler and coworkers at the Naval Ordnance Laboratory (NOL) discovered that nickel-titanium showed this shape memory effect [4]. Shape memory polymers (SMPs) are one type of shape memory materials defined as polymeric materials with the ability to sense and respond to external stimuli by returning to a predetermined shape. Polymer such as polynorbornene, trans-polyisoprene, styrene-butadiene copolymer, crystalline polyethylene, some block copolymer, ethylene-vinyl acetate copolymer and segmented polyurethane, etc., have been discovered with shape memory effect. Organic shape memory polymers have a lower recovery force than do shape memory alloys but offer easier processability, light weight, lower production costs, biocompatibility and color variation [5].

Compared with shape memory alloys, SMPs have better potentiality for textile and clothing, and related products. In the case of SMAs, mechanical properties can be adjusted only within a limited range and the maximum deformation that they can undergo is about 8%. In contrast, shape memory polymers (SMPs) have easy shaping, high shape stability, and adjustable transition temperature. Both the shape memory effect and the elasticity memory system effect of shape memory polymers make them a useful candidate for today's intelligent material systems and structures [6]. Here we present a review of temperature sensitive shape memory smart polymeric materials;

such materials can respond to changes in the temperature fields in their environment and make a desirable response for smart textiles.

7.2 A concept of smart materials

The term 'smart materials' is much used and more abused. All materials are responsive. Whether or not they are smart materials is a different question. The responsivity is often useful and forms the basis of transducer technology. Such material systems do exist and involve a spectrum of constituents. A simple pressure transducer that produces a voltage dependent upon the input pressure in a direct one-to-one relationship could never be regarded, as 'smart'. However, a pressure transmitter incorporating a thermocouple that measures both temperature and apparent pressure and corrects the apparent pressure taking due note of the sensor's temperature coefficient could be regarded as smart [7].

Figure 7.1 illustrates the functions of a range of transducer materials and indicates how their input/output characteristics may be summarized in a simplistic graphical form. However, returning to our original definition, there is no 'smartness' here since the output has the same, or sometimes a larger, information content than the input. To engender smartness we need responsivity to a second variable, illustrated conceptually in Fig. 7.2. If the

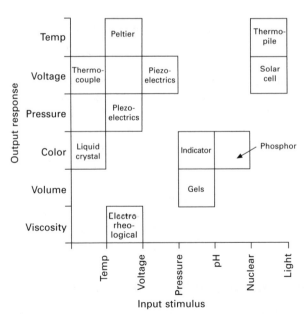

7.1 Response of some sensor materials. These materials are often referred to as 'smart', but are they? (Adapted with permission from Smart Structures and Materials, B. Culshaw, © 1996, Artech House Publisher [7]).

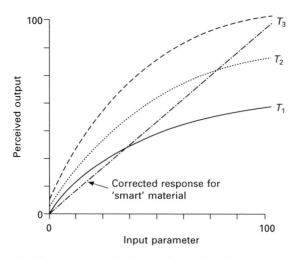

7.2 The response of a 'smart' material. The uncorrected response is a function of the parameter T, but after correction in the material, an unambiguous output is obtained, independent of T. (Adapted with permission from Smart Structures and Materials, B. Culshaw, © 1996, Artech House Publisher [7]).

material can be designed to produce a specific response to a combination of inputs, then it will fulfill the smartness requirements. The term 'smart' describes material that can sense changes in its environment and make a useful or optimal response by either changing its material properties, geometry, mechanical or electromagnetic properties. The smart materials are now one of the most studied fields, because of their ability to change their physical and thermomechanical properties when it is necessary. The concept of intelligent/smart materials implies substances or materials which are able to appropriately respond to changes of external factors by changing their structure and functions [8]. The basic concept of smart materials is that they have their own sensor which is able to change the characteristics of the materials. Smart materials are expected to adapt to their environment and provide a useful response to changes in the environment [9]. The environment may include fluid loading, material damping, the effect of the backing structure and changes in both the structure and the environment. In all cases, smart materials should sense the change and make a useful response. And overall, the intelligent materials will build themselves.

7.3 Shape memory polymer and smart materials

A material which changes one of its property coefficients in response to an external stimulus, and where this change in coefficient can be used to control the stimulus is called a smart material [10]. This definition is then consistent with the behavior of a shape memory polymer (SMP), with the potential for application as a smart material and also as a new functional material. For example, the well-known use as a control for knotting by a smart suture made with shape memory polymer which ties itself into a perfect knot where access is limited [11]. Its self-knotting action occurs when it is heated a few degrees above normal body temperature. At the start, a key characteristic of polymeric materials, which are good candidates for development into smart intelligent materials, is the ability to change properties, such as structure or composition, and function in a controlled response to a change in environment or operating conditions. The next level is the promise for development of some level of built-in intelligence, such as graded reaction to stimuli and ability to recognize or discriminate shapes or forms [12].

Stimuli sensitive polymers (SSPs) or shape memory polymers (SMPs) yield intelligent textiles that exhibit unique environmental responses. The molecular design of SSPs [13] facilitates phase change behavior in response to environmental stimuli and allows SSP textiles to change structures and properties. SSPs yield fabrics with such properties as air permeability, hydrophilicity, heat transfer, shape, and light reflectance that are responsive to such environmental stimuli as temperature, pH, moisture, light, and electricity. Shape memory polymers (SMPs) offer greater deformation capacities, easier shaping, and greater shape stability, and small changes to the chemical structure and composition of SMPs result in a wide variety of transition temperatures and mechanical properties.

7.3.1 Principle of temperature sensitive shape memory polymer

Shape-memory materials are stimuli-responsive materials. They have the capability of changing their shape upon application of an external stimulus. Shape memory may be triggered by heat, light, electric field, magnetic field, chemical, moisture, pH and other external stimuli [5–6]. Change in shape caused by a change in temperature is called a thermally induced shape-memory effect. These are materials which are stable at two or more temperature states. While in these different temperature states, they have the potential to be different shapes once their 'transformation temperatures' (Tx) have been reached. Shape memory alloys (SMAs) and Shape Memory Polymers (SMPs) are materials with very different shape changing characteristics. While exposed to their Tx, devices made from SMAs have the potential to provide force such as in the case of actuators. Devices made from SMPs in contrast, while exposed to their Tx, provide mechanical property loss as in the case with releasable fasteners. The shape memory polymers described in this chapter are all thermosensitive shape memory polymers.

Temperature sensitive shape memory polymers are a special class of adaptive materials which can convert thermal energy directly into mechanical work. This phenomenon, known as the shape-memory effect (SME) occurs when one of these special polymers is mechanically stretched at low temperatures, then heated above a critical transition temperature, which results in the restoration of the original shorter 'memory' shape of the specimen [14]. The proposed theory expresses well the thermomechanical properties of thermoplastic polymer, such as shape fixity, shape recovery, and recovery stress [1]. The mechanism of shape memory behavior with temperature stimuli can be shown as outlined in Fig. 7.3. These materials have two phase structures, namely, the fixing phase which remembers the initial shape and the reversible phase which shows a reversible soft and rigid transition with temperature.

At temperatures above the glass transition temperature (T_g), the polymer achieves a rubbery elastic state (Fig. 7.4) where it can be easily deformed without stress relaxation by applying external forces over a time frame $t < \tau$, where τ is a characteristic relaxation time. When the material is cooled below its T_g, the deformation is fixed and the deformed shape remains stable. The pre-deformation shape can be easily recovered by reheating the material

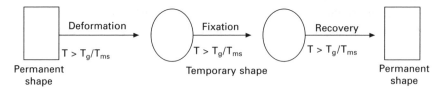

7.3 Typical temperature stimulating shape memory behaviors.

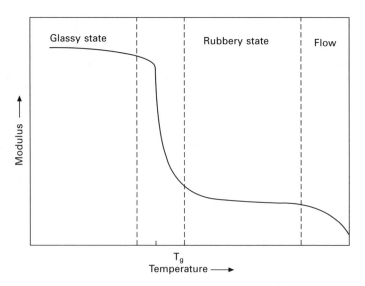

7.4 Temperature dependency elasticity of thermoplastic polymer.

to a temperature higher than the T_g [15]. Therefore, admirable shape memory behavior requires a sharp transition from glassy state to rubbery state, a long relaxation time, and a high ratio of glassy modulus to rubbery modulus. The micromorphology of SMPs strongly affects its mechanical properties. There are many factors that can influence these SMPs: chemical structure, composition, and sequence-length distribution of the hard and soft segment in segmented polymer, overall molecular weight and its distribution. An elastomer will exhibit a shape-memory functionality if the material can be stabilized in the deformed state in a temperature range that is relevant for the particular application. The shape is deformed under stress at a temperature near or above T_g or the segment crystal melting temperature, T_{ms}. The deformed shape is fixed by cooling below T_g or T_{ms}. The deformed form reverts to the original shape by heating the sample above T_g or T_{ms}.

7.3.2 Architecture of the shape memory polymer-network

Shape memory effect is not related to a specific material property of single polymers; it rather results from a combination of the polymer structure and the polymer morphology. Shape memory behavior can be observed for several polymers that may differ significantly in their chemical composition and structure. The shape memory polymers that show a thermally induced shape-memory effect are network structure. The netpoints, which can have either a chemical or physical nature, determine the permanent shape. The net chain shows a thermal transition in the temperature range which the shape-memory effect is supposed to trigger. Segmented polymer with shape-memory effect properties requires a minimum weight fraction of hard-segment-determining blocks to ensure that the respective domains act as physical netpoints [1]. Formation of a network structure with rigid point/segments (fixed phase) and amorphous/flexible segment/regions (reversible phase) are the two necessary conditions for their good shape memory effect [16]. In general, interpolymer chain interactions are so weak that one-dimensional polymer chains cannot keep a certain shape above Tg.

To maintain a stable shape, polymer chains should have a three-dimensional network structure. Interpolymer chain interactions useful for constructing the polymer network are crystal aggregate or glassy state formation, chemical cross-linking, and chain entanglement [5]. The latter two interactions are permanent and used for constructing the original shape. The other interactions are thermally reversible and used for maintaining the transient shapes. This can be reached by using the network chains as a kind of molecular switch. For this purpose the flexibility of the segments should be a function of the temperature. One possibility for a switch function is a thermal transition (T_{trans}) of the network chains in the temperature range of interest for the particular applications. The crystallites formed prevent the segments from spontaneously recovering the permanent shape that is defined by the netpoints. The permanent shape of the shape memory network structure is stabilized by covalent netpoints, whereas the permanent shape of the shape memory thermoplastics is fixed by the phase with the highest thermal transition at T_{perm} [1].

7.4 Some examples of shape memory polymer for textile applications

Many polymers have been discovered with shape memory effects. Some of them suitable for textiles applications are shown in Table 7.1 along with physical interactions used for memorizing the original and transient shapes of the polymers.

7.4.1 Polynorbornene

Norsorex is linear, amorphous polynorbornene **30** developed by the companies CdF Chemie/Nippon Zeon in the late 1970s. The molecular weight is as high as 3 million [20–21]. The polymer contains 70 to 80 mol% *trans*-linked norbornene units and has a glass transition temperature between 35 and 45 °C [17–18]. The shape memory effect of this amorphous polymer is based on the formation of a physical cross-linked network as a result of entanglements of the high molecular weight linear chains, and on the transition from the glassy state to the rubber-elastic state [19]. The long polymer chains entangle

Table 7.1 Shape memory polymer and mechanisms [1, 5]

Polymers	Physical interactions	
	Original shape	Transient shape
Polynorbornene	Chain entanglement	Glassy state
Polyurethane	Microcrystal	Glassy state
Polyethylene/nylon-6 graft copolymer	Crosslinking	Microcrystal
Styrene-1,4-butadiene block copolymer	Microcrystal/Glassy state of polystyrene	Microcrystal of poly(1,4-butadiene)
Ethylene oxide-ethylene terephthalate block copolymer	Microcrystal of PET	Microcrystal of PEO
Poly(methylene-1, 3-cyclopentane) polyethylene block copolymer	Microcrystal of PE	Glassy state/micro-crystal of PMCP

each other and a three-dimensional network is formed. The polymer network keeps the original shape even above Tg in the absence of stress. Under stress the shape is deformed and the deformed shape is fixed when cooled below Tg. Above the glass transition temperature polymers show rubber-like behavior. The material softens abruptly above the glass transition temperature Tg. If the chains are stretched quickly in this state and the material is rapidly cooled down again below the glass transition temperature the polynorbornene chains can neither slip over each other rapidly enough nor become disentangled. It is possible to freeze the induced elastic stress within the material by rapid cooling. The shape can be changed at will. In the glassy state the strain is frozen and the deformed shape is fixed. The decrease in the mobility of polymer chains in the glassy state maintains the transient shape in polynorbornene. The recovery of the material's original shape can be observed by heating again to a temperature above Tg. This occurs because of the thermally induced shape-memory effect [19]. The disadvantage of this polymer is the difficulty of processing because of its high molecular weight [5].

7.4.2 Segmented polyurethane

Shape memory polyurethane (SMPU) is a class of polyurethane that is different from conventional polyurethane in that these have a segmented structure and a wide range of glass transition temperature (Tg). Segmented polyurethane is composed of three basic starting raw materials, these are (i) long chain polyol, (ii) diisocyanate and (iii) chain extender. Diisocyanate and chain extender form a hard segment. On the other hand long chain polyol is soft segment. These types of polyurethanes are characterized by a segmented structure (block copolymer structure) and the morphology depends on chemical composition and the chain length of the soft segment (block). The SMPU has a microphase separated structure due to the thermodynamic incompatibility between the hard and soft segment. Hard segments can bind themselves via hydrogen bonding and crystallization, making the polymer solid below melting point temperature. Reverse phase transformation of soft segment is reported to be responsible for the shape memory effect.

The shape memory effect can be controlled via the molecular weight of the soft segment, mole ratio between soft and hard segment, and polymerization process [20]. If a SMPU is cooled from above Tg to a temperature below Tg, in the presence of a mechanical load, and after removal of load, significant deformation anywhere in the range of 10–200% becomes locked into the polymer. Phase-transition temperature of shape memory polyurethane is a little higher than the operating tempearure. A large drop of modulus and an enhanced micro-Brownian motion on heating through glass transition or soft segment crystal melting temperature can be used in the molecular design of the shape memory behavior [21–22]. Shape memory characteristics of the segmented polyurethanes having crystallizable soft segments are closely related to the temperature-dependent dynamic mechanical properties of the materials [21, 23]. A large glassy state modulus led to large shape recovery upon heating and standing at high temperature. On the other hand, high crystallinity of the soft segment regions [21, 24] at room temperature and the formation of stable hard segment domains acting as physical crosslinks in the temperature range above the melting temperature of the soft segment crystals are the two necessary conditions for a segmented copolymer with shape memory behavior.

The response temperature of shape memory is dependent on the melting temperature of the soft segment crystals. The final recovery rate and the recovery speed are mainly related to the stability of the hard segment domains under stretching and are dependent on the hard segment content. Control of hard segment content is important in determining the physical properties of shape memory polyurethane. The amount of hard segment rich phase would affect the ratio of the recovery, that is, the low content leads to the recovery of the deformed specimen being incomplete [25]. The recovery rate would be influenced by the modulus ratio and the size of the dispersed phase in the micromorphology.

Polyurethane with 20 or 25% of hard segment content could not show shape recovery due to weak interaction or physical cross-link. On the other hand with 50% hard segment did not show shape recovery due to the excess interaction among the hard segment and the resulting rigid structure. The maximum stress, tensile modulus and elongation at break increased significantly at 30% hard segment content, and the highest loss tangent was found typically at the same composition. Finally, 80–95% of shape recovery was obtained at 30–45 wt% of hard segment content [26]. The typical textile applications of SMPU are as fibre, coating, lamination, etc.

7.4.3 Polyethylene/nylon 6 graft copolymer

High density polyethylene ($\rho = 0.958$ g/cm^3) grafted with nylon-6 that has been produced in a reactive blending process of PE with nylon-6 by adding maleic anhydride (bridging agent) and dicumyl peroxide shows shape memory properties [27]. The nylon contents in the blends are in the range from 5 to 20 wt%. The maleated polyethylene/nylon 6 blend specimens are able to show good shape memory effect under normal experimental conditions. An elastic network structure is formed in these M-PE/nylon 6 blends, and the nylon domains (domain size less than 0.3 µm) dispersed in the PE chains in the matrix region. The high crystallinity of polyethylene segments at room temperature and the formation of a network structure in these specimens are the necessary conditions for their good shape memory effect. The nylon domains, which serve as physical cross links, play a predominant role in the

Table 7.2 Shape memory properties of HDPE-g-nylon-6

Samples	Strain fixation (%)	Strain recovery (%)	Recovery temperature (°C)
Ir-PE	96.2	94.4	100.5
PE with 20 wt% nylon-6	98.6	96.6	120.3
PE with 15 wt% nylon-6	98.9	96.0	121.0
PE with 10 wt% nylon-6	99.5	96.0	121.3
PE with 5 wt% nylon-6	99.8	95.0	118.8

(Adapted with permission from Li *et al. Polymer*, 39 (26), 6929, © 1998, Elsevier [27]).

formation of a stable network structure for the graft copolymer. All the M-PE/nylon blend specimens exhibit typical good shape memory behavior, having final recovery rates, R_f, more than 95% and high recovery speeds comparable to those of low-density PE cross-linked by reaction with ionizing radiation (Ir-PE), the commercial sample with a high degree of chemical cross-linking. The response temperature of blend samples, around 120 °C, is closely related with the melting temperature of the PE crystals in these specimens. Strain fixity rates of around 99% and strain recovery rates between 95 and 97% have been determined for these graft copolymers for an elongation 100% (Table 7.2). Shape memory properties of PE-g-nylon-6 with fixed contents of DCP (0.08 phr) and MAH (1.5 phr) along with Ir-PE are given in Table 7.2.

7.4.4 Block copolymers

Some block copolymers with phase-separated structures show the following shape memory properties.

Styrene-1,4-butadiene block copolymer

The crystal transformation of semicrystalline styrene-butadiene block copolymer attributes the shape memory properties [28–29]. Phase separated block copolymer contain 34 wt% polystyrene (PS) and 66 wt% poly(1,4-butadiene). The melting temperature of the poly(1,4-butadiene) crystallites (around 80 °C) represents the switching temperature for the thermally induced shape-memory effect. Aggregate or glassy state formation in polystyrene segment is used to memorize the original shape. Thus, polystyrene supplies hard domain. The high glass transition temperature and microcrystal structure of polystyrene segments hinders the polybutadiene chains from slipping off each other upon stretching. Below 40 °C the poly(1,4-butadiene) domain becomes crystallized and the deformed shape is fixed. The shape again returns to the original one upon heating at around 80 °C, at which microcrystal in the poly(1,4-butadiene) domain melts. A strain recovery of 80% is observed upon application of maximum of strain (ε_m) of 100%.

Ethylene oxide-ethylene terephthalate block copolymer

Microphase separated segmented copolymer based on poly(ethylene oxide) and poly(ethylene terephthalate) (EOET) shows shape memory behavior [30–32]. Polyethylene terephthalate (PET) domain formed hard segment phase, on the other hand poly(ethylene oxide) (PEO) domain formed soft segment rich phase. Thermally stimulated deformation recovery (R_f) depends on the stability of the physical cross-links formed by the hard segment, and at the same time, is influenced by the length of soft segment. A long PEO length could undergo a larger extension without dislocation from the anchor nodes of the PET-domain crystal that act as physical cross-links in the formation of the PEO-segment network. On the other hand, a higher crystallizability of longer PEO chains can impose a higher retardation of the relaxation of extended chains. Longer PEO segments would have a higher crystallinity, and a large number of PEO segments would crystallize rather than going into amorphous phase. The soft segment crystallization determines the thermally stimulated deformation recovery temperature T_r and T_M. R_f certainly also depends on the hard segment content and the molecular weight of the soft segment in the EOET segmented copolymers. Physical cross-linking formed by the hard segments are very well aggregated and not destroyed by stretching. With the same soft segment length, the higher the hard segment content, the better the aggregate formation and the corresponding deformation recovery is higher (Table 7.3).

Poly(methylene-1,3-cyclopentane) polyethylene block copolymer

New metallocene catalyst systems used to synthesize poly(methylene-1,3-cyclopentane) (PMCP), by the cyclopolymerization of 1,5-hexadiene [33].

Table 7.3 Shape memory properties of EOET copolymer

Block length of PEO (M_w)	HS (%)	T_r (°C)	T_M(°C)	R_f(%)
4000	27.6	45	45.2	84
4000	32.0	44	43.9	85
6000	21.2	48	47.3	90
6000	25.7	46	46.5	92
10,000	16.5	55	54.5	93
10,000	21.8	53	52.7	95

(Adapted with permission from Luo *et al. Journal of Applied Polymer Science*, 64, 2433, 1997, © 1997 John Wiley and Sons Ltd. [31]).

Block copolymer of PMCP and polyethylene obtained by addition polymerization of ethylene, i.e., after polymerization of 1,5-hexadiene shows a phase separated structure, where the PMCP domain act as a soft segment rich phase and the polyethylene domain as a hard segment phase. Crystal melting point of the hard segment was around 120 °C and that of the soft segment was nearly 64 °C. On the other hand, T_g of the soft segment was 5–10 °C. The sample was elongated at 45 °C in a rubbery state above T_g to 100% strain (ε_m), while maintaining the strain at ε_m, the sample was cooled to 25 °C and unloaded. Upon removing the constraint at 25 °C some recovery of the strain to ε_u occurs, because 25 °C is not far below but close to the T_g of the examined samples. The samples were heated to 85 °C, a temperature above the soft segment crystal melting temperature (T_{ms}), over a period of five minutes, and maintained at that temperature for the next ten minutes, allowing recovery from the strain. This completes one thermomechanical cycle (N = 1) leaving a residual strain ε_p where the next cycle (N = 2) starts. A shape fixity of more than 75% and shape recovery around 80% would obtained up to four thermomechanical cycles (Fig. 7.5).

7.5 Potential use of shape memory polymer in smart textiles

7.5.1 The use of the functional properties

Moisture permeability, volume expansion and refractive index vary significantly above and below T_g/T_{ms} according to the difference of the kinetic properties of molecular chains [34] between temperatures above and below T_g/T_{ms}. Temperature dependency of water vapor permeability is an important factor to be considered for the effective utilization of this type of smart materials. One of the applications of moisture permeability of a thin film is in sports wear. Moisture permeability is high above T_g/T_{ms} and low below T_g/T_{ms} (Fig. 7.6). For sports wear, the heat retaining property at low temperature and gas permeability at high temperature are excellent, yielding high quality for the sportswear. Imagine a coat that detected rain. It could change shape to keep the moisture out, but always be comfortable to wear. When the weather becomes cooler, it could change again to keep the wearer warm.

The shape-deformable matrix material contains an elastomeric, segmented block copolymer, such as a polyurethane elastomer or a polyether amide elastomer. The elastomeric polymer provides the force for dimensional change when the moisture-absorbing polymer softens and relaxes as a result of moisture absorption. The inclusion of a non-elastomeric polymer provides a degree of recovery when exposed to humidity. The activation process for the shape-deformable materials does not require substantially increasing the

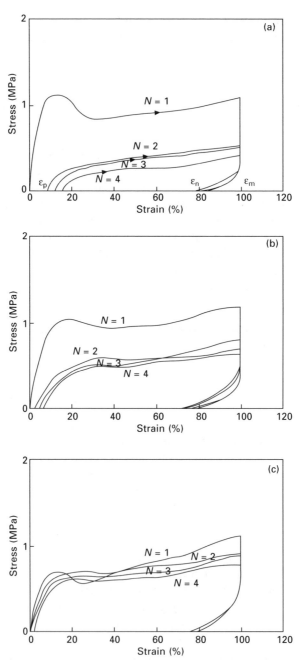

7.5 Cyclic tensile behavior of (a) PMCP, (b) PMCP-PE1, and (c) PMCP-PE2 (elongated at 45 °C and recovered at 85 °C), N represents number of cycles, ethylene segment in PE1 and PE2 are 8.1 and 17.0% (wt%) respectively. (Adapted with permission from Jeong *et al.*, Polymer International, 51, 275, 2002, (c) 2002 SCI, John Wiley and Sons Ltd. [33]).

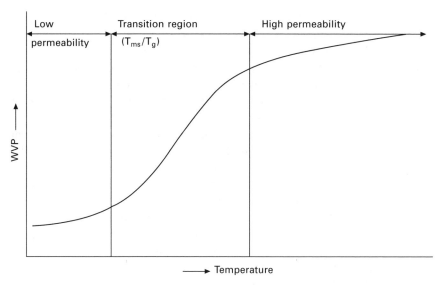

7.6 Temperature dependency water vapor permeability of shape memory polymers.

temperature of the materials, which helps in preventing leakage from the absorbent product. Also, by adjusting the chemical structure of the shape-deformable polymer, a specific polymer can be tailored to interact at a selected level of humidity. These kinds of shape memory materials would be applicable in disposable hygiene products, such as diapers, training pants, and incontinence products [35].

7.5.2 The use of changes of physical properties related to the phase transformation

Physical properties of materials are related to the crystal structure and microstructure of the involved phases. In spite of the diffusionless character of the phase transformation, the lattice parameter, crystal structure and microstructure of the parent phase and the reverse phase, differ significantly. As a consequence, during transformation, values of E-modulus, internal friction, electrical resistance, and hardness might change [36]. A typical example of changing phase change from temporary shape to permanent shape is suture. The technique of keyhole surgery minimizes scarring, speeds healing and reduces the risk of infection. However, it is extremely difficult to carry out delicate surgical procedures accurately in a confined space, such as implanting a bulky device or knotting a suture with the right amount of tension. In the latter case, if a knot is pulled too tight, necrosis of the surrounding tissue can occur, but if it is too loose, the incision will not heal properly and scar tissue develops [37].

Technology has developed a shape memory polymer that has been designed by MnemoScience to act as a smart suture that ties itself into a perfect knot. The suture's self-knotting action occurs when it is heated a few degrees above normal body temperature. The suture can therefore be used to seal difficult wounds where access is limited [11]. The material is a multi-block copolymer, in which block-building segments are linked together in linear chains. Specifically, the polymer they created contains a hard segment and a 'switching' segment, both with different thermal properties. One segment melts, or makes another kind of transition, at a higher temperature than the other. By manipulating the temperature and stress applied to the overall material, Langer and Leindlin end up with a material that forms a temporary shape at one temperature, and a permanent shape at a higher temperature. After increasing the temperature, the suture material shrunk, creating a knot with just the right amount of tension on the surrounding tissue. It is difficult to create such a knot in the confined spaces associated with endoscopic surgery. They demonstrated this by creating the first 'smart' degradable suture.

When SMPs change from glassy state to rubbery state, Young's modulus, tensile and elastic properties vary greatly, which would be useful for garments. For example, when the SMPs-based garments were washed at higher temperature or worn at body temperature, they could recover to the original state (wrinkle free). Shape memory fabrics regain the flatness of their appearance and retention of their creases after the fabrics have been immersed in hot water [6, 38].

7.6 General field of application

The recent advent of advanced 'intelligent' materials has opened the door for many useful applications in textile field. Although relatively few structures are presently built with such materials, the potential market for the application of these materials can be quite large. The unique properties of shape memory polymers resulting from thermally induced phase transformation has been exploited commercially to produce a variety of products. Due to the properties of shape memory polymers some suitable field of technical applications can be found. Shape memory polymers can be laminated, coated, foamed, and even straight converted to fibres. There are many possible end uses of these smart textiles, some of them are summarized in the following section.

7.6.1 Medical textiles

Smart fibre

The smart fiber made from the shape memory polymer designed by MnemoScience ties itself into a perfect knot. Features of the MnemoScience

[39] polymer material include its compatibility with body fluid, shape retention properties, retention of two shapes in memory, high shape fixity, high shape recovery, and its ability to form temporary and permanent shapes. The suture contracts to its permanent shape when heated and its self-knotting action occurs when it is heated a few degrees above normal body temperature. The suture therefore can be used to seal difficult wounds where access is limited. The other applications of smart fibers would be stents, and screws for holding bones together.

Surgical protective garments

Shape memory polymer coated or laminated materials can improve the thermophysiological comfort of surgical protective garments, bedding and incontinence products because of their temperature adaptive moisture management features.

Other applications

Kimberly-Clark company in US Patent 6 627 673 [40] disclosed humidity-activated shape-memory materials. The materials could find particular application in disposable hygiene products, such as diapers, training pants and feminine care and incontinence products. They may be used, for example, in products containing a gatherable or elastic part. Stretching the humidity-responsive material forms the latent deformation. The humidity-responsive material contains at least one shape-deformable matrix material.

7.6.2 Outdoor clothing

Films of shape memory polymer can be incorporated in multilayer garments, such as those that are often used in the protective clothing or leisurewear industry. The shape memory polymer reverts within a wide range temperatures. This offers great promise for making clothing with adaptable features [41]. Using a composite film of shape memory polymer as an interliner in multilayer garments, outdoor clothing could have adaptable thermal insulation and be used as protective clothing.

The United States Army Soldier Systems Center developed [42] a wet/dry suit with a shape memory polymer membrane and insulation materials that keep the wearer warm in marine environments but do not inhibit perspiration on land. Molecular pores open and close in response to air or water temperature to increase or minimize heat loss. Apparel could be made with shape memory fiber. Forming the shape at a high temperature provides creases and pleats in such apparel [43] as slacks and skirts. Other applications include fishing yarn, shirt neck bands, and cap edges.

7.6.3 Casual clothing

Permeability of SMP coated or laminated fabrics changes as the wearer's environment and body temperature change to form an ideal combination of thermal insulation and vapor permeability for underwear and outerwear. When the body temperature is low, the fabric remains less permeable and keeps the body warm. When the body is in a sweat condition, it allows the water vapor to escape into the environment because its moisture permeability becomes very high with increasing body temperature. This releases heat from the apparel. Using a composite film of shape memory polymers as an interlining provides [44] apparel systems with variable tog values to protect against a variety of weather conditions.

7.6.4 Sportswear

Sportswear should provide protection from wind and weather, dissipate perspiration, and have excellent stretch and recovery properties. Shape memory polyurethane fibers respond to external stimuli in a predetermined manner and are useful in sportswear.

7.7 Challenges and opportunities

The purpose in smart textile design is to utilize intelligent materials. Intelligent properties may be integrated in the material either in molecular or fibre level in the textile structure. The study of shape memory polymer has begun only in the last few years, and remains largely unexplored, and there are only a few marketable products. However, researchers are constantly finding combinations of technologies to increase avenues for commercialization. In general, thermally stimulated shape memory polymers show very large strains. There are many shape memory polymers that have been reported in the literature. Although these polymers have been proposed for a number of uses, their textile applications have been limited, due to the reversible temperature for the shape memory effects to be triggered is higher than the body temperature. Further investigations on the improvement of the induced strain magnitude, the stability of the strain characteristics with respect to temperature change, mechanical strength and durability after repeated driving are required to produce practical and reliable materials. The shape memory mechanism, the relation between structure and effect, especially their application to textiles and garments still have potentiality to develop. There are many challenges and opportunities in such an endeavor. Here, material design is very important for textile applications, especially their transition temperature to be triggered, and tailoring of hard and soft segment, as these all affect the shape recovery temperature, hand feel, etc. Compared with other types of shape memory materials such as shape memory alloys, the literature for understanding of the properties of shape memory polymers is still very fragmented and restricted. Investigation into their detailed characteristics and their potential is still required. Considering these factors, we think future research into the following will be interesting and challenging for the textile technologist:

- developing new shapes of polymer for textile applications
- shaping recovery to be triggered about room temperature/body temperature
- studying the relationship between shape memory polymer structure, ingredients and properties in function and processing
- adequately understanding the complexity of phase transformation systems and processes
- developing smart textiles with shape memory functions using shape memory polymer
- establishing the methods to evaluate the properties and performance of shape memory polymer and smart textiles.

7.8 Acknowledgement

The authors would like to acknowledge John Wiley and Sons Ltd., Elsevier Ltd., Artech House Publishers, for giving permission to reproduce their copyright material.

7.9 References

1. Lendlein A, and Kelch S, 'Shape-memory polymers', *Angew. Chem. Int. Ed.*, 2002, **41**, 2034–2057.
2. Okano T, and Kikuchi A, 'Intelligent biointerface: remote control for hydrophilic-hydrophobic property of the material surfaces by temperature', Proceeding of the third international conference on intelligent materials, *Third European conference on smart structures and materials*, Lyon, France, 3–5 June 1996, edited by Gobin P F and Tatibouët J, 34–41.
3. Koshizaki N *et al.*, 'Intelligent functionalities of composite materials', *Bulletin of Industrial Research Institute*, 1992, No 127, 99–128.
4. Srinivasan A V, and McFarland D M, 'Shape Memory Alloys', *Smart Structures*, Cambridge University Press, 2001, 26–72.
5. Irie M, 'Shape memory polymers', in *Shape memory materials*, edited by Otsuka K and Wayman C M, Cambridge University Press, 1998, 203–219.
6. Mondal S, Hu J L, Yang Z, Liu Y, and Szeto Y S, 'Shape memory polyurethane for smart garment', *Research Journal of Textile and Apparel*, **6** (2), (Nov 2002), 75–83.
7. Culshaw B, in *Smart structures and materials*, Artech House Publishers, 1996, 1–16.
8. Christophorov L N, 'Intelligent molecules: examples from biological charge transport', Proceedings of the third international conference on intelligent materials, *Third European conference on smart structures and materials*, Lyon, France, 3–5 June 1996, edited by Gobin P F and Tatibouët J, 58–65.

9. Varadan V V, Chin L C, and Varadan V K, 'Modeling integrated sensor/actuator functions in realistic environments', *Proceedings first European conference on smart structures and materials*, Glasgow, 12–14 May 1992, edited by Culshaw B, Gardiner P T and McDonach A, 1–6.
10. Melton K N, 'General applications of SMAs and smart materials', in *Shape memory materials*, edited by Otsuka K and Wayman C M, Cambridge University Press, 1998, 220–239.
11. 'Bard backs smart fibres for surgery', *Textile Month*, 2003 (DEC/JAN), 38–39.
12. Takagi T, 'A concept of intelligent materials and the current activities of intelligent materials in Japan', *Proceedings first European conference on smart structures and materials*, Glasgow, 12–14 May 1992, edited by Culshaw B, Gardiner P T and McDonach A, 13–18.
13. Vankar P, 'Challenges with Intelligent SSP in Fabrics', *Asian Textile Journal*, 9 (5), May 2000, 74.
14. Thrasher M A, Shahin A R, Meckl P H, and Jones J D, 'Thermal cycling of shape memory alloy wires using semiconductor heat pump modules', *Proceedings first European conference on smart structures and materials*, Glasgow, 12–14 May 1992, Edited by Culshaw B, Gardiner P T and McDonach A, 197–200.
15. Tobushi H, Hashimoto T, Hayashi S, and Yamada E, 'Thermomechanical constitutive modeling in shape memory polymer of polyurethane series', *J. Intelligent Materials Systems and Structures*, **8**(8), Aug. 1997, 711–718.
16. Hu J L, Ding X M, Tao X M, and Yu J, 'SMPs at work', *Textile Asia*, 2001, December, 377–381.
17. Sakurai K, and Takahashi T, 'Strain-induced crystallization in polynorbornene', *J. Appl. Polym. Sci.*, **34**, 1989, 1191–1194.
18. Sakurai K, Kashiwagi T, and Takahashi T, 'Crystal structure of polynorbornene', *J. Appl. Polym. Sci.*, **47**, 1993, 937–940.
19. Mather P T, Jeon H G, and Haddad T S, 'Strain recovery in POSS hybrid thermoplastic', *Polym. Prepr. Am. Chem. Soc. Div. Polym. Chem.*, 2000, 41, 528–529.
20. Kim B K, Lee S Y, Lee J S, Baek Y J C, Lee J O, and Xu M, 'Polyurethane ionomers having shape memory effects', *Polymer*, **39**(13), 1998, 2803–2808.
21. Li F, Zhang X, Hou J, Xu M, Luo X, Ma D, and Kim B K, 'Studies on thermally stimulated shape memory effect of segmented polyurethanes', *J. Appl. Polym. Sci.*, **64** (8), 1997, 1511–1516.
22. Li F, Hou J, Zhu W, Zhang X, Xu M, Luo X, Ma D, and Kim B K, 'Crystallinity and morphology of segmented polyurethane with different soft segment length', *J. Appl. Polym. Sci.*, **62** (4), 1996, 631–638.
23. Kim B K, Lee S Y, and Xu M, 'Polyurethanes having shape memory effects', *Polymer*, **37** (26), 1996, 5718–5793.
24. Kim B K, Shin Y J, Cho S M, and Jeong H M, 'Shape-memory behavior of segmented polyurethanes with an amorphous reversible phase: the effect of block length and content', *J. Polym. Sci.: Part B, Polym. Phys.*, **38**, 2000, 2652–2657.
25. Lee H Y, Jeong H M, Lee J S, and Kim B K, 'Study on the shape memory polyamides. Synthesis and thermomechanical properties of polycaprolactone-polyamide block copolymer', *Polym. J.*, **32** (1), 2000, 23–28.
26. Lee H S, Wang Y K, and Hsu S L, 'Spectroscopic analysis of phase separation behavior of model polyurethanes', *Macromolecules*, **20** (9), 1987, 2089–2095.
27. Li F, Chen Y, Zhu W, Zhang X, and Xu M, 'Shape memory effect of polyethylene/nylon 6 graft colpolymers', *Polymer*, **39** (26), 1998, 6929–6934.

28. Sakurai K, Tanaka H, Ogawa N, and Takahashi T, 'Shape-memorizable styrene-butadiene block copolymer 1. Thermal and mechanical behavior and structural change with deformation', *J. Macromolecular Sci. Phys*, **36**(6), 1997, 703–716.
29. Sakurai K, Shirakawa Y, Kashiwagi T, and Takahashi T, 'Crystal transformation of styrene-butadiene block copolymer', *Polymer*, **35**(19), 1994, 4238–4239.
30. Wang M, Luo X, and Ma D, 'Dynamic mechanical behavior in the ethylene terephthalate-ethylene oxide copolymer with long soft segment as a shape memory material', *European Polymer J.*, **34**(1), 1–5, 1998.
31. Luo X, Zhang X, Wang M, Ma D, Xu M, and Li F, 'Thermally stimulated shape-memory behavior of ethylene oxide-ethylene terephthalate copolymer', *J. Appl Polym Sci.*, **64**, 1997, 2433–2440.
32. Wang M, and Zhang L, 'Recovery as a measure of oriented crystalline structure in poly(ether ester)s based on poly(ethylene oxide) and poly(ethylene terephthalate) used as shape memory polymers', *J. Polym. Sci., Part B: Polym. Phys.*, **37**, 1999, 101–112.
33. Jeong H M, Song J H, Chi K W, Kim I, and Kim K T, 'Shape memory effect of poly(methylene-1,3-cyclopentane) and its copolymer with polyethylene', *Polym. Inter.*, **51** (4), 2002, 275–280.
34. Tobushi H, Hara H, Yamada E, and Hayashi S, 'Thermomechanical properties of shape memory polymer of polyurethane series and their applications', *Proceedings of the third international conference on intelligent materials*, Third European conference on smart structures and materials, Lyon, France, 3–5 June, 1996, 418–423.
35. 'Hygiene: Shape-memory materials', *Medical Textiles*, 2004, (April), 6–7.
36. Humbeeck J V, 'On the adaptivity of shape memory alloys for use in adaptive materials', *Proceedings of the third international conference on intelligent materials*, Third European conference on smart structures and materials, Lyon, France, 3–5 June 1996, edited by Gobin P F and Tatibouët J, 442–451.
37. Lendlein A, and Langer R, 'Biodegradable, elastic shape memory polymers for potential biomedical applications', *Science*, 296, 31 May, 2002, 1673–1676.
38. Li Y, Chung S, Chan L, and Hu J, 'Characterization of shape memory fabrics', *Textile Asia*, **35**(6), 2004, 32–37.
39. *Smart surgery, Future Materials* (through Textile Technology Index), Issue 1, 2004, 33–34.
40. Topolkaraev V A, and Soerens D A, 'Methods of making humidity activated materials having shape-memory', *USP 6627673*, September 30, 2003.
41. 'Potential uses of shape memory film in clothing', *Technical-Textiles-International* (through Textile Technology Index), 1999; 8(8), 17–19.
42. Draper D, 'It's Not Necessarily Magic, but.', *World Sports Activewear*, **7** (3), Autumn 2001, 22–26.
43. Russell D A, Hayashi S, and Yamada T, 'The potential use of memory film in clothing', *Techtextil Symposium – New Protective Textiles* (through Textile Technology Index database), April, 1999, 12–15.
44. Hayashi S, and Kondo S, 'Shape memory polymer yarn', *JP 2169713* (through Textile Technology Index database), June 29, 1990.

8
Development of shape memory alloy fabrics for composite structures

F BOUSSU, GEMTEX, France and
J-l PETITNIOT, ONERA, France

8.1 Introduction

Some smart materials have particular inner functions that allow them to be at the same time sensor and/or actuator.[1] They are able to instantly modify the physical properties (shape, conductivity, colour ...) in response to natural or artificial events (variation in temperature, electrical or magnetic fields, mechanical strains). To put it in a nutshell, the material 'reacts' to an external stimulus and 'adapts' its behaviour. In particular, actuators provide a mechanical force or change their aspect (deformation, change of colour or opacity) in order to prevent a modification of the environment or to generate an active response. Three main categories of smart material can be distinguished:

1. The piezoelectric materials: they generate an electrical tension when they are submitted to a strain or inversely an electrical tension can provide a strain. The range and the frequency of the signal are directly correlated to the mechanical buckling.
2. The magnetostrictive or electrostrictive materials: under a magnetic (or respectively electrical) field, they can change their shape.
3. The shape memory alloy (SMA) materials: during the variation of temperature, they modify their crystal lattice structure to reach one of the metallic phases, martensitic, austenitic or a blend of martensitic and austenitic.

This chapter deals with the last category of smart materials, the shape memory alloy. A complete and brief description of their different properties helps us to well understand their amazing capacities. Then, a review of the different kinds of application is also given with respect to the presented properties. This will be followed by focusing on the different applications in the textile field, especially the interesting damping capacity of SMA fabrics for ballistic applications. Different future trends and ideas for using SMAs in textile structures will conclude the chapter.

8.2 Definition and description of shape memory alloys

The shape memory effect was discovered in 1932 by Chang and Read on a gold-cadmium alloy. The same effect was also observed in 1938 on a zinc-copper alloy, lastly in 1963 on a nickel-titanium alloy. Afterwards, the first industrial applications were in the aeronautic field, for instance, a coupling system on F-14 aircraft in 1967. Since then these materials have found different applications in various industrial fields such as medical and biocompatibility, automotive, aeronautic and apparel.

Shape memory alloys have different mechanical behaviour from the usual materials. These are characterised by a specific state diagram of stress–temperature (see Fig. 8.1). As regards the temperature or the stress values, the material phase will be located in different areas as martensite, austenite or a blend of martensite and austenite. A complete and normalised definition of a shape memory alloy can be obtained in the French norm NF A51-080 (April 1991).[2]

In the same way as the martensitic changes in steels, the phase of 100% austenite in the material is obtained at high temperature and the phase of 100% martensite comes from the austenitic phase by cooling. The martensitic change presents four main characteristics:

1. the global motion of atoms occurs over very low distances
2. the transformation process occurs twice
   - first, change of the initial molecular structure to another
   - second, deformation of the crystal lattice

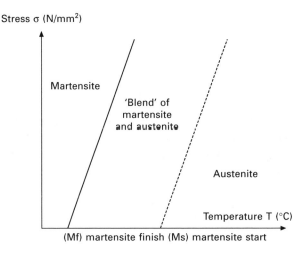

8.1 From austenite to martensite phases.

126   Intelligent textiles and clothing

3. the martensitic transformation is quasi-instantly (higher than the speed of sound)
4. the volume variation is very low, but a shearing of the structure appears in a determined direction.

During cooling, the buckling of the crystal lattice is homogeneous. It occurs by sliding and then a rotation. A martensitic crystal lattice is obtained at low temperature. If the decrease in temperature has been achieved without stress, different kinds of martensite are obtained. The final buckling depends on the number of these different obtained martensites. It can vary from 3 to 8% for a polycrystal to 10% for a monocrystal.

8.3 Interesting properties of shape memory alloys

According to the different compositions of shape memory alloys, five main effects can be, most of the time, observed:[3]

1. Superelasticity; the alloy is able to buckle in a reversible process under stress.
2. Single memory effect; the alloy is able, after mechanical buckling, to recover its initial shape under a thermal process.
3. Double memory effect; the alloy is able, after a 'training' process, to keep two different shapes at two different temperatures.
4. 'Rubbery' effect; the alloy subjected to a stress retains, after the stress is removed, a residual buckling. If the material is repeatedly subjected to stress the residual buckling will increase.
5. Damping effect; the alloy is able to absorb shocks or to reduce mechanical vibrations.

A complete and well referenced review of these different properties has been done by Wei *et al.*[4]

8.3.1 The superelasticity effect

The shape memory alloys having this characteristic may be subjected to a more extreme buckling than a conventional alloy such as steel. The process of superelasticity can be explained as follows. A stress is applied on the alloy as shown in Fig. 8.2. At a constant temperature T and greater than the temperature of martensite start (Ms), the stress values begin from σMs (martensite start stress) to σMf (martensite final stress).

As it takes the strain, three areas can be distinguished along the Y axis (stress applied on the material) as shown in Fig. 8.3:

1. from the origin to σMs: elasticity of the austenite

Development of shape memory alloy fabrics   127

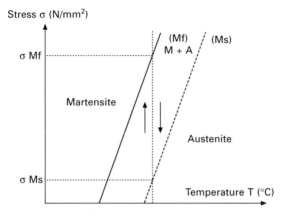

8.2 The modification steps from austenite to martensite phases.

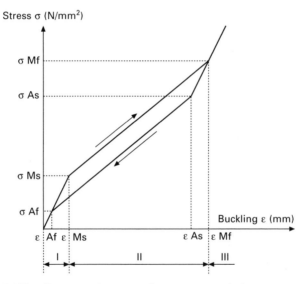

8.3 Tensile strength curve of a monocrystal shape memory alloy during change under stress at constant temperature.

2. from σMs to σMf: buckling due to the martensitic transformation
3. from σMf: elasticity of the martensite.

It can be also noticed that when the strain is removed, the profile of the curve is similar to the one when the strain is applied but shifted by hysteresis. This is due to the fact that stresses σAs and σAf, respectively at the beginning and the end of the inverse transformation, are different from the stresses σMs and σMf, respectively at the beginning and end of the direct transformation.

8.3.2 The single memory effect

This effect is observed when the alloy is submitted to a thermo-mechanical process corresponding to the following sequences (a) + (b) + (c) as shown in Fig. 8.4.

The single memory effect shown in Fig. 8.5 can be described as:

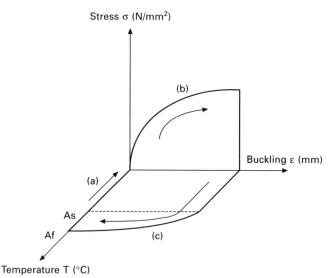

8.4 Thermo-mechanical process and single memory effect.

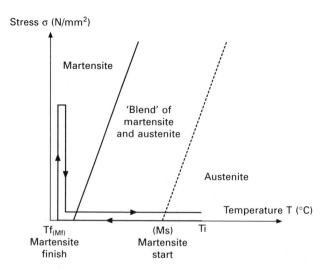

8.5 Sequences of the thermo-mechanical process allowing the single memory effect.

- Sequence (a): a cooling, without stress, from a Ti temperature greater than Ms to a Tf temperature less than Mf occurs. In this sequence, martensite is created in spite of the buckling of the material being null.
- Sequence (b): a stress is applied intermittently at a constant temperature, there is no phase change but a re-arrangement of the different kinds of martensite occurs during cooling.
- Sequence (c): heating is done up to a Ti temperature under a null stress. The alloy recovers its initial shape.

8.3.3 The double memory effect

In the double memory effect, shape modification occurs at two different temperatures, high for the austenite and low for the martensite. The obtained shapes are stable and are obtained after a 'training' period, without any stress, as shown in Fig. 8.6. A transformation buckling εM is observed during cooling as shown in Fig. 8.7. The double memory effect can be qualified as a 'super-thermal' effect in which the applied stresses are recovered by internal stresses coming from the 'training' process.

8.3.4 The 'rubbery' effect

This behaviour is linked to an internal mechanism at the martensite phase (see Fig. 8.8); it seems to be partially reversible. For a given stress (see Fig. 8.9), the (ε) reverse buckling obtained is obviously greater than the (εe)

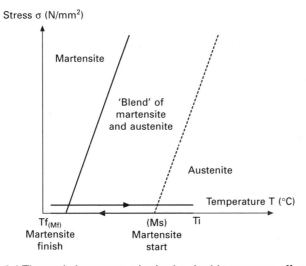

8.6 Thermal changes to obtain the double memory effect.

130   Intelligent textiles and clothing

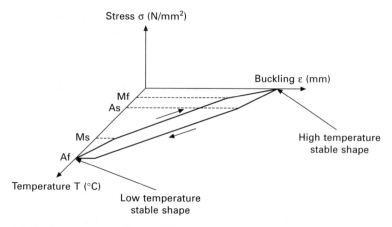

8.7 Double memory effect, without any external stress.

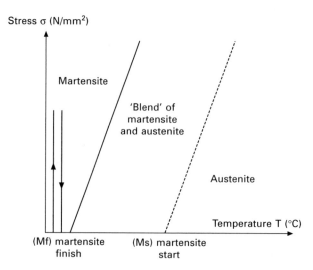

8.8 The 'rubbery' effect and the intermittent strain mechanical process.

usual elastic buckling. Thus, the modulus is much less than the elastic modulus which conducts to a 'rubbery' effect.

8.3.5 The damping effect

This effect is mainly linked to the transformation phase at the solid state for which an initial austenite phase gives birth to a martensite phase, in a reversible manner from a crystallographic point of view. This effect is also called internal friction and can be observed, for instance, during free mechanical

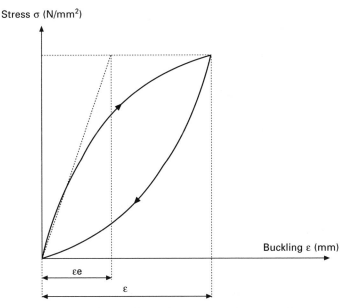

8.9 The 'rubbery' effect and the specific resulting buckling.

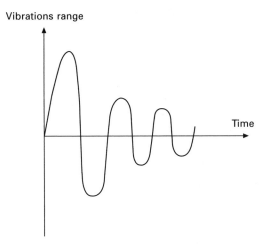

8.10 Decreasing of a vibration range due to internal friction of the material.

vibrations. A decreasing of the vibrations range with respect to the time occurs (see Fig. 8.10).

This friction is often characterised by a quality index Q also called the damping capability. According to the material phase and its buckling, three main areas can be observed (see Fig. 8.11) for which the shape memory alloy had different inner friction values:

132   Intelligent textiles and clothing

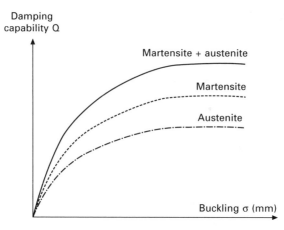

8.11 Variation of the inner friction with respect to the three main phases of SMA.

1. In the austenite phase, the inner friction is low.
2. In the martensite phase, the inner friction is associated with a reversible motion of the different martensite phases.
3. Friction is highest in the transition zone of the austenite and martensite phases.

Although many alloys had some shape memory properties and damping capacity, only two families are mainly used for the development of SMA, the nickel-titanium and the copper base alloys. These two families have been highlighted due to their mechanical, thermal and electrical properties, their shape making ability, their way of making and their cost.

8.4   Different kinds of alloys

8.4.1   Nickel-titanium base alloy

The nickel-titanium alloy, commonly named as Nitinol, is the most studied and used in spite of its relatively high cost. Good characteristics such as the single memory and the superelastic effects are mentioned in Table 8.1. It can be made in different diameters as a wire (from 0.1 to 2.5 mm), but we have to bear in mind that the cost of the yarn is increasing exponentially with respect to the diameter reduction. In addition, Nitinol has good corrosion index properties, good fatigue performance and good compatibility with the human body.

To improve some properties of the nickel-titanium base alloy, small quantities of other elements could be added as mentioned in Table 8.2; copper (Cu) and iron (Fe) reduce hysteresis and nobium (Nb) increases the As temperature.

Table 8.1 Properties of the main different SMAs

Properties	Unit	Ni-Ti	Cu-Zn-Al	Cu-Al-Ni	Cu-Al-Be
Composition		Nickel titanium	Copper zinc aluminium	Copper aluminium nickel	Copper aluminium beryllium
Melting point	°C	1260–1310	950–1020	1000–1050	970–990
Density	Kg/m3	6400–6500	7800–8000	7100–7200	7300
Young's modulus	Gpa	95	70–100	80–100	90
Tensile strength (in martensite domain)	Mpa	800–1000	800–900	1000	900–1000
Breaking strength elongation	%	30–50	15	8–10	15
Fatigue limit in austenite domain	Mpa	350	270	350	–
Transformation range	°C	–100 to 100	–100 to 100	–100 to 170	–200 to 150
Hysteresis (As–Mf)	°C	20–40	10–20	20–25	20–25
Spread (Af–As)	°C	30	10–20	20–30	15–20
Maximum buckling:					
Single memory effect	%	8	3–5	3–6	3–5
Double memory effect	%	5	2	3	2
Maximum temperature of use (1 hour)	°C	400	160	300	400
Maximum super elastic buckling					
Poly-crystal	%	4	2	2	3
Mono-crystal		10	10	10	10
Shock damping capacity (SDC)	%	15	30	10	–
Corrosion resistance		Excellent	Medium	Good	Medium
Bio-compatibility		Good	Bad	Bad	Bad

Table 8.2 Impact of an added element in nickel–titanium base alloy on different properties

	Hysteresis (°C)	Buckling (%)
Ni–Ti	20–40	6–8
Ni–Ti–Fe	2–3	1
Ni–Ti–Cu	10–15	4–5
Ni–Ti–Nb	60–100	6–8

8.4.2 Copper base alloy

As mentioned in Table 8.2, copper base alloys had less hysteresis than the nickel-titanium base alloys. Depending on their composition, wire diameters are also available from 0.1 mm to 2 mm. These different alloys are:

- Cu–Zn–Al is quite easy to produce at a moderate price but at a high temperature, its use is reduced due to a stabilisation of the martensite which increases the As temperature.
- Cu–Al–Ni, is difficult to produce but is not easily affected by stabilisation and ageing. This alloy has excellent properties in a wide range of temperatures (from 50 to 180 °C).
- Cu–Al–Be, has great thermal stability. The tiny quantity of beryllium allows the alloy to adjust its transformation temperatures from –200 to + 150 °C.

8.5 Different kinds of applications of shape memory alloys

The number of applications using SMAs is increasing as regards their specific properties, but limited to specific fields.[5] Principally, the limits of its use are governed by the cost of these materials and also by their low resistance, during their lifetime, to fatigue and ageing. At present, shape memory alloys can be found in the following.[6]

8.5.1 Bio-medical

Due to the bio-compatibility of nickel–titanium base alloys, many applications have occurred in the medical field. The following applications show the typical uses of SMA properties

- Medical staples are used inside the body. The main property required is the single memory effect. At the inner body temperature, the shape of the staple ensures the correct links which helps to fasten the fractured bones.[7]

Development of shape memory alloy fabrics   135

- A 'stent' looks like a sort of braid which reverts to a wider diameter at body temperature. The 'stent' is inserted in arteries to regulate blood pressure. The property used is the single memory effect.[5]
- Another successful medical application is nitinol's use as a guide for catheters through blood vessels.[8] The main property involved is the superelastic effect.
- When used as a blood-clot filter, nickel–titanium wire is shaped to anchor itself in a vein and catch passing blood clots. Cooling the part allows it to be inserted into the vein and body heat is enough to transform it to its functional shape. The main property required is the single memory effect.[9]
- Orthodontic wires reduce the need to retighten and adjust the wire. These wires also accelerate tooth motion as they revert to their original shapes. The main property involved is the superelastic effect.[10]
- Nitinol needle wire localisers are used to locate and mark breast tumours so that subsequent surgery can be more exact and less invasive. The main property involved is the superelastic effect.[8]

8.5.2 Aeronautic and aerospace

Mainly due to their capacity to transform a thermal process into a mechanical process without any added parts, SMAs had been identified, in some cases, as the best solution in aeronautic and aerospace applications. The following listed existing applications are good examples.

- Nitinol couplers have been used in F-14 fighter planes since the late 1960s. These couplers join hydraulic lines tightly and easily.[11] The real success of these couplers lies in the fact that the shape memory effect was the unique solution. Tighter connections and more efficient installations result from the use of shape memory alloys.[12]
- Cryofit hydraulic couplings are manufactured as cylindrical sleeves slightly smaller than the tubing they are to join. Their diameters are then expanded while martensitic, and when warmed to austenite, their diameters shrink and hold the tube ends. The tubes prevent the coupling from recovering its manufactured shape, and the stresses created as it attempts to do so create an extremely strong joint.[13]
- In the Betalloy coupling, the shape memory cylinder shrinks on heating and acts as a driver to squeeze a thin liner onto the tubes being joined.[14]
- For satellite solar panels, the shape memory alloy actuator, using the single memory effect, opens the solar panels by heating.[15]

8.5.3 Automotive

Future applications are envisioned to include engines in cars and aeroplanes and electrical generators utilising the mechanical energy resulting from shape transformations. Nitinol, with its shape memory property, is also envisioned for use as car frames.[7] Other possible automotive applications using SMA springs include engine cooling, carburettor and engine lubrication controls, and the control of radiator blinds.[16]

8.5.4 Apparel and spectacles

- Brassieres using SMA bones are more comfortable using the super elastic effect. Moreover, bones are not bent in washing machines. These bras, which are engineered to be both comfortable and durable, are already extremely successful in Japan. The superelastic effect is mainly used in this application.[17]
- Spectacle frames using superelastic SMAs, are much more resistant to breaking. Nitinol eyeglass frames can be bent totally out of shape and return to their parent shape upon warming.[9]

8.5.5 Miscellaneous

Other miscellaneous applications of shape memory alloys include use in household appliances, in structures, in robotics and in security devices as listed below.

- A deep fryer utilises thermal sensitivity by lowering the basket into the oil at the correct temperature.[18]
- Nitinol actuators used in engine mounts and suspensions can also control vibration. These actuators are helpful in preventing the destruction of buildings and bridges.[19]
- A fire-sprinkler with an SMA spring reacts to a given temperature and actuates the sprinkler. The main advantage of nitinol-based fire sprinklers is the decrease in response time.[20]
- Anti-scalding SMA valves can be used in taps and shower heads. At a certain temperature, the device automatically shuts off the water flow.[21]

8.5.6 Use of SMA in ballistics

Since 2000, a GEMTEX laboratory team has been working on SMA fabrics, especially with nitinol wires, and a weaving technique has been developed[22] taking into account the same ideas as in the Japanese patent JP8209488 in 1996.[23] Special measures during the warping, drawing-in and weaving processes have to be taken in order to keep the material in an austenite phase. Thus, thanks to the fabric comprised of 100% nitinol SMA, as well as in the warp and weft directions, different properties have been shown with respect to damping capability[24] and the superelastic effect.[25] It follows from this that several experiments have to be conducted with SMA fabrics coupled with polyparaphenylene terephthalamide (PPTA) and high tenacity polyethylene (PE) fabrics to make a composite structure improving the high-velocity impact resistance.

First let us look at the different patents and publications dealing with composite structures and SMA damping properties required for ballistic application.[26] In one patent a plain warp-weft weave structure is used for high-tenacity yarns.[27] This weave diagram does not use the unidirectional yarns structure commonly used in the composite material. It aims to keep the main characteristics of the yarn in the woven structure but is easier to manipulate. Specific weave diagrams can be used in the backing structure to make high-performance yarns as functional as possible. In a second patent, an armour material for protection against ballistic, flame and blast attack is presented, having the form of a wire mesh structure where parallel weft rods of hard metal such as tungsten, titanium or austenitic steel are linked by flattened helix wires of a yielding material such as mild steel.[28] This knitted structure permits some elasticity which can be convenient for the behaviour of the material during the impact. The blend of different types of metallic yarns alternately in the weft direction allows exploitation of the best property of each wire. Thus, different types of fabrics composed of different yarns can also be used in ballistic applications.

Particular attention must be given to tough materials such as S-glass, aramids, and high-performance polyethylene which behave differently at higher strain rates.[29] In another patent dealing with methods of protecting structures from impact, the components are interposed between the point of impact and a structure to be protected.[30] They comprise an SMA exhibiting pseudoelastic behaviour, and having a high strain to failure ratio. The patent includes, experimental results on damping behaviour of a beam composed of shape memory wires under mechanical stress.[31] It is observed that the damping increases significantly when the shape memory wires are stressed such that they lie within the pseudoelastic hysteresis loop. These results demonstrate that pseudoelasticity of shape memory wires can be used to augment passive damping significantly in structural systems. This indicates that the SMA yarns to be used in our ballistics application have to be in a transition phase depending on the stress and the temperature. The superelastic SMAs are shown to be effective at low velocities and may also be in high ballistic velocity applications.[32]

A previous study was conducted by Kiesling and it was demonstrated that an increase of 41% of the energy absorption can be obtained with only 6% of SMA inserted in volume.[33] Thus, an adjusted proportion of SMA fabrics will be used in our ballistic application with respect to the total volume and weight. Recently, in the thesis work of Roger Ellis,[34] just after Paine and Kiesling,[35] the concept of using high-strain SMA and ECPE hybrid components to improve the ballistic impact resistance of graphite composites has been studied. The following obtained results have to be highlighted:

- A relative improvement of 99% of the energy absorption is observed when the Spectra™ yarns are located at the back of the composite with an increase of only 12% of the total weight.
- Other and pure SMA fabrics must be tested by varying the yarns' diameters and the alloy compositions.
- At least, the recommended new structure to test tends to be knited rather than woven.

Finally, by taking into account all these previous results and recommendations, different backings have been realised made of pure aramid and blend of aramid and SMA fabrics corresponding to the NIJ Norm standard level 3 and 4.[36] At level 3, the armour protects against 7.62 mm full metal jacket bullets (US military designation M80), with nominal masses of 9.7 g impacting at a velocity of 838 m per second or less. Projectiles are fired six times at different locations. All the armours have been manufactured by MS Composite Company, during a final-year student project of six months. At the level 3, all the armours passed the ballistic tests. The mean resulted deformation after the impact for all these tests was very low. At the level 4, the armour protects against 30 calibre armour-piercing bullets (US military designation APM2), with nominal masses of 10.8 g impacting at a velocity of 868 m per second or less. Only one of five armours tested was failed at the level 4, mainly due to an excessive velocity of the projectile (measured at 890 m/s instead of 868 m/s). Faced with these promising results, we are engaged in a new campaign to develop new armour including a new kind of high-performance yarn. As a matter of fact, assuming a certain number of hypotheses and considering elementary computations on the buckling model of fabrics to impact, a new solution is proposed and will soon be tested. The new backing will integrate different fabrics with different type of yarns where each of them is used for its main property during impact. Its main composition is presented in Fig. 8.12.

The main properties of the four main blocks of the backing structure are detailed in Table 8.3. Thanks to a our previous tests and considering all the new recommendations, this new backing will succeed at different tests and especially reduce the blunt trauma.

8.6 Conclusion

The shape memory alloys have some different interesting properties than the other usual material such as steel. Various and numerous application fields of SMA demonstrate their advantages as regards these specific properties. In the ballistics domain, different ways put to the fore the main interest to make

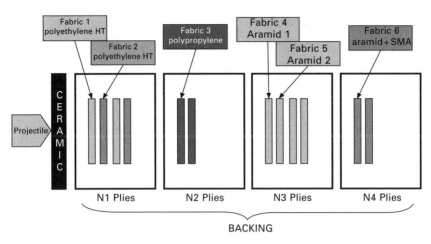

8.12 Description of future backing structures under test.

Table 8.3 Functions of different fabrics inside the backing

N1 plies	N2 plies	N3 plies	N4 plies
Fabrics 1 and 2 made of PEHT yarns (Dyneema™) with respectively a higher linear capacity for fabric 1 than fabric 2. Each of these fabrics has a specific weave diagram in order to be well coupled.	Fabric 3 made of polypropylene high linear density yarns.	Fabric 4 and 5 made of high tenacity aramid (Technora™) with respectively a higher linear density for fabric 4 than fabric 5. Each of these fabrics has a specific weave diagram in order to be well coupled.	Fabric 6 made of SMA and aramid yarns
Awaiting function	Awaiting function	Awaiting function	Awaiting function
Decreasing of shearing stress and absorption of the main part of energy during projectile penetration.	Absorption of the calorific energy by melting	Decreasing of shearing stress.	In the transition zone of the SMA, damping capacity at impact.

a blend of different yarns inside the composite structure to take the benefit of each of their thermomechanical properties. Different tests have allowed validating a certain number of hypotheses as regards the yarns' behaviour at high-velocity impacts and specifically the SMA wires. The obtained results are sufficient but can be improved by varying a certain number of parameters to aim at a final composite structure solution optimising all criteria including the total cost and weight.

8.7 Future trends

The immediate future aims are the achievement of a new series of tests using our new backing structures to validate levels 3 and 4 of the NIJ norm as regards high-velocity impact. The mid-term future aims are linked to the results of the previous immediate ones. However, two different ways can be explored:

- improvement of the fabric process using SMA wires, especially the warping and weaving operations
- optimisation of the 'training' process of SMA, especially for fabrics structure.

At last, long-term future efforts could be shared by a large community and deal with the knowledge of the fabric's behaviour made from high-performance yarns with respect to different thermomechanical tests (tensile, compression, shear, bend, fatigue and age strengths) at different velocities (low, medium and high speeds) and under different environments (low and high temperatures). This global knowledge will help to validate mathematical models of behaviour of the textile structure.

8.8 Internet links

SMA Providers

Memometal technologies, Bruz, France.
http://www.memometal.com/FR/Default.htm
Nimesis, Metz, France
http://www.nimesis.com/
NitiFrance, Lury sur Arnon, France.
http://www.nitifrance.com/fr/
Memry Corp. – Bethel, CT
http://www.memry.com/
MicroGroup, Inc. – Medway, MA
http://www.microgroup.com
Electropolishing Systems, Inc. – Plymouth, MA
http://www.electropolishingsystems.com/
New England Precision Grinding – Holliston, MA – Contact
http://www.nepg1.com/

SMA applications

http://www.nimesis.com/francais/technique/amf/applications.htm
http://aml.seas.ucla.edu/research/areas/shape-memory-alloys/index.htm

SMA properties

http://www.nimesis.com/francais/technique/amf/amf.htm
http://www.memry.com/nitinolfaq/nitinolfaq.html
http://aml.seas.ucla.edu/research/areas/shape-memory-alloys/overview.pdf

8.9 References

1. Patoor E., Berveiller M., *Technologie des alliages à mémoire de forme*, Hermès, 1994.
2. AFNOR, *Alliages à mémoire de forme (AMF), Vocabulaire et mesures, Shape memory alloys (SMA) – Vocabulary and measures*, NF A51-080, ISSN 0335–3931, Avril 1991.
3. ASTM, F2005 *Norm Standard Terminology for Nickel-Titanium Shape Memory Alloys*.
4. Wei Z.G., Sandstrom R., Miyazaki S., Shape memory materials and hybrid composites for smart systems, Part I, shape memory materials, *Journal of Materials Science*, 1998, **33** (15), 3743–3762.
5. Wu Ming H., Schetky L. McD., Industrial Applications for Shape Memory Alloys, *International Conference on Shape Memory and Superelastic Technologies (SMST)*, Pacific Grove, California, p.171–182, 2000.
6. Wei Z.G., Sandstrom R., Miyazaki S., Shape memory materials and hybrid composites for smart systems, Part II, shape memory hybrid composites, *Journal of Materials Science*, 1998, **33** (15), 3763–3783.
7. Memometal, Product Brochure, Rennes, FR.
8. Stoeckel D., Weikang Y., *Superelastic Nickel-Titanium Wires*, available from Raychem Corporation, Menlo Park, CA.
9. Simon M. *et al.*, *Radiology*, 1989, **172**, 99–103.
10. Yahia L., *Shape Memory Implants*, Springer-Verlag, ISBN 3-540-67229-X, 2000.
11. Kauffman G., Isaac M., Memory Metal, *Chem Matters*, October 1993, 4–7.
12. Borden T., Shape-Memory Alloys: Forming a Tight Fit, *Mechanical Engineering*, October 1991, 67–72.
13. Harisson J.D., Hodgson D.E., *Shape Memory Effects in Alloys*, Plenum Press, 1975.
14. Raychem Corporation, Product Brochure, Menlo Park, CA.
15. Benzaoui H., Lexcellent C., Chaillet N., Lang B., Bourjault A., Experimental study and modeling of a NiTi shape memory alloy wire actuator, *Journal of Intelligent Material Systems and Structures*, 1997, **8**, No. 7, 619–629.
16. Turner, J.D., Memory-metal Actuators for Automotive Applications, *Proceedings of the Institution of Mechanical Engineers*, 1994, **208**, 299–302.
17. Hiroyasu Funakubo, *Shape Memory Alloys*, Gordon and Breach Science Publishers, ISBN 2-88124-136-0, 1987.
18. Falcioni, John G., Shape Memory Alloys, *Mechanical Engineering*, April 1992, 114.
19. Baz A., Irman K., McCoy J., *Journal of Sound and Vibration*, 1990.
20. Memry Corporation, Product Brochure, Bethel, CT.
21. Wu M.H., Ewing W.A., SMST-94: The Proceedings of the First International Conference on Shape Memory and Superelastic Technologies, Pacific Grove, California, 311, 1994.
22. Boussu F., Petitniot J.L., Bailleul G., Vinchon H., The Interesting Properties of Shape Memory Alloy to Use in Technical Textiles, International Conference, 1st

AUTEX Conferences, Technical Textiles: Designing Textiles for Technical Applications, Guimaraes, Portugal, 27–28 June 2001.
23. Kitahira Takashi, Matsuo Akira, Takeuchi Masatoshi, Miyashita Seiichi, *Production of woven fabrics using wire of shape memory alloy*, Patent n°JP8209488, Publication date: 13 August 1996.
24. Joly D., Petitniot J-L., *Etude de l'effet amortissant de fils d'AMF dans des poutres en composite verre-époxy*, Août 1999, RT n°99/29, Onera.
25. Boussu F., Petitniot J.L., Development of Shape Memory Alloy fabrics for damping and shape control of composite structures, *World Textile Conference – 2nd AUTEX Conference, Textile Engineering at the dawn of a new millennium: an exciting challenge*, Bruges, Belgium, 1–3 July 2002.
26. Soroushian Parviz, *Metal matrix materials reinforced with shape memory fibers for enhanced ductility and energy absorption capacity, and method of manufacturing same*, United States Patent of DPD Inc., Lansing, Mich. (US), International Classification: B32B 5/12; Patent Number 6,025,080; Date of Patent: Feb. 15, 2000.
27. Cunningham David, Pritchard Laura, *Quasi-unidirectional fabric for ballistic applications*, Canadian Patent of Barrday Inc., Ontario (CA), International Patent classification: F41H 5/04, International Publication Number: WO 02/090866 A1, International publication date: 14 November 2002.
28. Macwatt David, *Yielding barriers*, Great Britain Patent of Passive Barriers Ltd, London (GB), International Patent classification: F41H 5/02, International Publication Number: WO 92/00496, International publication date: 9 January 1992.
29. Cuniff, P.M., An analysis of the system effects in woven fabrics under ballistic impact, *Textile Research Journal*, 1992, **62**, 9, 495–509.
30. Van Schoor M., Lengyel A., Masters B.M., Rodgers J.P, *Energy absorbing shape memory alloys*, United States patent of Mide Technology Inc., Medford, MA (US), patent Number: 0025985 A1, publication date: Feb 12, 2004.
31. Thomson P., Balas G.J., Leo P.H., The use of shape memory alloys for passive structural damping, *Smart Mater. Struct.*, March 1995, **4**, 1, 36–41.
32. Paine J.S.N., Rogers C.A., The response of SMA Hybrid composite materials to low velocity impact, *Journal of intelligent material systems and structures*, July 1994, **5**, 530–535.
33. Kiesling T.C., Chaudhry Z., Paine J.S.N., Rogers C.A., Impact failure modes of thin graphite epoxy composites embedded with superelastic Nitinol, *Proceedings of 37th AIAA/ASME/ASCE/AHS/ASC Structures, Structural Dynamics and Materials Conference*, Salt Lake City, Utah, 15–17 April 1996.
34. Roger L. Ellis, ballistic impact resistance of graphite epoxy composites with shape memory alloy and extended chain polyethylene Spectra^{TM} hybrid components, MSc. *Thesis, Mechanical Engineering*, Blacksburg, Virginia, December 1996.
35. Paine, J.S.N., Rogers, C.A., Observations of the Drop-weight impact response of composites with surface bonded Nitinol Layers, *Durability and damage Tolerance of Composites Symposium, ASME IMEC&E*, San Francisco, CA, Nov 12-18, 1995.
36. National Institute of Justice, Ballistic resistance of Police Body Armor, NIJ standard 0101.03, April 1987.

9
Study of shape memory polymer films for breathable textiles

J HU and S MONDAL, The Hong Kong Polytechnic University, Hong Kong

9.1 Introduction

Discussion of water vapor transfer will by nature focus on the ability of breathable textiles to reduce both heat stress and the uncomfortable sensation of wetness [1–2] whilst maintaining protection from wind and rain in comparison to totally impermeable fabrics [3]. In order to provide wearing comfort to the wearer, the garment should have a high capacity of water vapor transmission so that perspiration can evaporate and be transmitted from the body surface to the environment. Earlier fabrics coated/laminated with rubber or synthetic material, such as PVC, or acrylate were used as weatherproof clothing. These materials were watertight, however, their water vapor transmission was very low [4].

Today's modern technologies offer various possibilities for making waterproof and breathable textiles [5]. A well-known method to produce waterproof breathable textiles is to coat/laminate breathable nonporous films or laminate microporous films on a suitable base fabric's surface. Microporous film laminates 'breathe' due to their permanent, air-permeable pore structure. The basic principle consists of large numbers of micropores sufficiently large to allow the penetration of molecules of perspiration, but small enough to prevent water droplets going through the fabric [6]. Water vapor can easily escape from the body surface to the environment and prevent water droplet penetration. Because the diameter of a water vapor molecule is about 0.35 nm, while the diameter of a water droplet at standard pressure is bigger than 1 μm [7]. However, with the use of a porous film, deformation of films causes breaking or increases the size of pores that cause the delamination of coatings from the fabric surface and the water resistance becomes too poor. From the wearing point of view [8] the high water vapor permeability required for physiological comfort is important. Such moisture permeability cannot be obtained with a general nonporous polymer film. As the coated and laminated fabrics is composed of a thin layer of film and fabrics, so the film permeability plays an important role in order to make successful breathable textiles. In this chapter, we review the breathable textiles, followed by a study on microstructure and mass transfer properties of nonporous shape memory polyurethane films for breathable textiles described.

9.2 Breathability and clothing comfort

The movement of water vapor through a textile is an important factor in garment comfort. The human body attempts to maintain core body temperature around 98.6 °F. The balance between perspiration and heat production by the body and loss of the same is the comfort factor. The body would be in a state of comfort when the body temperature is about 35 °C and there is no moisture on the skin [9]. In order to maintain comfort, the primary functions of breathable textiles are:

1. to keep the wearer dry (from external water ingress and from internally generated condensation)
2. to provide protection against the cold
3. to protect against the wind chill factor, which is perhaps the bonus feature that the waterproof breathable textiles have due to their wind-resistant nature.

The total heat transfer through the clothing of the body to the environment, considering the thermal and evaporation resistance of the clothing, has been given by Woodcock [10]:

$$H = (T_s - T_a)/I + (P_s - P_a)/E \qquad 9.1$$

where H = total heat transfer, $T_s - T_a$ = temperature difference between skin and ambient, $P_s - P_a$ = water vapor pressure difference between skin and ambient, I = insulation of the clothing, and E = evaporation resistance of the clothing.

By incorporating thermal insulation products with waterproof, breathable textiles in the garment assembly, the effect of wind chill is greatly reduced. Through a combination of selected materials in the garment construction and suitable lining material, it is possible either to delay the formation of condensation by sweating and/or wick the condensed perspiration away during periods of activity in the inclement climates, thus increasing the wearer's comfort. It should be remembered that the body perspires to cool down during and after periods of strenuous activity and the fact that perspiring aids the return of body temperature to a comfortable balance. The perspiration rate of different activities are given in Table 9.1 [11].

Heat loss by evaporation is the only way to dissipate heat from the body when environmental temperature is greater than skin temperature. Liquid sweat is transformed into vapor at the skin surface, and it passes into the environment to cool. Evaporation of moisture from the skin surface is

Table 9.1 Approximate work and perspiration rates associated with various activities

Activity	Work rate (Watt)	Perspiration rate (g.day^{-1})	Limit of use			
Sleeping	60	2280				
Sitting	100	3800				
Gentle walking	200	7600	•			
Active walking	300	11,500	•			▶
with light pack	400	15,200		▲	■	▶
with heavy pack	500	19,000		▲	■	☐
in mountains	600–800	22,800–38,400		▲	■	☐
Very heavy work	1000–1200	38,000–45,600				

Note: • = coated; ▲ = hydrophilic membrane; ■ = microporous membrane; ▶ = woven cotton; ☐ = woven microfibres (adapted with permission from Holmes D A, *J. Coated Fabrics*, **29** (4), (April 2000), 306 © Technomic Publishing Co. Inc. [11]).

tremendously effective in disposing of body heat. The loss of heat through perspiration and heat-flux through fabrics is very important. Part of this heat loss is by moisture evaporation. The greater the rate of moisture evaporation, the greater will be the comfort. Mass diffusion may also result from a temperature gradient in a system; this is called thermal diffusion. Similarly, a concentration gradient can give rise to a temperature gradient and a consequent heat transfer. In sum, cold weather clothing, besides insulating, should ideally have three main features, it should be water vapor permeable, windproof and waterproof. Two types of fabrics are in use for foul weather clothing. They are impermeable fabrics and breathable fabrics. An impermeable fabric is both wind and waterproof but not water vapor permeable. A breathable fabric, on the other hand, meets all the features of foul weather clothing and is water vapor permeable [10].

9.3 Breathable fabrics

Textiles with good breathing properties have become indispensable. They protect from rain, snow and wind but allow water vapor (perspiration) to penetrate. This guarantees a high level of comfort, especially when the article is worn during physical effort in sports and during work [12]. Many textile products are waterproof but only a few provide 'breathability'. On the other hand, standard fabrics are 'breathable' but not waterproof. Some basic requirements of waterproof breathable textiles are summarized in Table 9.2.

9.3.1 Types of breathable fabrics

Breathable textiles can be categorized into four main types [10]:

Table 9.2 Basic requirements for waterproof breathable fabrics [13]

Waterproof breathable fabrics should, as indicated by the name, provide both waterproofness and breathability.

Windproofness

Abrasion resistance (wet and dry conditions)

Tear resistance

Strength of the coating/good adhesion of the membrane/film to the textile.

Easy care, wash resistance and washability

Lightness and packability (becoming increasingly popular for all outdoor activities)

Durability

Flexibility

Stretchability including stretch fabrics with thermal properties

Quietness

Handle, look, etc.

Other characteristics according to specific end uses: flame retardancy, chemical protection, high visibility, stain resistant and oil resistant

---

Note: besides these 'traditional' requirements the trend is clearly towards new functions and properties, such as 'clever fabrics', solar (UV) protection, sweat and moisture control (adapted with permission from Kramar L, *J. Coated Fabrics*, **28** (October), 1998, 107, © Technomic Publishing Co. Inc. [13]).

1. closely woven fabric with water-repellent treatment
2. microporous film laminates and coatings
3. hydrophilic film laminates and coatings
4. a combination of microporous coating with a hydrophilic top coat.

High-density-fabrics (HDF)

The fabrics can be woven with microfibers so densely that no interstices can be seen between the yarns. The microfibers are fibers that are less than 1 decitex per fiber. The fabrics made from microfibers are extremely soft and have a fine touch [5]. Water vapor permeability of such fabrics is high – a certain water resistance will be given by treatment with water-repellent agents such as flurocarbons and silicones. They provide for a higher watertightness as compared to the traditional textiles, but they do not give permanent protection in rain. Their water vapor permeability is excellent [4].

Examples are [5]:

1. Hoechst with its microfiber Trevira-Fineness (0.65 dpf) (polyester), which is used, for example, by Rotofil AG to weave their 'Climaguard'.
2. ICI with Tactel Micro/24 Carat (0.4 dpf) (polyamide) used, for example, by Finlayson in 'Microspirit'.
3. du Pont de Nemours fiber: 'Supplex' (0.9 dpf) (polyamide).

4. Burlington's: Versatech microfiber-based fabric.
5. Toray Ind. Inc. with 'Dyna-Bright' to weave their 'H$_2$Off'
6. Kuraray with 'Wramp' to weave 'Arcus'.

Microporous films, laminates and coatings

In the lamination technique a functional (water-resistant/breathable) barrier film is 'glued' to a suitable base fabric. Lamination of the film on a fabric uses special adhesives, sometimes even breathable ones. If non-breathable adhesives are used, care has to be taken not to cover the whole surface in the coating process. Microporous laminates/coatings 'breathe' due to their permanent, air-permeable pore structure. The basic principle consists of large number of micropores with a size sufficiently large to allow the penetration of molecules of perspiration, but small enough to prevent water droplets from going through the fabrics [13].

1. To render a film breathable, holes can be made in it: Teflon renders itself perfectly to this in a biaxial-stretching process, such as: Gore-Tex (W.L. Gore/USA) and Microtex (Nitto Elec. Ind./Japan).
2. Polyurethane/polyacrylate microporous film is also available, such as: Porelle film made by Porvair/GB in a coagulation process or Repel film (polyacrylate) made by Gelman Science/U.S.A. in a photopolymerization process.

Hydrophilic film laminates and coatings

The solid, or compact, structure of the product prevents penetration of water droplets whereas transmission of water vapor is provided by a molecular mechanism (absorption-diffusion-desorption) [13]. From the hydrophilic group containing polyurethane solution the solvent will be evaporated leaving a compact film behind [4]. Hydrophilic groups built into the polymer chains can absorb, diffuse, and desorb water vapor molecules through the film [14], examples are [5]:

1. Sympatex film (polyester), a product of Enka Glanzstoff (D)
2. Bion II film (polyurethane), a product of Toyo Cloth C^0/J
3. Excepor-U (polyamioacid/PU), a product of Mitsubishi-Kasei.

A combination of microporous coatings with hydrophilic top coat

The combination of microporous and hydrophilic layers is also possible [13]. A microporous coating or film, for example, can be further coated with a hydrophilic layer in order to increase waterproofness, and to seal the surface pores, reducing the possibility of contamination of the microporous layer by dust particles, detergents, pesticides, etc. A hydrophilic finish on a microporous structure is used to upgrade the water-resistance of microporous coatings. Care has to be taken to select a hydrophilic finish that does not impart an unacceptable decrease in breathability. Ucecoat NPU2307 finish on top of Ucecoat 2000 (S) microporous coating is an example [5].

9.3.2 Merits of nonporous over porous films for breathable laminates

Waterproof breathable textiles could be made by coating or laminating microporous film on a suitable base fabric. The water droplet penetration is effected via micro-fine pores which are produced by a special process during the production of the film or during coating. While perspiration can easily escape from the body surface to the environment very quickly, small drops of rain or spray cannot penetrate this breathable system, despite its pore structure because the diameter of water vapor molecules is 0.35 nm and of water droplets about 1 µm [7].

Besides the famous PTFE membranes (Gore-Tex, Gore), where the pores are achieved during production by stretching the membranes, there are also microporous hydrophobic polyurethane membranes on the market. The pores of these membranes are also produced by a special coagulation process [12]. The advantages of microporous membranes are [13]:

1. Better breathability: hydrophilic coatings/lamination is influenced by the thickness of the coating/laminates and number of hydrophilic groups present in the film structure.
2. Better handling: hydrophilic coating has a stiffer handle.
3. Hydrophilic coating tends to wrinkle in wet conditions.

Despite better breathability, the microporous films have several disadvantages, such as:

1. Pore sealing of micropore of coating or laminates occurs during usage and affects breathability. The pore-sealing micropores can be contaminated by a number of agencies, including particulate and air-borne dirt, pesticide residues, insect repellents, sun tan lotions, salts (marine environment), skin exudates, and detergents and surfactants used for laundering or drycleaning. All of these contaminants have been suspected of lowering the breathability [15].
2. During the use of microporous film, deformation of film causes films to break or an increase in the size of pores so that water proofness becomes too poor for practical use. Numerous small degraded loose segments can easily mix with human sweat, becoming an ideal culture medium in which bacteria and mold can grow. If the latter happens, the fabric can

easily be affected by all kinds of microorganism (e.g. bacteria and mildew) and deteriorate the films, thus affecting the waterproofness as well as fabric appearance.
3. Microporous film has a poor tearing strength compared to the solid structure of a nonporous one.

Compared to the microporus films the advantages of nonporous breathable films are as follows:
1. The film-making procedure is simple and the production speed is higher.
2. Having a solid structure and no holes the nonporous films are less sensitive to possible degradation.

Table 9.3 shows some features of nonporous and microporous films for textile lamination.

Having solid structures, nonporous films possess several advantages compared to microporous films. Since a certain amount of water vapor permeability is required in order to give the wearer comfort, how to improve the permeability of nonporous membranes is really a challenge for polymer chemists. Permeability of nonporous membranes could be enhanced by introducing hydrophilic groups into the polymer backbone, as the permeability occurs through the nonporous films by molecular mechanism (sorption–diffusion–desorption). However, too many hydrophilic groups causes swelling and weight loss of film increases during washing due to increasing solubility, and waterproofness continues to decrease after each wash [13]. The option may be in the shape memory polymer (SMP) films, the large change of mechanical and thermomechanical properties occurs across the glass transition temperature (T_g) or soft segment crystal melting point temperature (T_{ms}) [17].

In addition to the mechanical properties of SMP, it was found that SMP also undergoes a large change in moisture permeability above and below the T_g/T_{ms}. Based on the T_g/T_{ms} set at room temperature, the SMP has low moisture permeability below the T_g/T_{ms} and during the glassy state, and has high moisture permeability above T_g/T_{ms}, and during the rubbery state. This

Table 9.3 Features of microporous and nonporous films [16]

Nonporous	Microporous
Windproof	Windproof (arguments)
Waterproof and liquid proof	Waterproof and liquid resistance
Selective permeability	Non-selective permeability
High water entry pressure	Low water entry pressure
Good tearing strength	Low tearing strength
High water vapor transmission	High water vapor transmission

(Adapted with permission from Johnson L., and Samms, J., *J. Coated Fabrics*, 27 (July), (1997), 48 © Technolmic Publishing Co. Inc [16]).

9.3.3 Designing breathable textiles with nonporous films

Technical discussion of waterproof breathable fabrics usually raises the highly contentious subject of water vapor transmission properties. In this context, breathability refers to the capacity of a fabric layer, garment or clothing assembly to transmit water vapor, emitted from the body as perspiration, to the outside atmosphere. A discussion on the mechanism of water vapor permeability is useful in understanding the principle of designing breathable fabrics. Mass transfer through porous film occurs through the permanent pore structure of porous film. But non-porous films are dense, pinhole-free polymer membranes. These polymer films are also usually hydrophilic and absorb water very quickly. This is an important property that produces a 'wicking' action that actively attracts water vapor molecules. Non-porous film allows the transmission of water vapor through a process called active diffusion. This is the same transport mechanism by which helium escapes from a toy balloon. The permeant dissolves on the surface of the film on the side of the highest concentration, and then diffuses across the film. When the vapor arrives at the opposite surface, the permeant desorbs and typically enters the surrounding airspace as gas or vapor. Water vapor transfer through a nonporous membrane occurs in molecular mechanism, i.e., sorption-diffusion-desorption. A discussion of the mechanism of water vapor transfer through nonporous films follows.

Since in nonporous films, there are no micropores mass transfer can result from different phenomena [18]. There is a mass transfer associated with convection in that mass is transported from one place to another in the flow system. This type of mass transfer occurs on a macroscopic level and is usually treated by the subject of fluid mechanics. There will be a mass transfer on a microscopic level as the result of diffusion from regions of high concentration to regions of low concentration due to the concentration gradient. Higher concentration means there are more molecules per unit volume. Mass diffusion may also result from a temperature gradient in a system; this is called thermal diffusion. Similarly, a concentration gradient can give rise to a temperature gradient and a consequent heat transfer. Permeation is a collective process of diffusion and sorption and hence, the permeability of mass molecules in polymer depends both on diffusion and solubility [18]. Chemical structure and film thickness are the main determinants of permeability in a nonporous membrane [19].

Structural factors influencing film permeability

Not only the polymer-penetrant interaction but also the primary structure of the polymer itself are very important for an understanding of film functions such as sorption, diffusivity, and permeability of small molecules [20]. The cohesive energy of polymer membranes is determined by such factors as chain flexibility (internal rotation of repeating unit), coulombic interaction, van der Waals interaction, hydrogen bonding, and so on. A high permeability coefficient is generally attained when each factor except flexibility is low and is not affected by the penetrant. Therefore the polymer, which contains a monomer unit with a high charge density, dipole moment, and capability for hydrogen bonding, will give a low permeability coefficient. Cipriano *et al.* stated that the permeant fluxes increase with the increasing value of the prepolymers molecular weight [21]. This means that longer molecular chains originate larger polymer network holes and therefore higher water vapor fluxes.

Film thickness influencing permeability

Fick's law governs the rate of transport of the permeant through the film under the existing concentration gradient. If the diffusion coefficient (D) is not a function of concentration, integration across the film thickness l is given by [18]:

$$J = D(c_1 - c_2)/l \qquad 9.2$$

Where c_1 and c_2 are the concentrations of permeant at the high and low pressure faces of the film surface, and l is the thickness of the membrane. A linear relationship between the concentration of water vapor in equilibrium with the film and actual concentration of water vapor dissolved in the film is assumed by Henry's law as given in eqn 9.3, which holds for many polymers:

$$c = Sp \qquad 9.3$$

Substituting, values of c_1 and c_2 from eqn 9.3, eqn 9.2 becomes:

$$J = DS(p_1 - p_2)/l \qquad 9.4$$

Where p_1 and p_2 are the external partial pressure of the vapor on the high and low pressure sides of the membrane respectively. (DS) is termed as the permeability (P). From eqn 9.4, it is clear that for ideal systems, the permeation rate of permeant is directly proportional to the pressure gradient and inversely proportional to membrane thickness.

9.4 Water vapor permeability (WVP) through shape memory polyurethane

9.4.1 Glass transition temperature as transition point for WVP [22]

Experimental

Shape memory polyurethane (SMP), MS-4510, with a solids content of 30%, was supplied by Mitsubishi Heavy Industries Ltd., Japan. N,N'-dimethyl formamide (DMF) was obtained from BDH laboratory, England. The solution of SMP was dissolved to form 15% by weight solution. The film was cast on a glass plate by doctor blade. Three casting temperatures were chosen; 70, 120 and 150 °C, and the resultant films were coded as SMP-70, SMP-120 and SMP-150, respectively. DSC measurements were carried out over the temperature range of –50 to 220 °C using a Perkin Elmer 7 Series DSC, purging with N_2 and chilled with liquid N_2, about 10 mg of sample scanned at a heating rate of 10 °C/min.

Dynamic mechanical properties were measured with a dynamic mechanical thermal analyzer (Rheometry Scientific DMTA MK 3). The samples ($10 \times 3 \times 0.03$ mm) were investigated in the temperature range from –50 to 120 °C, using the tensile mode at the heating rate of 5 °C min^{-1} and a frequency of 1 Hz under N_2 gas purging. A length to thickness ratio of samples is larger than 10 for neglecting the DMA's dependence on the Poisson ratio. The maximum peak of the tan (delta) curve is considered as glass transition temperature.

The shape memory effect was examined by bending mode [23]. The samples were deformed to an angle θ_i (~90°) at a bending temperature (T_b = 100 °C), well above T_g of the samples, and kept the bending time (t_b = 2 min). Then the deformed samples were quenched to 0 °C for about 1 min and then the external force released. The deformed samples were heated at a constant heating rate and recorded the data of the angle θ_f and the corresponding temperature. The recovery ratio was defined as $(\theta_i - \theta_f)/\theta_i \times 100\%$.

Water vapor permeability (WVP) was measured according to ASTM D1653-93. That is, an open cup containing distilled water was sealed with the cast film of SMP, and the assembly was placed in the test chamber with a controlled temperature (25 °C, 35 °C, 40 °C, 50 °C and 60 °C) and humidity (relative humidity (RH) 80%, 65%, 40%). The steady water vapor flow was measured by plotting the weight change of the cup containing the water against time.

Results and discussion

DSC curves and thermal data of shape memory polyurethane samples are shown in Fig. 9.1 and Table 9.4. The glass transition temperature of the soft

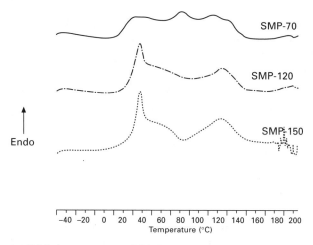

9.1 DSC thermogram of SMP (adapted with permission of ref. 22).

Table 9.4 Thermal data of DSC testing (Adopted with permission of ref. 22)

Samples	T_g, °C	ΔH_1, J/g	T_{mh}, °C	ΔH_2, J/g
SMP-70	31	–	125.8	7.9
SMP-120	39.2	16.8	136.3	27.5
SMP-150	41.5	20.9	133.3	51.2

segment (T_g) is 30–42 °C in DSC curves, and the stress relaxation peak (ΔH_1) appears at the glass transition domain both in SMP-120 and SMP-150 because the hard segment restricts the mobility of the soft segment due to the increased crystallinity of the hard segment. The second transition (T_{mh}) is an endotherm (ΔH_2) in the range of 135 °C, indicating the melting of hard segment crystals. These results showed that SMP-120 and SMP-150 are phase separated into an amorphous soft segment domain and a partially crystalline hard segment domain. However, the transition in SMP-70 is very weak and two small endothermic peaks appear at temperatures of about 90 and 125 °C due to the dissolution of the hard segment in the soft domain and the lower degree of segment separation. The DSC curves confirm that the film preparation at lower temperatures, such as sample SMP-70, are not favorable for crystallization of the hard segment and therefore segment separation.

The shape memory effect of SMP at different temperatures, is shown in Fig. 9.2. The recoverable ratio of the specimens was less than 10% at a low temperature range (0 to 30 °C). The SMP-120 recovered deformation rapidly when it was heated to a high temperature, and little residual deformation remained; the SMP-70 showed a wider temperature range for recovering and

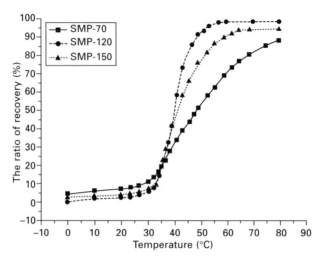

9.2 Shape memory behavior of SMP films (adapted with permission of ref. 22).

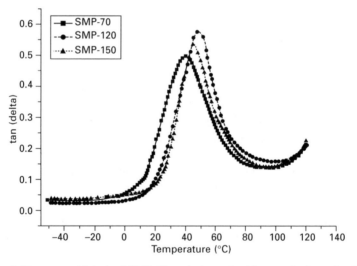

9.3 Loss tan (delta) of SMP films (adapted with permission of ref. 22).

a smaller recovering ratio than the others. These results demonstrated that the shape memory behavior of SMP is influenced by the morphology of the soft and hard segment phase domains [24]. Their recovering ratio is directly related to their storage modulus ratio, that is, the SMP films could show better shape memory behavior if their storage modulus ratio were high Fig. 9.3 and Table 9.5. Consequently, the SMP-120, with a high modulus ratio, shows better behavior of deformation recovery than the others. The crystallinity in the hard segment domain is good for keeping the deformation at a temperature

Table 9.5 DMTA data of SMP films (Adopted with permission of ref. 22)

Samples	Glass transition temperature (T_g)	Storage modulus ratio ($E'_{T_g-20°C}/E'_{T_g+20°C}$)
SMP-70	24	137.7
SMP-120	31	293.9
SMP-150	34	194.0

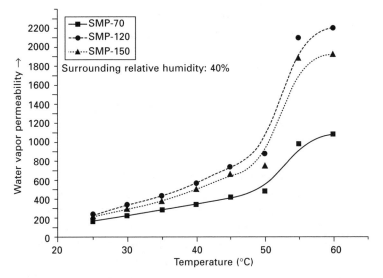

9.4 Water vapor permeability of SMP films (adapted with permission of ref. 22).

range lower than the T_g of the soft segment and recovery to the original shape with the heating process.

The water vapor permeabilities variations of shape memory polyurethane films at different temperatures are shown in Fig. 9.4. Water vapor permeability increases appreciably above the T_g of the soft segment in all samples. This shows that moderate crystallinity is more suitable for better water vapor permeability than low and very high crystallinity according to the thermal data of all samples (Table 9.4). We can explain this as follows: generally, the permeability of small molecules through the polymer membranes is enhanced when their diffusivity increases with increasing temperatures [25–26]. According to the concept of free volume in polymers, the glass transition occurs in the polymers when the fractional free volume (FFV, the ratio of the free volume and specific volume in polymers) reaches the standard value of $f_g = 0.025$. Above T_g, that is, in the rubbery state, FFV increases linearly with temperature:

$$\text{FFV} = F_g + (\alpha_1 - \alpha_2)(T - T_g) \quad\quad 9.5$$

where α_1 and α_2 are thermal expansion coefficients in the rubbery and glass states, respectively.

This increases the free volume in the polymer, and the micro-Brownian motion of the soft segment obviously increases to make the intermolecular gap large enough to allow water vapor molecules to be transmitted through the SMP film. That is, the diffusivity of water vapor molecules in the SMP film increases with increasing temperature. Therefore, large changes in moisture vapor permeability above and below the T_g of the soft segment are observed. However, the glassy state of the soft segment at low temperature plays the role of water vapor barrier, so it decreases the water vapor permeability and provides a waterproof barrier at low temperature.

9.4.2 Soft segment crystal melting temperature as transition point for WVP [27]

Experimental

The shape memory polyurethane (SMPU) used for this study was obtained from Mitsubishi Heavy Industries. Differential scanning calorimetry (DSC) was carried out over a temperature range from −40 °C to 220 °C using a Perkin Elmer DSC7. The samples were scanned at a heating rate of 10 °C/min and sample weight was 9.6 mg. After the first scan, melted specimen was quenched to −40 °C at a cooling rate of 20 °C/min. The sample was again scanned at 10 °C/min. SMPU film was sealed over the open mouth of a test dish which contained water, and the assembly placed in a controlled atmosphere. After keeping at 10 °C, 20 °C, 30 °C, 40 °C for 24 hours, the dish was weighed and the rate of water vapour permeation through the film was determined.

Results and discussion

The DSC thermogram of SMPU is shown in Fig. 9.5. DSC curve 1 was obtained over a temperature range from −40 to 220 °C at a heating rate of 10 °C/min. It showed that the endothermic peak began from about 10 °C, and the highest point of the peak was at 50 °C, which was caused by crystal melting. DSC curve 2 was obtained over the same temperature range at the same heating rate as curve 1 after quenching to −40 °C. There was only one exothermic peak at −10°C and one endothermic peak at 50 °C, which was related to re-crystallization and crystal melting, respectively. This result could support the curve 1 findings. With the temperature rising further, we saw no distinct endothermic peak from curve 2, indicating that only soft segments formed crystal structure in the SMPU.

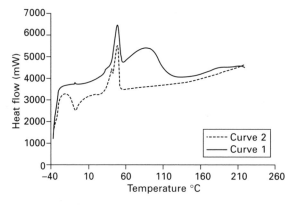

9.5 DSC curve of SMPU (adapted with permission of ref. 27).

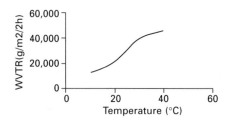

9.6 Temperature-dependent water vapor permeability of SMPU films (adapted with permission of ref. 27).

Figure 9.6 shows the relationship between water vapor permeability of SMPU film and temperature. When the temperature rises from 10 to 40 °C, i.e., in the soft segment crystal melting range, WVP increases significantly. In theory, the phase transition of a polymer from the crystalline to the amorphous phase will result in an increased amorphous area, which will also lead to increased free volume. It is well known that for dense samples, water vapor transport proceeds by diffusion through the film by free volume theory, driven by vapor concentration difference [28]. When the experimental temperature reaches crystal melting point temperature, the relative amount of amorphous area increases, which leads to increased free volume, therefore the film can provide more paths for water vapor permeation, and thus the WVP increases significantly.

9.4.3 Soft segment crystal melting temperature as transition point at room temperature

Experimental

Polytetramethylene glycol (M_n = 2000 g mol^{-1}, PTMG 2000)-based temperature-sensitive shape memory polyurethanes (TSPU) were synthesized

Table 9.6 Composition of TSPU [29]

Sample code	Feed (×10⁻³ mol)				
	PTMG2000	PEG	MDI	1,4-BDO	PEG (wt %)
S_6	14.75	22.5 (PEG-200)	44	7.15	9.86
S_7	14.75	2.25 (PEG-2000)	37	20	9.98
S_8	14.75	1.33 (PEG-3400)	36.08	20	10.03
S_9	9	–	38	28.79	–

(Adapted with permission from Hu J. L., and Mondal, S., *Polym. Inter.*, in press, © 2005, John Wiley and Sons Ltd. [29]).

with three different block lengths of hydrophilic segment, polyethylene glycol (M_n = 200, 2000 and 3400 g mol⁻¹, PEG 200, PEG 2000 and PEG 3400), by a two step polymerization process [28]. The detailed compositions are given in Table 9.6.

Membranes were cast from diluted PU solution (concentration about 5% w/v) in N,N-dimethyl formamide (DMF) on Teflon coated steel plate. In order to obtain pinhole free membrane, solvent was evaporated slowly at 60 °C for 12 h and final residual solvents were removed under vacuum at 80 °C for another 12 h. Then the Teflon plates were removed from the vacuum oven and kept at room temperature for 2 h. After 2 h membranes were removed from the Teflon plate. The thickness of the membranes was 45–60 μm for mass transfer.

Perkin-Elmer DSC 7 was used to measure the heat of fusion (ΔH), T_m, etc. Each sample having a weight from 5 to 10 mg was scanned from –50 to 120 °C at a scanning rate of 10 °C min⁻¹ under dry nitrogen purge. In order to find the role of PTMG in the PU, the DSC and WAXD testing for pure PTMG was carried out. The water vapor flux (WVF) was measured according to ASTM method E 96-80B. Round-mouth conical plastic cups with a diameter of 60 mm and a height of 90 mm were filled with deionized water. Membranes were placed over the top of the cups, securing perfect sealing between cup and membranes. The gap between the membranes and water surface was about 4 mm. The cups were placed in a constant temperature chamber at the desired temperature (12, 18, 25, 35 or 45 °C). During all WVF measurements air surrounding the membranes had a constant temperature and 70% relative humidity. An average of three different samples was used for each WVF measurement, which are expressed in the units g m⁻² d⁻¹, where d is a day (24 h).

Results and discussions

DSC results are shown in Fig. 9.7 and Table 9.7. From Fig. 9.7, it can be seen that no endothermic peak was observed for the sample S_9 containing no

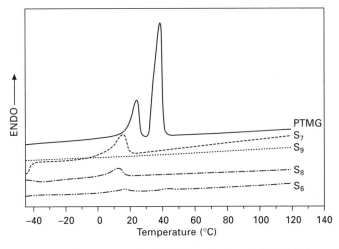

9.7 Heating thermogram of PTMG and related TSPU (adapted with permission from Hu J. L., and Mondal, S., *Polym. Inter.*, in press, © 2005, John Wiley and Sons Ltd. [29]).

Table 9.7 DSC data of TSPU [29]

Samples	ΔH_1^a	T_{ms1}^b	ΔH_2^a	T_{ms2}^b
S_6	0.66	15.50	0.32	42.83
S_7	28.37	15.67	–	–
S_8	22.52	12.83	–	–
S_9	–	–	–	–
PTMG	32.34	24.50	81.13	39.00

ΔH is heat of fusion, T_{ms} crystal melting temperature, T_g glass transition temperature, ^{a}data are in g J^{-1} or ^{b} are in °C (adapted with permission from Hu J. L., and Mondal, S., *Polym. Inter.*, in press, © 2005, John Wiley and Sons Ltd. [29]).

hydrophilic segments. That may be due to the flexible nature of the PTMG soft matrix, where hard segments act as a reinforcing filler and prevent the crystallization of the soft matrix. On the other hand, introducing the hydrophilic segment in the PU enhances the crystallization of the soft matrix, that may be due to the PEG segment increasing the mobility of the polymer molecule, which facilitates the crystallization process. With low molecular weight PEG-200, the actual percentage crystallinity is very low compared to the calculated percentage crystallinity from polyol weight fraction, and may be due to the plasticization effect of PEG-200, that would not make for favorable conditions for soft segment crystallization. The percentage crystallinity is highest with PEG-2000 as compared with PEG-200 and PEG-3400. This results from the fact that crystallization in polymers involves the steps of (primary) nucleation and relatively rapid spherulitic growth, followed by a slow, kinetically difficult improvement in crystal perfection [30]. The molecule must undergo a considerable degree of motion during crystallization. The motion may be optimum with PEG-2000, because this molecular weight is comparable with polyol (PTMG) molecular weight, 2000 g mol^{-1}, and enhanced intermolecular packing of small crystalline domains. The decreased percentage crystallinity with PEG-3400, may be due to chain entanglement, which hinders the crystallization process of the soft domain.

The water vapor flux (WVF) data are shown in Table 9.8 and Fig. 9.8. From Table 9.8 we can see that in all cases WVF increases with temperature, due to two reasons. First of all as the temperature increases the difference in the saturation vapor pressure of water in the cup and surroundings also increases, which would also increase permeability. The second reason is the

Table 9.8 Water vapor flux data of TSPU [29]

Sample	WVF (g m^{-2} d^{-1})				
	12 °C	18 °C	25 °C	35 °C	45 °C
S_6	170	210	310	410	680
S_7	276	359	460	660	1080
S_8	280	365	520	750	1220
S_9	96	124	210	310	480

(Adapted with permission from Hu J. L., and Mondal, S., *Polym. Inter.*, in press, © 2005, John Wiley and Sons Ltd. [29]).

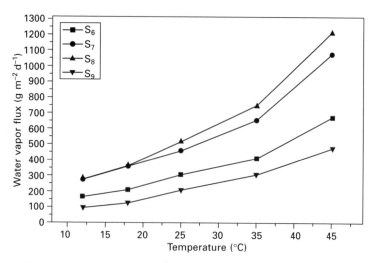

9.8 WVF results of TSPU at different experimental temperatures (adapted with permission from Hu J. L., and Mondal, S., *Polym. Inter.*, in press, © 2005, John Wiley and Sons Ltd. [29]).

structural change with increasing temperature. In all cases first factor remains constant, so the differences of water vapor permeability at a particular temperature is due to the change in morphological structure.

The permeability of small molecules through the nonporous polymer membrane is enhanced when their solubility and diffusivity in the polymer increased [31]. The fractional free volume increases with temperature according to eqn 9.5, which provides more paths for water molecules to pass through the membrane. The increase of free volume in the polymer, and the micro-Brownian motion of the soft segment obviously increases the intermolecular gap enough to allow water vapor molecules to pass through the membrane [22]. With increasing the block length of the PEG component in the polymer, the WVP also increases (Fig. 9.9), due to the increased flexibility of the polymer and increase of hydrophilicity [22], which increases the interaction between water molecules and polymer chain segments (Fig. 9.10), and increases the permeability because permeability of non-porous membranes follows the law: sorption–diffusion–desorption. On the other hand, longer polymer chains of PEG originate larger polymer network holes that will also enable the water vapor molecules to pass through the membranes.

Moreover, when the experimental temperature reaches the soft segment crystal melting point, discontinuous density changes occur, which take advantage of micro-Brownian motion (thermal vibration). Micro-Brownian motion occurs within the membrane when the temperature rises above a predetermined activation point. The activation energy can be considered as the energy to 'loosen' the polymer structure, which is related to the change in thermal expansivity. An increase in temperature provides energy to increase segmental mobility, which increases the penetrant diffusion rate. As a result of this motion, more micropores are created in the polymer membrane which

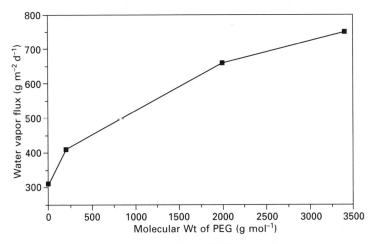

9.9 Effect of block length of PEG on WVF at 35 °C.

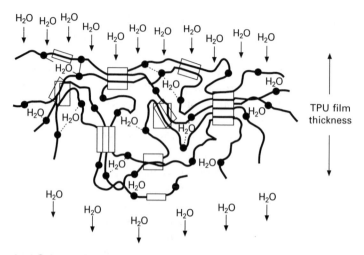

9.10 Schematic representation of water molecule transport through a non-porous hydrophilic membrane (adapted with permission from Johnson L., and Samms, J., J. *Coated Fabrics*, 27 (July), (1997), 48 © Technolmic Publishing Co. Inc [16]).

allow water to escape. That is why the WVF curves for S_7 and S_8 increase sharply (Fig. 9.8), because of the confluence of hydrophilicity, increase of free volumes, and micro-Brownian motion in the membranes at the soft segment crystal melting point. But in the case of S_6 the increase of WVF is due to the confluence of hydrophilicity, and increase of free volume. On the other hand in the case of samples S_9, without hydrophilic segments, the WVF is very low as compared to S_7 and S_8, even as compared to S_6, because in these samples, increase of WVF with temperature is due to the increase of free volume and increase of saturation vapor pressure, which is the same in all cases.

So, in summarizing the experimental results and discussion of water vapor permeability of shape memory polyurethane films, we can say that a large change in water vapor permeability occurs at the transition point (glass transition temperature/crystal melting point temperature) due to the morphological change of the polymer membrane. In addition, the presence of hydrophilic groups in the polymer structure also enhances the permeability due to the increasing solubility of water vapor molecules in the membrane.

9.5 Future trends

The property of breathability is in great demand by today's breathable textile industry. The inherent desire is to create fabrics that will provide protection from water droplets while allowing water vapor to escape. The use of breathable textiles will undoubtedly continue to expand into apparel applications. Some commercial breathable textile products are already available in the market. Mitsubishi is testing a new breathable fabric in the United States. The fabric is called Dream Cloth [32]. It is designed to keep people warm in cool temperatures, and cool in warm temperatures. The fabric achieves this through a polymer coating that acts like human skin. The molecular structure of the fabric coating opens when the wearer becomes hot, thus allowing body heat and vapor to escape. When the temperature drops, the molecular structure closes to provide insulation against the cold.

With the advantages that shape memory polymers can offer in coating/lamination products, unparallel hand and drape will emerge and have important implications for sports and leisure-wear in the apparel markets. Forecasting future trends is always speculative, but we are confident for the future of shape memory polymer films in breathable textiles.

9.6 Acknowledgement

Experimental results presented in this chapter have been reproduced with permission from *Polymer International*, © 2005, SCI, John Wiley & Sons Ltd. and *Textile Research Journal*, © 2003 & 2004 Sage Publications. The Author also would like to acknowledge Technomic Publishing Co. Inc. for giving permission to use their copyright materials.

9.7 References

1. Pause B, 'Measuring the water vapor permeability of coated fabrics and laminates', *J Coated Fabrics*, **25**(Apr.), 1996, 311–320.
2. Ruckman J E, 'An Analysis of Simultaneous Heat and Water Vapour Transfer through Waterproof Breathable Fabrics', *J Coated Fabrics*, **26**(Apr.), 1997, 293–307.
3. Keighley J H, 'Breathable fabrics and comfort in clothing' *J Coated Fabrics*, **15**(Oct.), 1985, 89–104.
4. Gottwald L, 'Water vapor permeable PUR membranes for weatherproof laminates', *J Coated Fabrics*, **25**(Jan.), 1996, 169–175.
5. Roey M V, 'Water-resistant breathable fabrics', *J Coated Fabrics*, **21**(Jul.), (1991) 21–31.
6. Lomax G R, 'Design of waterproof, water-vapour-permeable fabrics', *J Coated Fabrics*, **15**(Jul.), 1985, 40–66.
7. Lomax G R, 'Intelligent polyurethane for interactive clothing', *Textile Asia*, Sept., 2001, 39–50.
8. EU Patent No 83305387.9 (date of filing 14.09.83), 'Moisture-permeable waterproof fabric', *J. Coated Fabrics*, **14**, 1985, 148–164.
9. Shishoo R L, 'Technology for comfort', *Textile Asia*, **19** (6), 1998 (June) 93–110.
10. Sen A K, 'Coated Textiles: Principles and Applications', tech editor Damewood J, Technomic Publishing Co., USA, 2001, 133–154.
11. Holmes D A, 'Performance characteristics of waterproof breathable fabrics', *J. Coated Fabrics*, **29** (4), (April 2000), 306–316.

12. Kubin I, 'Functional and fashion coating for apparel', *Melliand International*, 7(June), 2001, 134–138.
13. Kramar L, 'Recent and future trends for high performance fabrics providing breathability and waterproofness', *J. Coated Fabrics*, **28**(October), 1998, 107–115.
14. Lomax G R, 'Hydrophilic polyurethane coatings', *J. Coated Fabrics*, **20** (October), 1990, 88–107.
15. Mooney C L and Schwartz P, 'Effect of salt spray on the rate of water vapour transmission in microporous fabric', *Textile Res. J.*, **55** (8), 1985, 449–452.
16. Johnson L and Samms, J, 'Thermoplastic Polyurethane Technologies for the Textile Industry', *J. Coated Fabrics*, 27 (July), (1997), 48.
17. Hayashi S, Ishikawa N and Giordano C, 'High moisture permeability polyurethane for textiles applications', *J. Coated Fabrics*, **23**, July, 1993, 74–83.
18. Cussler E L, *Diffusion: Mass Transfer in Fluid Systems*, (Cambridge University Press, Cambridge), 1st edn, Chap. 15 (1997).
19. Baker W R, *Membrane Technology and Applications*, (McGraw-Hill), Chap 2, (2000).
20. Strathmann H, *Synthetic Membranes and Their Preparation, Handbook of Industrial Membrane Technology*, edited by Porter M C, (Noyes Publications, New Jersey), (1989) 1–60.
21. Cipriano M M, Diogo A and Pinho M N D, 'Polyurethane structure design for pervaporation membranes', *J Membr Sci*, **61** (1991) 65–72.
22. Hu J L, Zeng Y M and Yan H J, 'Influence of processing conditions on the microstructure and properties of shape memory polyurethane membranes', *Text. Res. J.*, **73** (2), 172–178 (2003).
23. Lin J R and Chen L W, 'Shape-memorized crosslinked ester-type polyurethane and its mechanical viscoelastic model', *J. Appl. Polym. Sci.*, **73** (7), 1305–1319, (1999).
24. Kim B K and Lee S Y, 'Polyurethane having shape memory effects', *Polymer*, **37** (26), 5781–5793, (1996).
25. Jeong H M, Ahn B and Kim B. K, 'Temperature sensitive water vapor permeability and shape memory effect of polyurethane with crystalline reversible phase and hydrophilic segments', *Polym. Int.*, **49**, 1714–1721, (2000).
26. Jeong H M, Ah B, Cho S M, et. al. Water vapor permeability of shape memory polyurethane with amorphous reversible phase', *J. Polym. Sci., Part B, Polym. Phys.* **38**, 3009–3017, (2000).
27. Ding X, Hu J L and Tao X M, 'Effect of crystal structure of shape memory polyurethane film on water vapor permeability', *Text. Res. J.*, **74** (1), 39–43, (2004).
28. Duda J and Zielinski J M, *Free-volume theory, in diffusion in polymers*, P. Neogi, ed., Marcel Dekker, NY, 1996.
29. Hu J L and Mondal S, 'Structural characterization and mass transfer properties of segmented polyurethane: Influence of block length of hydrophilic segment, *Polym. Inter.*, 54, 764–771 (2005).
30. Kim B K, Shin Y J, Cho S M and Jeong H M, 'Shape-memory behavior of segmented polyurethanes with an amorphous reversible phase: the effect of block length and content', *J. Polym. Sci.: Part B, Polym. Phys.*, **38**, 2000, 2652–2657.
31. Van K D W, 'Properties of polymers', 3rd edn, Elsevier, New York, 1990, p. 535–583.
32. Maycumber S G, 'Fabric that gets cool when you get hot', DNR: *Daily News Record* (through Textile Technology Index database), May 6, Vol. 23 Issue 86, 1993.

10
Engineering textile and clothing aesthetics using shape changing materials

G K S T Y L I O S, Heriot-Watt University, UK

10.1 Introduction

This chapter focuses on the Design/Technology interface (Stylios, 2003) for developing SMART textile materials. Textiles are capable of unique shape changing behaviour by incorporating shape formed alloys and/or polymers incorporated within their structure. These are called in the first case shape memory alloys (SMA) and in the second case shape memory polymers (SMP), and belong to the family of shape memory materials (SMM).

Other properties of SMMs include pseudoelasticity or recoverable stroke (strain), high damping capacity and adaptive properties which are due to their ability to reverse the transformation during phase transitions. SMMs can sense physical changes in their environment, such as thermal, mechanical, magnetic or electric. These physical stimuli make them to respond by transforming their shape, position, strain, stiffness, natural frequency, damping, friction and other static and dynamic characteristics.

This chapter will show how the shape changing behaviour of textiles is being achieved by programming SMA, or by spinning SMP and ultimately by developing them into yarns for incorporation into woven, knitted or other textile structures. Their shape changing ability and the possibility of engineering this shape change for SMART applications have inspired the aesthetics of new designs for interior textiles and knitted garments, and have contributed to new hybrid mood changing textiles for smart ambience.

10.2 Innovative design concepts in textiles and clothing

Consumers are increasingly becoming more adventurous in curiosity and taste and with increasing disposable income, are driving the design industry in different endeavours. There is an increasing interest in products that are not only visually and aesthetically well made, but that can also excite, surprise and entertain the user/wearer. Lifestyles are changing too, with less time on our hands, accessories and products that facilitate ordinary chores, or help us to save time become attractive. Blinds or screens that open and close by themselves to accommodate the amount of light or heat in an office or room, upholstery that changes colour and shape depending on our mood, garments that can self-iron are examples that illustrate the self-managing capabilities of some products that have raised interest from consumers and industry alike.

Technological advances are making these innovations possible. In recent years, there has been considerable interest in non-static designs, in textiles, in fashion as well as in textile art. With the advent of new materials such as chromatic dyes and pigments, which change colour under a given stimulus, phase change materials, which change temperature within their environment and shape memory alloys and polymers, design concepts can now be more dynamic and interactive with the user or wearer. We are in an era of 'SMART TEXTILES', which at the flick of a button, or with an environmental change, can change shape, colour, texture, pattern, drape and handle. This concept has been explored in the fashion community, e.g., with colour-changing garments and with designer fashion incorporating shape memory wires (Marks, 2001).

Dynamic and interactive designs are the result of increasing multidisciplinary collaborations, in particular between designers and technologists (Stylios *et al.,* 2005; Chan *et al.*, 2002). Corpo Nove (Italy) has already attempted the use of a shape memory alloy in a designer shirt which rolls up its sleeves when the body gets warm. Increasingly, multinational companies such as Philips and Siemens, which originally were more technology rather than design-based are investing more and more in design/technology products. Designers are gradually changing from being traditionally 'users of technology' to now having a fundamental understanding of many of the technologies that they use. This leads to design now being used for specific functions in addition to being a visual and aesthetic attribute. Photochromic patterns that visually warn the wearer of excessive UV exposure, texture and morphological changes in an interior textile office partition, which not only creates different aesthetic effects, but also provides a function such as increasing the airflow or reducing sound and noise are other recent examples of functional design technology.

10.3 The principles of shape changing materials and their end-uses

Shape memory materials are able to 'remember' a shape, and return to it when stimulated, e.g., with temperature, electrical current, UV light, etc. The most common types of such materials are shape memory alloys and polymers, but ceramics and gels have also been developed.

10.3.1 Shape memory alloys

The first shape memory effect was observed in a gold-cadmium alloy in the early 1930s, and it was only in the 1960s that nickel-titanium alloy, a cheaper non-toxic alternative, was discovered. To date, nickel-titanium is still the most popular SMA and has been used in numerous functional engineering applications. Other shape memory alloys include copper-based or iron-based type variations.

The shape memory mechanism in alloys is generally caused by two distinct structural states: an austenite phase (highly ordered phase at higher temperature, also called the parent phase) and a martensite (less ordered, lower temperature, deformable) phase. In general, heat or mechanically induced stress is able to cause a change in phase type, e.g., from martensite to austenite, or between variants of martensitic phases (de-twinning), which generates the shape memory effect (Panoskaltsis *et al.*, 2004). With heat for example, the material is able to change from the martensite to the parent phase, through diffusionless transformation, leading to a shape 'recovery'. Hence, provided that the material has been 'fixed' into a specific physical form in its high-temperature parent phase, at a lower temperature, it can be distorted, but will 'remember' the original form when reheated. Figure 10.1 illustrates the shape memory recovery process of a SMA spring.

An attractive feature of SMAs is that they enable a two-way shape memory effect, also known as the all-round shape memory effect (Otsuka and Ren, 2005). This allows for repeated cyclic applications as the material is able to remember a shape at two different temperatures. Alloys are able to recover a large proportion of their deformation (up to 100% of the original programmed shape). However, they exhibit low strain ranges (up to 8%) compared to polymers (Shaw and Kyriakides, 1995).

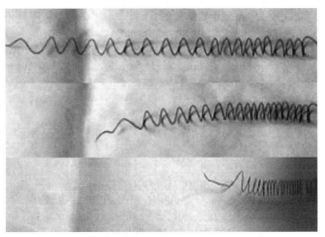

10.1 Shape memory recovery of SMA spring with time when T > 50 °C.

10.3.2 Shape memory polymers

The shape memory mechanism and effects in polymers are unique and different from those occurring in metals and ceramics (Liu *et al.*, 2004). The mechanism relies on a cross-linked structure at a molecular level. In the case of temperature-sensitive shape memory polymer, below glass transition temperature T_g the polymer is a stiff cross-linked network of chains, but above T_g, the network lends itself to a rubbery plateau state. The shape memory effect can be created by forming a shape while the polymer is in the rubbery state, and 'freezing' or 'fixing' this entropic state at a low temperature (Liu *et al.*, 2004; Ohki *et al.*, 2004). This frozen form can then be released only at a higher temperature.

Recently, the development of a UV-sensitive shape memory polymer has been reported (Borchardt, 2005). In this case, the mechanism is dependent upon the grafting of photosensitive groups into a polymer network shaped as required. The polymer is exposed to UV light, which causes the photosensitive groups to cross-link and fixes the new shape. Subsequent exposure to light of a different wavelength cleaves the cross-links and brings back the original shape.

The mechanism of shape recovery of SMP is dependent on the combination of a partially crystalline hard segment and a soft amorphous segment at the transition temperature T_g. Above T_g, the permanent shape can be deformed by the application of an external stress. After cooling below T_g, the amorphous segment is 'frozen' in a glassy non-crystalline state of high elastic modulus and hence obtains the temporary shape (Metcalfe Annick *et al.*, 2003). The material recovers to its permanent shape upon heating to $T > T_g$.

Shape memory polymers have some advantages over their alloy counterparts. They are lightweight, are able to withstand larger strains (up to 400%), possess a wide range of recovery temperatures, have low manufacturing costs and better processability (Yang *et al.*, 2005; Ohki *et al.*, 2004). For textile applications, when spun they are more flexible and can blend easily with other conventional yarns and woven and/or knitted structures (Chan and Stylios *et al.*, 2003b). Examples of shape memory polymers include segmented polyurethanes, poly(cyclooctene), poly(lactic acid) and poly(vinylacetate) blends (Liu *et al.*, 2002). A recent study of SMPs has been carried out by Hn (forthcoming).

10.3.3 Shape memory coatings

The area of shape memory coating (SMC) materials is new and reported developments are scarce. One promising SMC is reported by Stylios and Wan (2006) who have developed a highly adhesive resin coating solution by dissolving polyurethane SMP in dimethylacetamide.

10.3.4 Colour changing materials

Thermochromic materials change colour reversibly with changes in temperature. They can be made as semi-conductor compounds, from liquid crystals or metal compounds. The change in colour occurs at a pre-determined temperature, which can be varied. Current research involves the development of temperature-sensitive polymer-based pigments that visually and reversibly change colour at a prescribed temperature in the region of 15 to 35 °C. The temperature of the colour change (called the thermochromic transition) can be altered by the structure of the polymer-based pigment used and can be tailor made by chemical modification. In effect, thermochromic coated polymer films are thermal sensors that detect change of temperature with visual transformation.

With different constitutions of thermochromic and generic colour pigment, numerous colour variations can be produced. The thermochromic pigment can be incorporated into a coating solution for the film formulation or directly as paint with a special binder (such as PEG) for texture surface effects. Materials containing 0.1–1.0% by weight of thermochromic pigments in a host polymer have a visually retrievable, reversible thermochromic transition. The combination of SMM and thermochromic coating is an interesting area which produces shape and colour changes of the textile material at the same time.

10.3.5 Typical end-uses

Until recently, end-uses for SMMs were principally in the technical and functional engineering fields. SMPs and alloys have been used as heat-shrinkable devices, biomedical devices such as stents (Keiji, 2003) or other implantables and deployable structures. In the last ten years or so various companies developed new techniques for making SM textiles with some success (Kobayashi and Hayashi, 1990; Yoshida, 1991; Kitahira *et al.*, 1996; Butera *et al.*, 2004) and more recently, SMMs have started to find use in the fashion and clothing sector, with one of the first examples being the Corpo Nove shirt (Marks, 2001). There is also significant interest in using shape memory polymers as coatings for crease recovery and easy care apparel textiles, and for upholstery and other interior applications.

10.4 Technical requirements for shape changing textiles and clothing

10.4.1 Shape changing under stimuli

Many man-made and natural materials are naturally shape changing. A classic example is cotton, which expands when exposed to humidity and shrinks back when dried. Such behaviour has not been used for aesthetic effects because the changes, though physical, are in general not noticeable to the naked eye. In the case of shape changing textiles and clothing, a key requirement is to have a noticeable, visible shape changing effect, be it surface-bound, or form-bound. This consequently leads to two essential criteria: the efficacy of the material and the ability for the effect to be triggered as and when required.

In a woven, knitted or non-woven textile, the interactions between yarns and fibres contribute to the strength and mechanical properties of the fabric. With blended SMMs, these interactions can also act as a resisting force, to constrict the effect of the SMM. Hence for example, Chan and Stylios (2003a) found that in the case of wrap-spun shape memory polymer yarns, the twist of conventional fibres around a core shape memory filament prevented the core from performing its shape-changing function. The situation was further exacerbated when the yarn was turned into fabric. It is therefore essential that the shape changing 'force' is higher than the resisting forces in the textile structure.

The second criterion is the ability to be stimulated to the right level when required, or under normal conditions of use. Temperature-sensitive shape memory alloys and polymers can be programmed to respond at specific temperatures, which falls within a broad range. Programming however requires that the maximum effect be achieved under the temperatures of use. Other stimuli can also be used, e.g., electrical currents, magnetic fields or UV lights. Wang *et al.* (2004) found that the shape change in nickel-titanium alloys largely depended on the magnitude of the electrical current. For SMPs, the challenges of making the polymer conducting current are still to be met, but work is already in progress (Yang *et al.*, 2005). As for magnetic shape memory responses in alloys, they have been found to be highly variable, therefore posing a challenge for their applications (Heczko, 2005).

10.4.2 Problems associated with processability

Incorporating shape memory materials into textiles has been tried for over 15 years now, but many of the difficulties of turning the materials into textiles and clothing still exist. Alloys, due to their low strains, are difficult to weave and knit, because a certain amount of stretch in the material is required in order to interweave and inter-loop. With respect to aesthetic requirements, some novel woven or knitted structures require added flexibility for ease of production and suitability. Polymers, which have higher strains and lend themselves more to textile processing, are easier to turn into yarns, fabrics and clothing (Oohira, 1990; Terada, 1990). The textile assembly consisting of the shape memory materials must be able to be programmed, i.e., treated to train the material to remember a shape. If high temperatures are required (e.g. for some alloys), it may not be possible to treat the whole textile assembly, so the alloy has to be programmed prior to its inclusion into the textile structure. This can pose problems associated with the creation of specific effects. Chan and Stylios (2003a) have successfully pre-programmed and used SMA in textile fabrics as well as spun and programmed SMP into various textile woven structures for the first time.

10.4.3 Durability of shape changing effects

Durability of effects during use includes requirements for resistance to abrasion caused by ageing, wear and tear, but also good performance in cyclic repetition of the shape memory effect over a period of time. It is well known that shape memory alloys such as Nitinol can undergo deformation and shape reversing many times, but much less is known in the area of SMPs. Hysteresis and fatigue in cyclic loading of SMMs have been studied to some extent for metal alloys (Eggeler *et al.*, 2004). Among other parameters, temperature, microstructure and surface quality of the material inevitably affect their fatigue behaviour. Bhuniya *et al.* (2005) reported that the addition of small amounts of titanium to SMAs reduces their microstructural degradation caused by ageing. This is of interest for textile applications, where shape memory responses and reactions must be efficient for at least the estimated lifespan of the product in order to justify the costs.

10.4.4 Aesthetic degradation

Aside from performance, aesthetically, shape changing textiles should aim at having minimum visual and tactile degradation over time. This is particularly the case for products that will undergo harsher environmental conditions, e.g., interior textiles such as window blinds exposed to strong sunlight and environmental pollution. Nickel-titanium alloys are generally resistant to the environment and will not undergo corrosion. In the case of shape memory polymers, little is known about their long-term visual and tactile effects. Many polymers often experience brittleness, stiffness and colour change with ageing, in particular if exposed to harsh environmental conditions. This could impact on the visual effects, handle, comfort and drape of the textile characteristics which are particularly important for clothing.

10.4.5 Requirements for fashion and clothing

The requirements for fashion and clothing are slightly higher than for non-wearable textiles, in particular if the material is to be worn next to the skin and has to be washed. Chan and Stylios (2003a) and Winchester and Stylios (2003) highlighted the relatively harsh handle of the alloy and bulkiness of the shape memory polymer filaments, but also showed that this is not in itself a deterrent to the use of such material in textiles, as they can be used sparsely, but with significant visual and functional impact. They have also shown how blending with conventional or specialised yarns like elastane, can improve handle, tactile and recovery properties. Alloys pose a particular challenge for designers as they will provide the most dramatic shape memory effects, but also adversely affect drape, handle, touch and comfort.

10.5 Engineering textile and clothing aesthetics with shape memory materials

10.5.1 Extrusion of SMP

In a recent research project, a polyurethane-based SMP was used. The polymer pellets were dried prior to processing and extruded using an ESL Labspin 892 pilot-scale screw extruder into continuous monofilaments and multifilaments. The T_g of the polymer was 25 °C. The raw resin pellets were dried for eight hours in a hopper circulation oven at 80 °C until moisture was less than 0.03%. Without drying the resin, its viscosity becomes too low when melted, causing deformation by foaming, flashing and dropping at the nozzle. The temperature profiles of the machine suitable for processing of SMP yarn of 0.4 mm to 0.6 mm diameter using a die diameter of 1 mm, are as follows:

rear (feed zone): 170–180 °C
centre (compression): 175–185 °C
front (metering zone): 170–180 °C.

The key of this operation is to control the viscosity of SMP in the nozzle at the extrusion machine, while assuring uniform melting of the polymer. The viscosity of SMP is more temperature-dependent than traditional polymers, requiring stricter temperature and processing controls for extrusion. In order to control the diameter of the SMP yarn, the extrusion rate of yarn has also to be regulated.

Table 10.1 shows the characteristics of the fibre.

It should be noticed that the recoverable force of the pre-deformed 'frozen' SMP itself is ascertained as weak since the soft state of SMP is caused by the

Table 10.1 Fibre characteristics

Fibre diameter	0.10–0.34 mm
Tensile stress	0.1–0.8 kN/mm^2
Elongation at break	260–980%

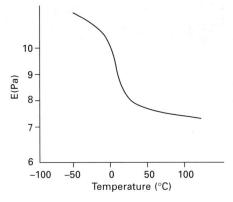

10.2 Dependence of elasticity modulus of SMP on temperature.

rise in temperature. Several physical properties of SMPs other than the SME are also significantly altered due to external variation in temperature, particularly at the glass transition temperature of the soft segment. These properties include elastic modulus, hardness and flexibility. As an example, temperature dependence of elasticity modulus E' of SMP is illustrated in Fig. 10.2, in which the elastic modulus of an SMP is changed dramatically when heated above the glass transition temperature of the soft segment.

10.5.2 Yarn and fabric formation

Using a Gemmel and Dunsmore Fancy Wrap Spinner, extruded SMP fibres were used as:

- core filaments in wrap-spun yarns
- blend material with other fibres (polyester, viscose and Lurex)
- filaments in unblended yarns.

The yarns were woven and knitted into three-dimensionally changing structures. Knitting and weaving of the material did not raise difficulties, as in the case of knitting the SMA. A range of SMP yarns and complex fabric structures made therefrom were designed and made up, as shown in Fig. 10.3.

10.5.3 Shape programming

Unlike SMA, which has to be programmed before incorporation in the textile structure because of the high treatment temperature, SMPs are treated after their incorporation into yarns and fabrics. It is therefore possible to program the shape of the whole fabric for the shape memory effect. The treatment in this particular case was performed at a temperature of 50 °C. The shape

174   Intelligent textiles and clothing

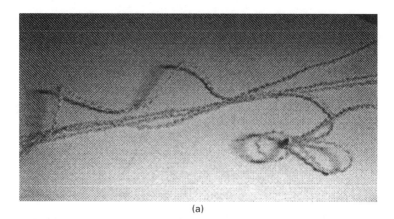

(a)

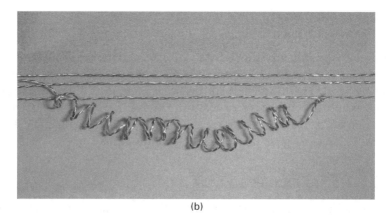

(b)

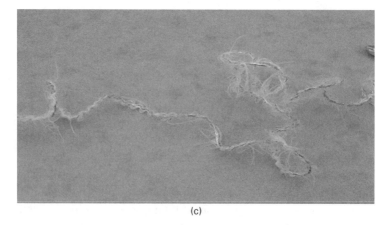

(c)

10.3 SMART yarns; various yarn composite blends with SMP.

memory effect on the fabric was clearly visible upon application of an external heat source such as a fan heater or a hair dryer.

10.5.4 SMP effects in fabrics

The main visual effect of the fabrics comes from the fact that areas containing the SMP are highly deformable, but upon heating, contract back to their original shape (Chan and Stylios, 2003b). Visual textural and structural effects that were explored from this deformation were:

- opening and closing of gaps within the structure of the fabric
- surface movements, including extended floats
- honeycomb structures
- double cloth elevated effects.

It was found that the effect of SMP as a core component in wrap-spun yarns was less pronounced, possibly as a result of the conventional fibres restricting the recovery movement of the polymer.

Figure 10.4 illustrates shape memory performance of SMP textile samples with uniform and densely woven SMP yarn of 0.4 mm diameter in the fabric structure. To observe the SME in the textured fabrics, the SMP composite is covered with two sheets of aluminium foil and deformed into the shape illustrated in Fig. 10.4(a) under higher temperature. The sample is then placed in a fridge under a mechanical constraint. The SMP filament recovers to its original shape of being flattened at a high temperature from being deformed at a low temperature upon heating to $T > T_g$, which allows the shape of the fabric sample to vary with environmental temperature. This is because the straightened state of SMP yarn at high temperature originates from the extrusion state when leaving the nozzle of the extrusion machine without any further shape memory training. Below the transformation temperature T_g, the bend shape of this textile is hibernated to provide a

10.4 Shape memory recovery of SMP composite woven uniformly and densely of SMP yarn at 50 °C with recovery time (a) 0 sec, (b) 15 sec, (c) 30 sec.

support force for the deformed state. After the specimen is heated up over the transformation temperature T_g, the hibernated SMP filament becomes soft and recovers to the original flattened shape. However, the shape memory effect in the SMP filament is mono-directional and it is difficult to perform an invert shape variation procedure. The problem is solved by adding some reinforcement yarn of high elastic modulus to the SMP matrix.

Conventional yarns of different material performance can be blended and woven with SMP yarn as illustrated in the samples in Figs 10.5, 10.6 and 10.7. The SMP filament was woven spaciously and loosely along the weft to allow room for the SME to take place. In contrast with the sample shown in

10.5 Shape memory recovery of SMP composite loosely woven fabric with SMP yarn at 50 °C with recovery time (a) 0 sec, (b) 30 sec, (b) 60 sec.

10.6 Shape memory recovery of SMP composite loosely woven fabric at 50 °C with recovery time (a) 0 sec, (b) 30 sec.

Engineering textile and clothing aesthetics 177

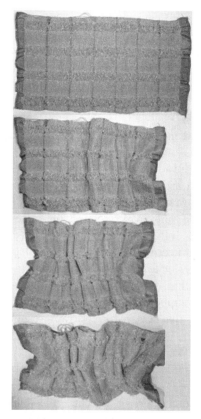

10.7 Shape memory recovery of SMP composite loosely woven fabric with flexible yarn at 50 °C with recovery time (a) 0 sec, (b) 30 sec, (c) 60 sec.

Fig. 10.3, the composite structure freezes the fabric at the flattened state. The initial flat shape of these SMP composites is fixed by exerting an external stretch force when the sample is in a freeze state. The SME occurs similarly from its original flattened state at low temperature to an embossed matrix with convex edges state, when being contracted at a high temperature. In this case, complete contraction occurs because the elastic modulus of the SMP yarn decreases dramatically when the environmental temperature is over the T_g temperature.

The recovery process may also be described as metamorphic in which the polymer exhibits a gradual shape variation during transformation. However, fabric designs based on SMP yarn blended with various kinds of flexible and light yarns can show interesting and aesthetically appealing effects, as shown in Fig. 10.7. The change of fabric shape depends on fabric design and SMP specific training at a given external temperature. It is apparent that shape memory design and training can create a number of aesthetic appeals with different texture yarns, showing significant shape change in response to environmental variation. More work is expected to be carried out in fabric design with reinforcement of high elastic modulus incorporated into SMP yarn to improve recovery.

10.5.5 Potential applications and limitations

Trained correctly, SMP can be used in a number of textile applications as SMART materials. The high levels of deformation and 'stretch' possible with SMP, in combination with their lower bending rigidity, make the material suitable for textile processing such as knitting and weaving to form fully flexible structures with good texture and handle. Also, the handle of the SMP, being softer and more flexible than SMAs, renders the material ideal in applications where comfort and drape are important, e.g., fashion and clothing, upholstery, sportswear, protective clothing or medical garments. Other application areas include interior textiles (e.g. blinds, partitions and curtains that can open or close depending on temperature) and medical textiles (e.g. responsive wound dressings or supporting materials). However, designs have to take into account the fact that SMPs have slower response than SMAs.

10.5.6 Programming SMAs

SMAs are normally 'trained' to remember one or two particular shapes while they are in the austenite phase. In recent work reported here, nickel-titanium (Ni-Ti) SMA wires of 0.1–0.3 mm (transformation temperatures $A_s = 25.5$, $A_f = 46.5$, $M_s = 10$ and $M_f = -14.5$ °C measured by DSC) were trained and programmed by a thermomechanical process involving heating the 'shaped' alloy for up to four hours at 650 °C in an inert atmosphere, followed by quenching in water. It is demonstrated that a solution treatment at 650 °C/60 minutes and ageing treatment at 380 °C/100 minutes yields an M_s of about 14 °C, while ageing treatment at 480 °C/100 minutes yields an M_s of about 20 °C. The variation is consistent with the formation of lenticular Ti_3Ni_4 precipitates. When the specimen is annealed at a temperature lower than 400 °C, the Ti_3Ni_4 precipitate particle is fine and the dispersion density is high, so the precipitate Ti_3Ni_4 has great coherence with the matrix. On the contrary, when annealed at a temperature higher than 400 °C, the precipitate Ti_3Ni_4 grows up and the low dispersion density destroys the coherence between Ti_3Ni_4 and the matrix (Nishida and Wayman, 1984; Scherngell and Kneissl, 1999). As conventional textile materials are not able to withstand the high temperatures required for the programming, use of SMAs in textiles normally requires programming of the alloy before yarn spinning, weaving or knitting.

Engineering textile and clothing aesthetics   179

The shape memory recovery of a textile that contains a trained SMA spring with varied temperature, in which the shape of the fabric changes under the influence of the trained SMA spring incorporated into the fabric structure has been investigated. In this case, the fabric can display a two-way shape memory effect when the spring is trained in such a way. Since the training of two-way SME springs is carried out before the spring wire is woven into a fabric structure, various fabric structures with different yarn textures, have been developed. The trained SMA wire may be engineered to enhance the aesthetic appeal of fabrics or clothes, showing the potential for significant shape change in response to environmental variation. An example of a potential application is the creation of an intelligent window curtain with self-regulating structures changing under a range of temperatures.

10.5.7 SMAs in yarns

Yarns can be designed with trained Ni-Ti SMA wires as core component, wrapped with conventional fibres such as polyester, viscose and polyamide. A range of yarns with different twist levels and fibre content have been produced in a recent work using a Gemmel and Dunsmore Fancy Wrap spinner, shown in Fig. 10.8. The level of coverage was found to be important in order to prevent the alloy, as the core structure, protruding from the yarn during the development of the shape memory effect. Yarns were optimised for maximum stability during mechanical deformation by altering the yarn specifications particularly their twist level.

10.5.8 Fabric development

Experimental fabrics have been made of SMA wire and/or of SMA wrapped-spun yarns. Woven structures consisting solely of untrained SMA wires are programmed after weaving and trained as a whole structure. Knitting a structure consisting entirely of SMA wires is more complex due to the low extensibility of the wire which creates difficulties in loop formation, stability and regularity. In the case of knitting wrap-spun SMA/fibre yarns, it was found that due to high stiffness and low tensile properties of the yarns, the process is significantly affected by the properties of the core SMA and hence complex structures were not possible. This was further prevented by the fact that the balance of the core/wrap structure could be easily disrupted during loop formation. Knitted structures consisting of selected areas of the core-wrapped yarns were found to be more stable. For woven fabrics, the wrap-spun yarns lend themselves for handloom weaving more easily, and various interesting structures have been produced.

180 Intelligent textiles and clothing

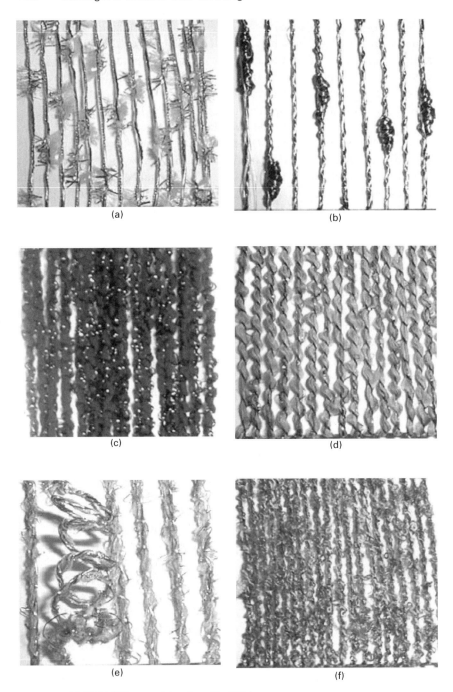

10.8 SMART yarn blends with SMA.

10.5.9 Utilising the SMAs aesthetic effects

A range of textured and sculptured woven and knitted fabrics have been designed and produced. The following three-dimensional concepts have been developed using the alloy's shape memory properties as the basis for imparting physical changes in the fabric structure:

- movement in the fabric or on the fabric surface, e.g., visual wave effects and other textural changes on the surface of the fabric
- opening and closing of apertures in the fabric structure to alter properties such as opacity, insulation, diffusion of light, and air flow
- accordion-style opening and closing of the fabric structure, to create structures that can open and close, e.g., for interior screens.

By training the alloy to remember a specific shape in the deformable martensite phase, shape memory fabrics can be produced. When temperatures increase, the alloy changes from the martensite phase to the austenite one, and reverts to the shape that it was trained to remember. This creates structural and textural changes in the yarns and in the fabric. The changed structure will retain its shape until mechanically deformed again when fabric cooling takes place (Chan and Stylios, 2003b).

10.5.10 Potential exploitation and limitations

The exploitation of SMAs in SMART textiles relies on two main factors: the ability of SMAs to change shape upon the application of a stimulus such as heat or an electric current, and the flexible structure of woven and knitted fabrics. In the past, SMAs have been used in a number of non-textile applications, ranging from simple day-to-day items such as flexible frames for eyewear and coffee-pot thermostats, to high-performance devices such as vascular stents, medical anchors and guides and orthodontic braces. With the possibility now to incorporate them within a flexible textile structure, SMAs can find applications in the home and furnishing sector, clothing and sportswear sector, geotextiles, etc., where the shape changing effect can be triggered either by a change in ambient temperature, material temperature, or on the application of a small voltage.

Using alloys as composite textile structures, shape changes can be facilitated, particularly because of the high recovery properties. On the other hand, although SMAs can be used in selected areas only, they still add weight to the structure, and because of their stiffness, can have a detrimental effect on drape and comfort.

10.5.11 Shape memory coatings

SMCs pose most interesting possibilities due to their ability to be coated on existing textiles without the need for knitting or weaving them into the

182 Intelligent textiles and clothing

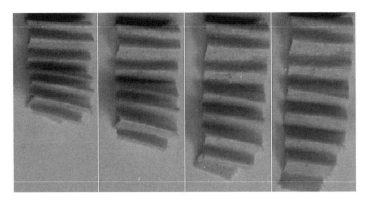

10.9 Shape memory recovery of fabrics coated with SMP when T > T_g (a) 0 sec, (b) 15 sec, (c) 30 sec, (d) 45 sec.

structure. Figure 10.9 shows shape memory recovery of fabric samples coated with SMP polyurethane. The SME may be trained from being flat at high temperature to being bent at low temperature, tailored to suit specific ranges of environmental temperatures. Observations revealed that SMP layered fabric have stronger shape recoverability compared with single layer shape memory woven fabrics, and therefore permit more flexibility in shape memory design, especially for SMP layers coated on stiffer fabrics (Wan and Stylios, 2004). The increment of shape memory recovery force in the textiles is due to a strong cross-linking of the fabric matrix that consists of high elastic modulus yarns. SMP reinforced with the matrix fabric having higher elastic modulus exhibits higher recoverability with the application of heat, which could be of great benefit to SMART textiles. In addition, when combined with reversible colour change pigment, intelligent hybrid coatings may be created to demonstrate a number of SMART functions, such as shape and colour change, in response to variations in environment temperature.

10.6 Aesthetic interactive applications of shape changing smart textiles

10.6.1 Textile art

One of the main areas where the characteristics of SMMs can be used purely for aesthetic and interactive purposes is in the area of textile art and design. Displays, 3D structures, flat panels and other forms of textiles including SMPs and alloys have a strong element of interaction between the viewer or the environment and the artwork. Krähenbühl for example utilised SMAs to create sculptures that react to wind and temperature (Gotthardt and Krähenbühl, 2005). Similar concepts can also be used in flexible textile art.

10.6.2 Fashion

In the fashion world, a number of aesthetic interactive effects can be incorporated in clothing and accessories, including shoes, bags, gloves, and the like. Corpo Nove has already illustrated the use of SMA in a shirt that automatically rolls up its sleeves when it is warm (Marks, 2001). Although this is, strictly speaking, not aesthetically driven, it does illustrate the potential for visual impacts. The development of 'mouldable' or 'reformable' fashion items that can be folded, twisted and changed into a desired shape when worn, and that will spring back to their original shape when stimulated is an interesting concept for the personalisation and customisation markets. Likewise, the developments of wearable items that, upon a flick of a button, can change shape, length, size, or other dimensional feature are interesting concepts to explore. The sportswear industry has already expressed interest in shape memory shirts that can open or close (macroscopically or microscopically) with body temperature in order to act as a heat and moisture management system, and also have an aesthetic visual effect. However, in addition to aesthetic and functional properties, SMMs can also be engineered for sensory and psychological effects. As an example, Winchester and Stylios (2003) demonstrated the use of shape memory alloys and polymers to create fabrics that can have interesting visual surface movements for knitwear, as shown in Fig. 10.10.

10.6.3 Interiors

Another area where the shape changing characteristics can be capitalised is in the interior textile industry, where again, the duality of the properties (aesthetic and functional) can be explored (Chan and Stylios, 2003a). Partitions, wall hangings and panels, window blinds, etc. with the ability to change

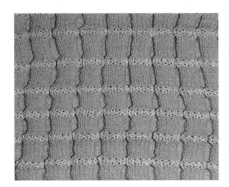 

10.10 SMART knitted and woven structures for garments.

184 Intelligent textiles and clothing

10.11 SMART interiors.

shape, texture and structure can create functional effects (more light, more air, more sound, etc.), but can also be used as decorative elements to interact visually and sensorily with people, Fig. 10.11. As illustrated by Chan and Stylios (2003a), the combination of woven 3-dimensional effects with shape memory properties creates exciting interactive visual effects with a huge potential for interior products.

10.7 The concept of mood changing textiles for SMART ambience

10.7.1 Introduction

SMART materials have been used for their functionality in the medical, aerospace, automotive fields and the like. Nowadays SMART material technologies (including SMMs, chromatic dyes and pigments, thermoelectric films and wearable electronics) are being optimised to explore the concept of SMART ambience, where an interior environment can respond interactively to the needs of users. A flexible composite matrix material integrating aesthetics and functionality is currently being developed by Stylios (2005). Focusing particularly on enhancing the functionality and aesthetics of SMART materials, this new field will bring a strong element of innovation to the interior and technical textiles industry.

Characteristics, properties and responses are being optimised for interior environment control both following moods and feelings through audio signal conversion technology in the first instance for interactivity. Based on the end-use requirements, this explores and combines several categories of SMART materials to create innovative hybrid architectures of flexible matrices. It tries therefore to create SMART structures, systems and prototypes with tailor-made functionality and aesthetics (Fig. 10.12) combining mood-changing technologies with SMART technologies for the development of a SMART ambience.

Engineering textile and clothing aesthetics   185

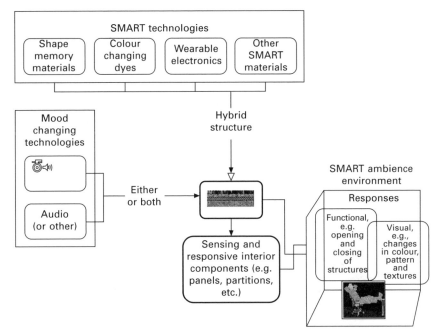

10.12 Combining mood-changing technologies with smart technologies for the development of a SMART ambience.

10.7.2 Technology approach

A new generation of hybrid structure(s) that can be used for interior components such as panels, partitions, and the like is being developed. The structure(s) are based on SMART technologies, integrated with mood changing technologies, Fig. 10.12, so that a new concept of a SMART ambience environment can be established.

Responses include both visual and functional changes in the structure, as a result of mood/physiological changes, or on command. For example a new hybrid composite matrix architecture, shown in Fig. 10.13 or structures that can change colour, pattern and/or texture can be created with the combined use of chromatic materials and SMMs. Colour and pattern effects are being imparted by chromatic materials. Texture and shape changes, e.g., changes in the openness and closeness of a structure, are being developed using shape memory polymers or alloys. Beside visual effects, functionally is possibly by changes of the shape memory which in turn can change the degree of air flow (hence temperature), improve privacy between partitioned areas and sound-absorption properties.

Upon command changes to the structure can be triggered by sending a signal to the structure, which then triggers these changes. Mood-dependent changes can be triggered in several ways, and are now being explored. The

186 Intelligent textiles and clothing

10.13 SMART fabric architecture.

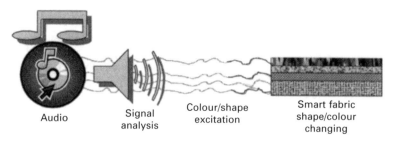

10.14 SMART fabric architecture mood selectable (SFAMS); mood changing SMART fabric matrix capable of changing shape and colour depending on personal moods.

first approach is by integrating audio output such as vocals directly to the structure through voltage by breaking the signal into primary colours and shapes Fig. 10.14. Another approach is by a webcam positioned on an office computer that can pick up facial expressions and detect typical moods (stress, happiness, sadness, etc.). Upon detection of these moods, specific changes in the SMART hybrid structure can be programmed to take place. In another approach, an inconspicuous wearable device (e.g. a wristband) with small sensors wirelessly connected to a central processing system can be used to detect physiological changes (body temperature, skin resistivity, heart rate, etc.) related to moods. The device can then transmit commands to the SMART hybrid structure to change. Finally and simply the environment can change at will by a switch or a computer selection by the user.

10.8 Summary

Recent advances in shape memory alloys (SMAs) and shape memory polymers (SMPs) have inspired people to create SMART textiles with self-regulating structures and performance in response to environmental variations, which constitutes a new field in the scientific frontier of SMART materials. By appropriate shape memory training of SMA and SMP, fabric designs can be engineered to enhance their functionality and particularly the aesthetic appeal of new textiles and clothes by significant shape change in response to environmental variation. Main technologies include the training procedures for shape memory effect (SMEs) in SMAs and SMPs, the regulation of phase transformation temperatures in SMAs, the manufacture of SMP yarns and SMP coatings for the requirements of new fabric design. Further attention is also paid to the combination of SMP or SMA with conventional yarn blends engineered to promote a given SME. Finally a hybrid SMART fabric architecture has been described and discussed for the next generation of mood changing textiles applied to SMART ambience.

10.9 Acknowledgement

I wish to acknowledge staff and students who have made the fascinating world of shape changing textiles possible; Dr T Wan, Dr Y Chan Vili, Dr S Lam Po Tang, Mr A McCullough and Mr S Wallace.

10.10 References

Bhuniya, A. K., Chattopadhyay, P. P., Datta, S. and Banerjee, M. K., 'On the Degradation of Shape Memory Effect in Trace Ti-added Cu–Zn–Al Alloy', *Materials Science and Engineering*, Vol. A 393, pp. 125–132, 2005.

Borchardt, J. K., 'Shape-Memory Polymers See the Light', *Materials Today*, Vol. 8, Issue 6, p. 15, June 2005.

Butera, F. and Alacqua, S., 'Fabric for composite materials comprising active threads, and method for obtaining said fabric', European Patent No EP1420094, 19 May 2004.

Chan, Y. Y. F. and Stylios, G. K., 'Designing Aesthetic Attributes with Shape Memory Alloy for Woven Interior Textiles' in *INTEDEC 2003, Fibrous Assemblies at the Design and Engineering Interface*, Edinburgh, UK, September 2003(a).

Chan, Y. Y. F. and Stylios G. K., 'Engineering the Design Attributes of Woven Interior Textiles Using Shape Memory Polymer', Advanced Flexible Materials and Structures: Engineering with fibres, The Fibre Society 2003 Spring Symposium, June 30–July 2, 2003(b), Loughborough, UK.

Chan, Y. Y. F., Winchester, R. C. C., Wan, T. and Stylios, G. K., 'The Concept of Aesthetic Intelligence of Textile Fabrics and their Application for Interior and Apparel', *IFFTI 2002*, Hong Kong, November 2002.

Eggeler, G., Hornbogen, E., Yawny, A., Heckmann, A. and Wagner, M., 'Structural and Functional Fatigue of NiTi Shape Memory Alloys', *Materials Science and Engineering*, Vol. A 378, pp. 24–33, 2004.

Gotthardt, R. and Krähenbühl, E., 'Science & Sculpture: The Fabulous Discoveries of 8 Years of Collaboration', Materials Research Society (MRS) Spring Meeting, San Francisco, USA, March-April 2005.

Heczko, O., 'Magnetic shape memory effect and magnetization reversal', *Journal of Magnetism and Magnetic Materials*, No. 290–291, pp. 787–794, 2005.

Horie, H., 'Shape Memory Woven Fabric', Japanese Patent No JP2000345444, 12 December 2000.
Hu, J., *Shape Memory Polymers and Textiles*, Woodhead Publishing Limited, forthcoming, ISBN No. 1-84569-047-8.
Keiji, I., 'Method for Manufacturing Yarn for Vessel Stent', US Patent No US2003055488, 20 March 2003.
Kitahira, T., Matsuo, A., Takeuchi, M. and Miyashita, M., 'Production of Woven Fabrics using Wire of Shape Memory Alloy', Japanese Patent No JP8209488, 13 August 1996.
Kobayashi, K. and Hayashi, S., 'Woven Cloth from Shape-Memory Polymer', Japanese Patent No JP2112433, 25 April 1990.
Liu, C. *et al.*, 'Tailored Shape Memory Polymers: Not all SMPs are Created Equal' in *First World Congress on Biomimetics*. 2002. Albuberque, New Mexico.
Liu, Y., Gall, K., Dunn, M. L., Greenberg, A. R. and Diani, J., 'Thermomechanics of Shape Memory Polymers: Uniaxial Experiments and Constitutive Modeling', *International Journal of Plasticity*, Vol. 36, No. 10, pp. 929–940, 2004.
Marks, P., 'Sleeves Up', *New Scientist*, Issue 2301, p. 24, 28 July 2001.
Metcalfe Annick, Desfaits Anne-Cecile and Salazkin Igor, 'Cold Hibernated Elastic Memory Foams for Endovascular Interventions', *Biomaterial*, Volume 24, issue 3, pp. 491–497, February 2003.
Nishida M. and Wayman C. M., *Scr. Metall.* 18, pp. 1389–1394, 1984.
Ohki, T., Ni, Q-Q., Ohsako, N. and Iwamoto, M., 'Mechanical and Shape Memory Behavior of Composites with Shape Memory Polymer', *Composites: Part A* 35, pp. 1065–1073, 2004.
Oohira, M., 'Weaving Process using Shape Memory Filament Yarn and Apparatus Therefor', Japanese Patent No JP2221435, 4 September 1990.
Otsuka, K. and Ren, X., 'Physical Metallurgy of Ti–Ni-based Shape Memory Alloys', *Progress in Materials Science*, Vol. 50, pp. 511–678, 2005.
Panoskaltsis, V. P., Bahuguna, S. and Soldatos, D., "On the Thermomechanical Modeling of Shape Memory Alloys", *International Journal of Non-Linear Mechanics*, Vol. 39, pp. 709–722, 2004
Scherngell, H. and Kneissl, A. C., *Mat. Sci. Eng.* A273-275, p. 400, 1999.
Shaw, J. A. and Kyriakides, S., 'Thermomechanical Aspects of NiTi', *Journal of Mechanical and Physical Solids*, Vol. 43, No. 8, pp. 1243–1281, 1995.
Stylios, G. K., 'Fibrous Assemblies at the Design and Engineering Interface', Book of Proceeding, ITEDEC 2003, Heriot-Watt University, ISBN No. 0-9546162-0.
Stylios, G. K., 'The Concept of Programmable Fibrous Assemblies for SMART Ambience', Keynote Paper, AUTEX 2005, 27–29 June 2005, Portoroz, Slovenia.
Stylios, G. K., Luo, L, Chan Vili, Y. Y. F. and Lam Po Tang S., 'The Concept of SMART Textiles at the Design/Technology Interface,' 5th International Instanbul Textile Conference, Recent Advances in Textiles and Clothing", Turkey, 19–21 May, 2005.
Stylios, G. K. and Wan, T., 'Shape Changing SMART Fabrics', Transactions of the Institute of Measurement, in print, 2006.
Terada, F., 'Fabric', Japanese Patent No JP2289142, 29 November 1990.
Wan, T. and Stylios, G. K., 'Investigating Shape Memories Technologies for SMART Fabric', 2nd International Textile Conference of the North India Section of the Textile Institute, New Delhi, 2–3 December, 2004.
Wang, Z. G., Zu, X. T., Feng, X. D., Zhu, S., Bao, J. W. and Wang, L. M., 'Characteristics of Two-way Shape Memory TiNi Springs Driven by Electrical Current', *Materials & Design*, Vol. 25, pp. 699–703, 2004.

Winchester, R. C. C. and Stylios, G. K., 'Designing Knitted Apparel by Engineering the Attributes of Shape Memory Alloy', *International Journal of Clothing Science and Technology*, Vol. 15, No. 5, pp. 359–366, 2003.

Yang, B., Huang, W. M., Li, C. and Chor, J. H., 'Effects of Moisture on the Glass Transition Temperature of Polyurethane Shape Memory Polymer Filled with Nano-Carbon Powder', *European Polymer Journal*, Vol. 41, Issue 5, pp. 1123–1128, May 2005.

Yoshida, M., 'Special Fabric', Japanese Patent No JP3130147, 3 June 1991.

Part III

Chromic and conductive materials

11
Introduction to chromic materials

P TALVENMAA, Tampere University of Technology, Finland

11.1 Introduction

Chromic materials are the general term referring to materials, which change, radiate or erase colour. Chromism as a suffix, means reversible change of colour and by extension, a reversible change of other physical properties. Due to colour changing properties, chromic materials also are called chameleon materials. This colour changing phenomenon is caused by the external stimulus and chromic materials can be classified depending on the external stimulus of induction:[1]

- photochromic: stimulus is light
- thermochromic: stimulus is heat
- electrochromic: stimulus is electricity
- piezochromic: stimulus is pressure
- solvatechromic: stimulus is liquid
- carsolchromic: stimulus is an electron beam.

Chromism has been studied since before the 1900s and the main applications are in the areas of photochromism, thermochromism and electrochromism such as paints, inks, tiles, eyeglasses, windows and many optical applications.[3] The most applied chromic solutions in textiles and fibre materials are the two first groups, photochromic and thermochromic.

Photochromic materials change their colour by light and these materials are both organic and inorganic, but the most studied are organic photochromic materials. There are also photochoromic materials called heliochromic compounds, which are activated by unfiltered sunlight and deactivated under diffuse daylight conditions. Therefore they are suitable for sun lens applications.[7]

Thermochromic materials change their colour as a result of reaction to heat, especially through the application of thermochromic dyes whose colours change at particular temperatures. Two types of thermochromic systems have been used successfully in textiles, the liquid crystal type and the molecular rearrangement type.[1]

Electrochromic materials are capable of changing their optical properties (transmittance and/or reflectance) under applied electric potentials; when that potential is stopped or it goes into reverse, these materials retain or return to their original optical state in a reversible way. The variation of the optical properties is caused by insertion/extraction of cations in the electrochromic film.[4]

Piezochromism is the phenomenon where crystals undergo a major change of colour due to mechanical grinding. The induced colour reverts to the original colour when the fractured crystals are kept in the dark or dissolved in an organic solvent.[7]

Solvatechromism is the phenomenon, where colour changes when it makes contact with a liquid, for example, water. Materials that respond to water by changing colour are also called hydrochromic and this kind of textile material can be used, e.g., for swimsuits.[1]

11.2 Photochromic materials

Photochromic materials change colour reversibly with changes in light intensity. Usually they are colourless in a dark place and when sunlight or UV-radiation is applied the molecular structure of the material changes and it exhibits colour. When the relevant light source is removed the colour disappears. Changes from one colour to another colour are possible by mixing photochromic colours with base colours.[2]

In photochromic pigments the structure changes when it is exposed to sunlight or ultraviolet radiation, causing a colour change. The reversible change can be colourless to colour or, by a combination of photochromic conventional dyes, one colour to another. The lifetime of these compositions is rather short for industrial applications or building applications.

Photochromic materials are used in lenses, paints, inks and mixed to mould or cast materials for different applications.[2] Photochromic compounds in textiles are used mainly for decorative effects in jacquard fabrics, embroideries and prints in different garments.

11.2.1 Definitions of photochromism

Photochromic compounds are chemical species having photochromic properties. There are two major classes of photochromic materials, inorganic and organic; most inorganic materials are based on silver particles. Photochromism is a reversible transformation of a chemical species induced in one or both directions by absorption of electromagnetic radiation between two forms, A and B, having different absorption spectra (Fig. 11.1).[7] The

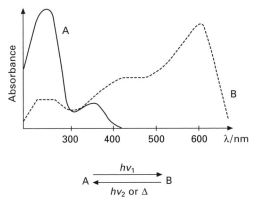

11.1

thermodynamically stable form A is transformed by irradiation into form B. The back reaction can occur thermally (photochromism of type T) or photochemically (photochromism of type P).

The bistability of organic photochromic compounds is of special interest for optical data memory systems (photon-mode erasable optical recording media). Numerous other uses have also been explored with encouraging results; applications concerning single molecules, polymers and biological molecules. In the case of optical storage memory systems, the interest in photochromic compounds is based on the fact that photon-mode recording has certain advantages over the heat-mode, regarding resolution speed of writing and multiplex recording capability. Also fatigue due to material movement is eliminated. Despite these advantages, the photochromic molecules still await a wide range of practical applications. The reason is that they must fulfil many other requirements, e.g., can be cycled many times without loss of performance and thermal stability.[8]

The fact that some chemical species can undergo reversible photochemical reactions goes beyond the domain of variable optical transmission and includes a number of reversible physical phenomena such as optical memories and switches, variable electrical current, ion transport through membranes, variable wettability, etc. For this purpose, organic photochromic compounds are often incorporated in polymers, liquid crystalline materials, or other matrices.[7]

11.2.2 Photochromic applications

Photochromism is simply defined as a light-induced reversible change of colour; a most common example of that phenomenon is photochromic spectacles that darken in the sun and recover their transparency in diffused light. The first commercial spectacles used glass lenses impregnated with inorganic salts (mainly silver) but in recent years organic photochromic lenses are replacing the inorganic ones.[7]

Photochromic phenomena are presently also being used in many textile applications like T-shirts, handbags and caps. There is, e.g., nightwear, which indoors is white while in daylight the colour changes to blue, green, purple or whatever. The California-based company SolarActive International manufactures a wide variety of photochromic products including colour-changing textured yarns for textile knitting and weaving and embroidery threads stimulated by UV-light.[6] Embroidery threads are made by using mass pigmentation of polypropylene in the molten state with different photochromic compounds. Polypropylene fibre is then produced by melt spinning to serve as the basis of solar-active thread. The thread appears white on fabric when indoors and/or away from UV-light. Outside, rainy or sunny, the UV radiation penetrates the thread, activating the photochromic compounds which change the thread to specific colours. The threads return to white when removed from UV-light within about one and a half minutes. The colour-changing effect has been tested and proven to last for more than 2000 alternations, in and out of UV-light, which is equivalent to the life of a garment. Since the photochromic material is not just a coating but a part of the polypropylene, the material does not wash off and remains with the fibre for the life of the product.[5]

The Swedish Interactive Institute has developed a curtain made from a colour-changing material sensitive to UV-light. A computer-controlled UV lamp dynamically illuminates various parts of the curtain thereby creating a dynamic textile pattern depending on a computational interpretation of certain given information. The textile can thus be connected to a computer as a kind of display.[9]

One commercial application of photochromic textiles includes the fabrics of Taiwan company Super textile Corp. These high-tech photochromic fabrics utilise a special microcapsule dye technology that changes colour upon absorbing sunlight and UV radiation. The fabrics can be combined with other coating materials and dyestuffs to help produce a colourful effect. These products are non-hazardous and they can be used in general applications such as knapsacks, dolls and warning signs.[10]

11.3 Thermochromic materials

Thermochromic materials change colour reversibly with changes in temperature. The materials can be made as semi-conductor compounds, from liquid crystals or using metal compounds. The change in colour occurs at a predetermined temperature called the thermochromic transition. This can be altered by adding doping agents to the material.[3] Thermochromic materials using encapsulated dyes were developed in the 1970s. The technology was patented by Japanese companies and subsequently used mainly in the textile and novelty industries.[11] Thermochromic materials are used in a variety of commercial products including textiles and novelty items, toys and baby spoons.

Most thermochromic commercial products are derived from liquid crystals or complicated mixtures of organic dyes. Both of these methods require the materials to be microencapsulated. This results in a variety of limitations including poor thermal stability and difficulty in processing. In addition, some materials that are currently available are toxic and easily extracted from most plastics and therefore not approved for use in articles that come into contact with food products.[12] Thermochromic materials are used as well for fun purposes and for many useful applications like instrumentation, warning, information and indicating data, e.g., body thermometers.

11.3.1 Definitions of thermochromism

Thermochromism is the reversible change in the colour of a compound when it is heated or cooled. The thermochromic colour change is distinguished by being quite noticeable, often dramatic and occurring over a small or sharp temperature interval. The mechanics of thermochromism have been investigated since the beginning of 1970. There are different mechanisms according to which materials are used. The following four materials are used in thermochromism:[13]

1. organic compounds
2. inorganic compound
3. polymers
4. sol-gels.

Organic compound thermochromism

Organic compound thermochromism has various applications for fibres, optical sciences, photo-storage instruments and optical sensors. The mechanism responsible for thermochromism varies with moleculer structure. It may be due to an equilibrium between two molecular species, acid-base, keto-enol, lactim-lactam, or between stereoisomers or between crystal structures. We can divide this into three parts on the basis of materials:[13]

1. variation in crystal structure
2. stereoisomer
3. molecular rearrangement.

The advantages of thermochromism of these organic compounds are that colour change takes place sharply and that there are many factors to control temperature easily. Two types of these thermochromic systems have been used successfully in textiles, the liquid crystal type and the molecular rearrangement type.[1]

Liquid crystal type

Thermochromic liquid crystals show different colours at different temperatures because of selective reflection of specific wavelengths of light from their structure. In an appropriate temperature range intermediate between a low-temperature crystalline phase and a high-temperature isotropic liquid phase, these materials form a cholesteric liquid crystal. In a cholesteric liquid crystal, changes in temperature result in thermal expansion, which leads to a change in layer spacing and hence pitch and therefore the colour change observed will vary with temperature.[13] These materials are used in the manufacture of thermochromic printing ink. Thermochromic liquid crystals have the advantage that they express a finely coloured image, but they have the disadvantage of high cost and low colour density.

RO—⟨C6H4⟩—C(=O)—O—⟨C6H4⟩—CH$_2$CH(CH$_3$)CH$_2$CH$_3$

11.2

Molecular rearrangement type

The molecular rearrangement of an organic compound that arises from tautomerisation, such as that which occurs as a result of, for example, an acid-base, keto-enol or lactim-lactam equilibrium, can lead to an increase in the conjugation of the molecule and formation of a new chromophore. Such molecular rearrangement can be effected by a change in temperature or by alteration of the polarity of the solvent and/or the pH of the system. Because of the pH-dependent chromism, the temperature-dependence of the acid-base equilibrium means that pH sensitivity can result in thermochromic behaviour.[13]

11.3

According to the definition of thermochromism given earlier, crystal violet lactone does not exhibit thermochromism, as the equilibrium is pH- rather than temperature-dependent; indeed, the ring-opening reaction does not occur, even on heating low pH values (below approximately pH4) the coloured form predominates, while an increase in pH results in the equilibrium moving to the left, with the effect that the system becomes colourless.

Inorganic thermochromism

Many metals and inorganic compounds are known to exhibit thermochromic behaviour either as solids or in solution. Inorganic thermochromic systems have not enjoyed widespread textile applications as the observed change in colour often occurs in solution at high temperature. The majority of textile end-uses demand a reversible, solid system that is suitable for application by printing and that exhibits distinct colour changes over a small temperature range.[13]

11.3.2 Thermochromic applications

Colorant manufacturers and academic research organisations have worked for many years to develop fast colouring matters by researching dyes and pigments that are chemically inert and physically unresponsive once they have been applied to a substrate.[14] Thermochromic organic colorants have been developed for use in producing novel coloration effects in textiles as well as other applications. In the textile sector most thermochromic applications are the same as photochromic ones, like T-shirts and caps, which use thermochromic inks for decoration purposes. An increase in temperature produces a reversible change in colour and the heat from direct sunlight is enough to produce the change.[15]

The diverse range of non-textile applications to which thermochromic materials have been put include thermometers and temperature indicators for special purposes like security printing and non-destructive testing. They can indicate the proper temperature, e.g., cold enough for storage or hot enough for serving.[12] Compared with other markets, the textile sector generally lags behind in the exploitation of thermochromic materials. One reason is that the colour changing temperature of the majority of these materials lies well above room temperature, typically over 100–200 °C and have not so much use for producing thermochromic effects on textiles.

There are, however, thermochromic materials based on a special class of liquid crystalline substances and by using these it is possible to achieve significant changes in colour appearance over narrow temperature ranges, 5–15 °C, and also to detect small variations in temperature, < 1 °C.[14] The Japanese company Pilot Ink Co. Ltd have worked for a long period with thermochromic materials and have also a significant amount of patents. Pilot Ink Co. is the leading company developing especially thermosensitive materials for different mouldings like toys and hoses. Thermosensitive material is coloured by mixing the three primary colours so it can display virtually any hue from white to black. The thermosensitivity is functional at temperatures ranging from –10 °C to 60 °C, within the range of ordinary manufacturing processes.[15]

11.4 Colour-changing inks

The two major groups of colour-changing inks are thermochromic, which change colour in response to temperature fluctuations, and photochromic, which respond to variations in exposure to UV light (primarily sunlight). Both materials are reversible and will change colour over and over again with the appropriate exposure.[17]

11.4.1 Temperature-sensitive inks

The two types of thermochromic inks are liquid crystals and leucodyes.[t] The most famous thermochromic application ever, the 'mood ring', was a liquid crystal. Today liquid crystals are used in many products, including aquarium thermometers, stress testers, and forehead thermometers. Unfortunately, liquid crystal thermochromics are very difficult to work with and require highly specialised printing and handling techniques.

The other type of thermochromic ink, the leucodye, is used in wide range of applications, e.g., for textiles, security printing, novelty stickers, product labels and advertising specialities. Some products printed with leucodye thermochromic inks change from one colour to another, rather than transitioning from coloured to clear. This is achieved with an ink that combines a leucodye with a permanent-colored ink formulation. For example, the ink manufacturer may formulate a green ink by adding a blue leucodye to a yellow ink. In its cool state the printed ink layer is green and once warmed reverts to yellow as the leucodye becomes clear or translucent. Leucodyes can be designed to change colour at various temperature ranges, from as low as –25 °C up to 66 °C. A wide range of colours is also available.

In order to function, a leucodye requires a combination of chemicals working together in a system. This special system of materials needs to be protected from the components of the ink to which it is being added, so it is microencapsulated. The microencapsulation process takes a small droplet of the leucodye and coats a protective wall around it. The leucodye microcapsules contain the complete colour-changing thermochromic system, which, when added to inks, give them their colour-changing properties.

Under normal conditions, the thermochromic leucodye inks have a shelf life of six months or more. After they are printed, they function or continue to change colour for years. The post-print functionality can, however, be adversely affected by UV light, temperatures in excess of 121 °C and aggressive solvents. Thermochromic textile inks will withstand about 20 washings before showing a significant deterioration and can last even longer when dried without heat. Chlorine bleach is not recommended, because this will degrade the performance of the printed ink on textiles.[17]

11.4.2 Light-sensitive inks

Photochromic inks change from clear when indoors to coloured when taken outdoors. UV-light changes the chemical structure of the photochromic material and makes it absorb colour like a dye. It then reverts to a clear state when the UV source is removed. The colour change can occur thousands of times, depending on the application. A photochromic dye can also change from one colour to another when it is combined with a permanent-pigment ink. These inks are also available in a full range of colours, including a four-colour process system. One of the most popular uses for photochromic inks is on screen-printed garments such as T-skirts.

In their pure state, photochromics are powdered crystals that must be dissolved in the inks to which they are added. Some manufacturers microencapsulate the photochromics in their own system, as with leucodye microcapsules. Microencapsulating photochromic systems enables them to be used in inks that cannot dissolve them, such as water-based systems.

Photochromic material is inherently unstable and actually changes its chemical structure when exposed to UV light. Because the dye is so vulnerable in its excited state, stabilisation is the prime challenge for photochromic ink manufacturers. Without stabilisation most photochromic inks would not last even a few days in sunshine and may even expire before being printed.

A properly stabilised photochromic ink will last for years in store before printing, but even the best of them will withstand only a few months of outdoor exposure after printing. The best photochromic textile inks will withstand about 20 washings after printing and are more susceptible to the negative effects of chlorine bleach than their thermochromic counterparts. Bleach must never be used on garments printed with photochromic inks.[17]

11.5 Electrochromic materials

Electrochromic materials are materials that change colour upon application of an electrical potential. Electrochromic systems have been used successfully in the past in mirrors and windows for anti-glare and anti-reflective applications. However, other potential applications, including large area privacy/security glazing and high-contrast angle-independent dynamic displays have not yet reached the market. This is due to the limitations of existing electrochromic materials.[18]

Collaborative research efforts in different companies, universities and research centres have concentrated on studying electrochromic applications in fibres and textiles. One example of this kind of research work is carried out by Clemson University, Georgia Institute of Technology and Furman University in the USA and the goal is to create colour tunable fibres and fibre composite structures. This is being accomplished by incorporating into or onto fibres, molecular or oligomeric organic chromophoric devices capable of colour change over the visible portion of the electromagnetic spectrum by the application of a static or dynamic electric field. Anticipated applications for this type of material will include wall and floor coverings, camouflage materials and fabric based electronic displays which can change colour or information display content based on the electric field applied. Such fields will be low-voltage fields (3–5 volts). One very interesting effect found with this work is the possibility of information storage and retrieval from organic fibres based on molecular recognition. This may lead to microdevices encapsulated in fibres for a variety of technical applications.[19]

Electrochromic materials, which change their optical properties in response to an electric field and can be returned to their original state by a field reversal, have major advantages:[20]

- a small switching voltage (1–5 V)
- they show specular reflection
- they possess a grey scale
- they require power only during switching
- they exhibit adjustable memory, up to 12–48 hours.

11.5.1 Definitions of electrochromism

Electrochromic material changes colour in a persistent but reversible manner by an electrochemical reaction and the phenomenon is called electrochromism. Electrochromism is the reversible and visible change in transmittance and/or reflectance that is associated with an electrochemically induced oxidation-reduction reaction. It results from the generation of different visible region electronic absorption bands on switching between redox states. The colour change is commonly between a transparent ('bleached') state and a coloured state, or between two coloured states. When more than two redox states are electrochemically available, the electrochromic material may exhibit several colours and may be termed as polyelectrochromic or can be said to possess multicolour-electrochromism. This optical change is effected by a small electric current at a low d.c. potential of the order of a fraction of a volt to a few volts.[21]

11.5.2 Electrochromic applications

Electrochromic devices are the most popular technology for large area switching devices. Much of this technology is being developed for building and automotive windows, mirrors and glazing applications. Electrochromic materials are finding applications as specialised displays, electronic books, paper-like, and banner displays. Many chromic materials can be fabricated on plastic substrates, which is an advantage for future display aplications.[20]

One example of an innovation is a paper-thin electrochromic display developed by the Siemens company. The display is controlled by a printed circuit and can be powered by a very thin printable battery or a photovoltaic cell. The goal is to be able to create the entire device – the display and its power source – using the same printing method, so that manufacturing costs would be as low as possible. Siemens expects to achieve this by 2007.[22]

Textile integrated display applications have been developed by France Telecom: a flexible, battery-powered optical fibre screen is woven into clothing. Each plastic fiber-optic thread is illuminated by tiny LEDs that are fixed along the edge of the display panel and controlled by a microchip. The threads are set up so that certain portions are light when the LEDS are switched on, while other sections remain dark. These light and dark patches essentially act as pixels for the display screen.[23]

IFM's patent Electric Plaid™ is a combination of textile display technology and design material. The novel fabric contains interwoven electronic circuits from stainless steel yarns and thermochromic colour-changing ink, which are connected to drive electronics. Flexible wall hangings can then be programmed to change colour in response to heat from the conducting wires.[23] Electrochromic inks have some applications in textiles, e.g., jacket, which could be black and every now and then it displays a pictures.[20]

11.6 Conclusion

The majority of applications for chromic materials in the textile sector today are in the fashion and design area, in leisure and sports garments. In workwear and the furnishing sector a variety of studies and investigations are in the process by industrial companies, universities and research centres. There is a large number of patents, especially in Japan and in the USA, in different ranges of chromic materials, but many innovations still await their commercial applications.

Most photochromic materials are based on organic materials or silver particles. The lifetime of these compositions is rather short for industrial applications. Most thermochromic materials are organic materials, having a short lifetime and a limited range of temperature. For various applications, especially for textile applications, there is a need to develop the wide range of temperature and lifetime properties of these chromic materials. Electrochromic materials in textiles have many promising applications. There is, however, a demand for a considerable volume of scientific study in different areas of expertise before these can be put into large-scale production.

Chromic materials are one of the challenging material groups when thinking about future textiles. Colour changing textiles are interesting, not only in fashion, where colour changing phenomena will exploit for fun all the rainbow colours, but also in useful and significant applications in workwear and in technical and medical textiles.

11.7 References

1. Intelligent textiles http://www.tut.fi/units/ms/teva/projects/intelligenttextiles/index5.htm / 07.06.2005.
2. Dag Lukkassen, Annette Meidell, *Advanced Materials and Structures and their Fabrication Process*, October 13, 2003, third edition, 86–87.
3. Michael Connley: Chromism /05/05/05) http://howard.engr.siu.edu/mech/faculty/hippo/Papers.htm / 03.08.2005.
4. Cho-han Fan., Electronical Theses Heap of NSYSU, Study of NiO Electrochromic Films Prepared by Liquid Phase Deposition, http://www.scirus.com / 15.07.2005.
5. H.C. Dunn, J.N. Etters, University of Georgia, Athens, 'Solar Active Products that Change Color on Exposure to UV', *Textile Chemist and Colorist & American Dyestuff Reporter*, Vol. 32, March 2000, 20–21.
6. SolarActive International Color Change Products http://www.solaractiveintl.com/newsrelease.htm /05.08.2005.
7. Henri Bouas-Laurent, Heinz Dürr, Organic Photochromism, IUPAC Technical Report, *Pure Appl. Chem.*, Vol 73, no. 4, pp 639–665, 2001. www.iupac.org/publications/pac/2001/pdf/7304x0639.pdf /05.08.2005.
8. G.M. Tsivgoulis, New Photochromic Materials http://www.mariecurie.org./annal/volume1/tsigvoulis.pdf /05.08.2005.
9. Lars Hällnäs, Linda Melin, Johan Redström, A design Research Program for textiles and Computational Technology, *Nordic Textile Journal* 1/2002, s.56–62
10. Products Focus http://www.taiwantrade.com.tw/tp/2004web/2/pl.htm /24.05.2004.
11. Lyle Small, Thermochromic Inks in Brand & Document Security Applications, Chromatic Technologies Inc. http://theprintdirectory.co.uk/hp/Technical%20Topics/thermochromic.htm /10.06.2005.
12. Thermochromic Pigments for Rapid Visual Assesment of Temperature, Sensors and Surface Technology Thermochromic Pigments Research Area http://bilbo.chm.uri.edu/SST/thermochromic.html /04.10.2005 /10.6.2005.
13. LED-english, Thermochromism http://infochem.hanyang.ac.kr/researches/researches_thermo_english.html /24.05.2005.
14. Andy Town, The heat is on for new colours, *JSDC* Volume 115 July/August 1999 www.sdc.org.uk/pdf_features/JA99F1.PDF.
15. http://www.designinsite.dk/htmsider/k1506.htm /13.05.2005.
16. News Releases 2005 http://www.kuraray.co.jp/en/release/2005/050623.html / 04.10.2005.
17. J. Homola, *Colour Changing Inks*, Color Change Corp. / 29.01.2003 http://www.screenweb.com /04.08.2005.

18. Openloop.ElectroChromic. http://open.loop.ph/twiki/bin/view/Openloop/ElectroChromic /04.08.2005.
19. R.V. Gregory, R.J. Samuels, T. Hanks, Chameleon Fibers: Dynamic Color Change From Tunaable Molecular and Oligomeric Devices., *National Textile Center Annual Report*: November 1999, M98-C01.
20. Carl M. Lampert, Chromogenic smart materials, *Materials Today* March 2004.
21. P.R. Somani, S. Radhakrishnan, Electrochromic materials and devices: present and future, *Materials Chemistry and Physics*, Volume 77, Issue 1, 2 January 2003, p.117–133. http://www.sciencedirect.com
22. http://docbug.com/blog/archives/00460.html /04.10.2005.
23. Paula Gould, Textiles gain intelligence, *Materials Today*, October 2003.
24. Steinbeis Europa Centrum, http://www.steinbeis-europa.de /05.08.2005.
25. Maureen Byko, From electric Corsets to Self-Cleaning Pants:The Materials Science and Engineering of Clothing, http://www.tms.org/pubs/journals/jom/0507/byko-0507.html /05.10.2005.
26. Andrew McWilliams, GP-309 Smart and Interactive Textiles http://www.bccresearch.com/advmat/GB309.html / 05.08.2005.
27. Sci.chem FAQ-Part 5 of 7, http://www.faqs.org/faqs/Sci/chem-faq/part5/ /05.08.2005.
28. *The Times*, London, December 4, 2004.
29. P.R. Somani, S. Radhakrishnan, Electrochromic materials and devices: present and future, http://www.ingentaconnect.com/ 09.10.2005.
30. The Society of Dyers and Colorists, technical, Archive news, http://www.sdc.org.uk/technical/news-archive.htm
31. H.L. Schreuder-Gibson, M.L. Realff, Advanced fabrics, http://www.mrsorg./publication/bulletin

12
Solar textiles: production and distribution of electricity coming from solar radiation applications

R R MATHER and J I B WILSON,
Heriot-Watt University, UK

12.1 Introduction

This chapter is devoted to the incorporation of solar cells into textile fabrics. It reflects the increasing commercial interest in solar cells and the versatility which their integration into textiles would afford. After the background to the theme of the chapter has been set, a brief outline of the design and mode of action of solar cells is given. We then discuss the specifications that textile fabrics have to meet to be practical substrates for solar cells, and we also discuss the challenges in meeting these specifications. Suitable textile constructions are considered, together with a few commercial examples. An important aspect of a textile-based solar cell is the conferment of electrical conductivity on the textile, and strategies that have been devised for achieving conductivity are given. The final section outlines the wide variety of potential applications for solar cells in textiles.

12.2 Background

There is increasing concern worldwide about the huge dependence on oil as a source of energy. Although coal resources are immense, their extraction and use tend to potentially damaging environmental problems. There are concerns too about nuclear fission as a source of energy, and controlled fusion has yet to be proven feasible. Consequently, there is growing interest in harnessing energy by other means that do not rely on the consumption of reserves, as with oil and coal, nor involve dangers (real and perceived) with utilising nuclear fission. Total renewable resources available are more than adequate for human needs, but are unevenly distributed across the globe and generally have large diurnal and seasonal variations. They are often more appropriate for distributed power supplies rather than for centralised power stations and electrical energy distribution.

These alternative energy sources are numerous, and include geothermal, wind and waves, renewable natural products (biomass) and ocean tides.

However, the source of direct energy which arguably arouses the greatest potential is the sun itself. There are numerous examples of attempts to utilise solar energy for storage of chemical energy (mimicking natural photosynthesis) and for the storage of electrical energy in batteries. In particular, devices in which sunlight is absorbed through a dye on titanium dioxide particles, Grätzel cells, may either give an electrical current through a liquid electrolyte, or in other configurations may produce hydrogen by electrolysis of the liquid thus storing the energy they provide from sunlight rather than directly powering an electrical load.[1] Nevertheless, greatest success has been achieved in the direct conversion of solar power into electrical energy, with the aid of photovoltaic devices, 'solar cells'. Apart from using an 'endless' source of energy, the application of photovoltaic devices offers a number of other advantages: the technology is clean and noiseless, maintenance costs are very small, and the technology is attractive for remote areas, which are difficult to supply by conventional means from a grid.

12.3 Solar cells

Sunlight may be converted into electricity by heating water until it vaporises, and then passing the steam through a conventional turbine/generator set. However, this triple conversion of energy (optical to thermal to mechanical to electrical) has losses at each step and it should be more efficient to convert optical energy directly to electrical energy. This is the function of photovoltaic cells, which are solid-state devices in which photons release valence electrons from the constituent atoms. Thermodynamics limits the efficiency for conversion of solar radiation by a simple photovoltaic cell to ~32%, and today's crystalline cells can approach this (Si cells have achieved ~24% conversion efficiency); multiple junction cells have a thermodynamic limiting efficiency of ~66% and triple junction cells have achieved ~32%.

To see how this works, and the implications for cells on flexible materials, we need a little semiconductor science. In a semiconductor, valence electrons are freed into the solid, to travel to the contacts, when light below a wavelength specific to that solid is absorbed (i.e. the electron binding energy). This threshold wavelength must be selected to provide an optimum match to the spectrum of the light source; too short a wavelength will allow much of the radiation to pass through unabsorbed, too long a wavelength will extract only part of the energy of the shorter-wavelength photons (the part equivalent to the threshold wavelength). The typical solar spectrum requires a threshold absorption wavelength of ~800 nm, in the near infra-red, to provide the maximum electrical conversion, but this still loses some 40% of the solar radiation. Further losses occur through incomplete absorption or reflection of even this higher energy radiation. The current that is delivered by a solar cell depends directly on the number of photons that are absorbed:

the closer the match to the solar spectrum and the larger the cell area the better.

As well as a current (i.e. number of electrons per second), electrical energy has a potential associated with it (i.e. the voltage) and this is determined by the internal construction of the cell, although it is ultimately limited by the threshold for electron release. In conventional silicon solar cells (as illustrated schematically in Fig. 12.1), there is an electrical field built into the cell by the addition of minute amounts of intentional impurities ('dopants') to the two cell halves; this produces what is known as a PN junction. The junction separates the photon-generated electrons from the effective positive charges (known as 'holes') that pull them back and in so doing, generates a potential of ~0.5 volts. This potential depends on the amounts of dopants, and on the actual semiconductor itself. Multiple junction cells have integrated layers stacked together during manufacture, producing greater voltages than single junctions.

The power developed by a solar cell depends on the product of current and voltage. To operate any load will require a certain voltage and current, and this is tailored by adding cells in series and/or parallel, just as with conventional dry-cell batteries. Some power is lost before it reaches the intended load, by too high a resistance in the cells, their contacts and leads. Note that at least one of the electrical contacts must be semitransparent to allow light through, without interfering with the electrical current. Thus optimising the design of a solar cell array involves optics, materials science, and electronics. A historical view of photovoltaic technology was given by Goetzberger et al. in 2003.[2]

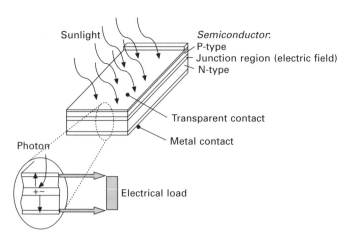

12.1 Schematic of a photovoltaic cell.

12.4 Textiles as substrates

Despite the many attractions of solar cells as vehicles for providing energy, the way in which they are constructed provides problems in application. Typically, solar cells are either encased between glass plates, which are rigid and heavy, or the cells are covered by glass. Glass plates are fragile, and so care has to be taken with their storage and transport. The rigid nature of solar cells requires their attachment to flat surfaces and, since they are used outdoors, they have to be protected from any atmospheric pollution and adverse weather.

It is, therefore, not surprising that increasing attention has been turned to the construction of lighter, flexible cells, which can still withstand unfavourable environments, yet nevertheless maintain the durability required. Several examples have recently appeared of solar cells in plastic films. Examples are the 'power plastic' developed by Konarka Technologies Inc.[3] in which the cell is apparently coated or printed onto the plastic surface, and an integrated flexible solar cell developed at Sky Station International Inc. for high-altitude and stratospheric applications.[4] Iowa Thin Technologies Inc. have claimed the development of a roll-to-roll manufacturing method for integrating solar modules on plastic.[5]

However, whilst the successful incorporation of solar cells into plastics represents an important step in the expansion of solar cell technology, still greater expansion of the technology would be achieved through their incorporation into textile fabrics, especially if the cells are fully integrated into the fabrics. Textiles are materials with a huge range of applications and markets (some established for thousands of years!), and they can be produced by a number of fabrication processes, all of which offer enormous versatility for tailoring fabric shape and properties. The different types of woven, knitted and nonwoven constructions that can nowadays be achieved seem almost infinite, and indeed technical uses are beginning to be found for crocheted and embroidered constructions.[6,7,8]

Some examples of the application of solar cells in textiles have now been reported. Architect, Nicholas Goldsmith, has designed a solar tensile pavilion which, while providing shade and shelter, can also capture sunlight, and transform it into electrical power.[9] The skin of the tent consists of amorphous silicon cells, encapsulated and laminated to contoured panels of the woven fabric. Zabetakis et al. have recently described the design of an awning, which provides protection from the sun and at the same time converts the incident sunlight into electricity.[10] A type of sail has been designed in which attached to the sailcloth are strips of flexible solar cells.[11] The incorporation of the solar cells has been designed to allow for expansion of the sailcloth when wind blows on it.

Textiles offer further advantages too. Fabrics can be wound and rewound. Thus, Warema Renkhoff GmbH in Germany has developed a retractable woven canopy,[12] in which energy is collected from the solar cells integrated into the canopy, and then stored. The stored energy is applied to the operation of the winding drive motor. In addition, textile fabrics can be readily installed into structures with complex geometries, a feature which opens up a number of potential applications (discussed later).

12.5 Technological specifications

Conventional single crystal silicon cells may produce 0.5 V when not connected to a load, but this falls according to the current that is drawn, until it reaches zero for a short-circuited load. The current in bright sunlight may be a few amps from a square metre of cells but this will fall towards zero if the resistance of a connected load is steadily increased. Optimum performance (and hence highest conversion efficiency) of a solar cell or array therefore requires the electrical load and the current vs. voltage characteristic of the solar cells to be matched. This load matching may be a severe limitation on provision of power to a desired load, if the available area is insufficient. On the other hand, when the cell area is very large there is ample provision for selecting an appropriate combination of current and voltage. One potential advantage of a low-cost flexible array is the ability to tailor the interconnections between individual cells at the manufacturing stage, and perhaps with some contact designs to enable this to be done by the user.

As was noted above, the current delivered by a solar cell depends not only on the illumination but also directly on the illuminated cell area. This is a fundamental barrier to the applications that can be addressed by a restricted area of cells, such as would be available on clothing, for instance. The voltage from a solar cell does not depend on cell area, but on cell construction, and there are ways to increase this voltage above that given by the simple 'single junction' devices, by stacking cells on top of each other, usually no more than three in total.

This added complexity generally reduces the production yield, and requires careful optimisation of each cell. Since the same current must flow through each cell in the stack, each cell must generate exactly the same current. Therefore an increasing thickness is required for each cell into the stack if it is to produce current from a successively weaker flux of light. This is more readily achieved by making the topmost cells absorb the shortest wavelength light, and successive cells the longer wavelengths, by adjusting their compositions to match the desired spectral bands. In amorphous silicon cells, this adjustment is achieved by alloying other elements into the semiconductor. The addition of carbon makes the cells more transparent to long wavelengths and the addition of germanium increases the long wavelength absorption.

It is therefore apparent that providing a series of standard solar cell arrays, usually labelled by their 'peak-watt' output (that given in a stated sunlight intensity), allows the user to make up an electrical supply to meet particular needs, as long as the space available is not too restricted. Large-scale users can readily couple any cell array to either an active load-matching circuit or to a buffer energy store (e.g. electrochemical battery), but small-scale users would probably need more choice of array size and shape. Flexible textile cells would address this shortcoming of traditional cells and arrays.

12.6 Challenges to be met

The active cell components, the semiconducting layers, therefore have a great part to play in determining the cell performance, not only from an efficiency point of view (i.e. matching the illumination spectrum) but also from the load matching perspective. The separate improvement of both output current and voltage demands effective optical absorption and efficient photo-generated charge separation. There may be unavoidable trade-offs in meeting both of these simultaneously. Textile substrates offer new ways of improving the optical absorption in thin layers by scattering light that passes through the layer back into the layer for a second chance of absorption. Rough surfaces in conventional silicon cells are sometimes a source of electrical problems but substrate texturing has always been seen as a possible tool for advanced thin-film cell design.

Transferring some of the best features of amorphous silicon cell technology to a flexible textile substrate device[13] would provide more rapid evolution of the new cells than a separate development based on single crystal cell technology, given the processing temperature restraints of the preferred textiles. There are still plenty of challenges in ensuring efficient charge collection from a stack of thin layers on a non-planar base.

Finally the word 'textile' still often suggests applications in clothing, but users would expect to have the same performance from a solar cloth as from a passive cloth, with respect to folding, wear, washing, abrasion, etc. Few of these properties have yet been tested with coatings of the type used in solar cells, nor have solar cells been assessed in such conditions, even if hermetically sealed with a thin polymer layer.

12.7 Suitable textile constructions

The successful integration of solar cells into textiles has to take into account the type of fibre used and the method of textile fabrication. The selection of a type of fibre is strongly influenced by its ability to withstand prolonged irradiation by ultraviolet (UV) light. Fibre selection is also governed by the temperatures required to lay down the thin films comprising the solar cell, although it has been shown that nanocrystalline silicon thin films, an improved form of amorphous silicon, can be successfully deposited at temperatures as low as 200 °C, and under the appropriate conditions even single crystal silicon may be grown, albeit epitaxially on silicon wafers.[14] These two factors of UV resistance and maximum temperature restrict the choice of commodity fibres. Commercial polyolefin fibres melt below 200 °C. Cotton, wool, silk and acrylic fibres start to decompose below this temperature. Polyamide fibres are likely to be too susceptible to UV radiation.

Polyethylene terephthalate (PET) fibres, however, are viable substrates. They melt at 260–270 °C and exhibit good stability to UV light.[15] They are commercially attractive too because of their existing widespread use. Thus, fabrics composed of a large variety of PET grades are currently available. PET fibres possess good mechanical properties and are resistant to most forms of chemical attack.[15] They are less resistant to alkalis, but a solar cell would not normally find use in an alkaline environment. Fabrics constructed from PET fibres should, therefore, be suitable as substrates for solar cells, whilst also possessing flexibility and conformability to any desired shape.

Fabrics composed of glass fibres produced from E-glass formulations[16] could also be used. One advantage could lie in their transparency, as with plate glass in conventional solar cells. Moreover, the price of E-glass is similar to that of PET.[17] E-glass fibres, however, suffer from poor resistance to acids and alkalis and are prone to flexural rupture. Other glasses are more stable to environments of extreme pH. S-glass is used for specialist sports equipment and aerospace components, but its price is approximately five times that of E-glass.[16]

Many high-performance fibres, which readily withstand temperatures up to 300–400 °C, could also be considered, although aramids would not be sufficiently stable to UV radiation. Examples of suitable high-performance fibres could be polybenzimidazole (PBI) fibres, polyimide (PI) fibres and polyetheretherketone (PEEK) fibres. However, these fibres are expensive. There are clear commercial attractions, therefore, in adopting PET fibres.

The type of textile fabric construction is also important to the performance of the solar cell. The type of construction affects the physical and mechanical properties of a fabric, and also its effectiveness as an electrical conductor. Where conduction in textile fabrics is required, woven fabrics are generally considered to be best,[18,19] in that they possess good dimensional stability and can be constructed to give desired flexibilities and conformations. Moreover, the yarn paths in woven structures are well ordered, which allows the design of complex woven fabric-based electrical circuits.[18]

Knitted structures, on the other hand, do not retain their shapes so well, and the rupture of a yarn may cause laddering. These problems are heightened if the shape of the fabric is continually changing, as in apparel usage. Nonwoven fabrics do not, as yet, generally possess the strength and dimensional stability of woven fabrics. More significantly, the construction of electrical circuits in them is limited, because their yarn paths are highly unoriented. Embroidery, however, may offer an opportunity for circuit design.

12.8 Conductive layers for PVs

Solar cells require two contacts in a sandwich configuration, with the active semiconductor between them. At least one of the contacts must be transparent to solar radiation, over the waveband which the semiconductor absorbs. Conventional crystalline silicon cells have a thick metal contact on the back surface, and a gridded metal contact on the front surface. They do not require complete coverage of the front surface because the top layer of silicon is sufficiently conducting ('heavily doped') to deliver the photo-current without significant resistance losses. Amorphous silicon cells, like most thin-film cells, have a more insulating top semiconductor layer and so require the whole surface to be contacted. This is achieved without blocking the incoming light, by a layer of transparent conducting oxide (TCO) such as ZnO or Indium tin oxide (ITO), often with other elements added to enhance the conductivity without reducing the transparency. In addition, a fine gridded contact is often superimposed to further improve the current collection efficiency.

It is also possible to build the cell with the light entering through the substrate: this is known as the superstrate configuration. Typically a glass sheet, coated with TCO is then covered with thin-film semiconductor, which in turn is covered by an opaque metal layer. Textile substrates are unlikely to be sufficiently transparent that this construction may be used, although it is not necessary for the material to be transparent enough to show an image through it, only to be translucent. Hence the textile material is likely to be coated with a conducting layer that may be opaque. Alternatively, the textile may itself form part of the active cell structure, either enabling the conduction of photo-current generated within an imposed semiconductor, or in future materials, enabling photo-currents to be generated within the fabric (probably by semiconducting polymers). Organic solar cells, or polymer solar cells still have a long process of development before they can be considered for this application, perhaps being a more intractable problem than the inverse application of light-emitting polymers.

Unless the textile can be made from conducting fibres, an additional material is needed, that should not degrade the other desirable textile properties. It is possible, of course, to incorporate metal fibres within a fabric, but it is questionable whether these can make sufficiently good contact to the whole thin-film cell area. They may be helpful in addition to a thinner, continuous conductor. Although polycrystalline silicon can be sufficiently conducting to enable only partial coverage by the contacting layer, it generally requires much higher processing temperatures, before or after its deposition, than most textiles can withstand.

We therefore require a continuous conducting layer to be placed over the whole textile surface before adding the semiconductor. The choice of materials (probably metals) is determined by electrical conductance (lower conductance requires thicker layers), compatibility with the substrate (chemically, physically, and during processing), and compatibility with the semiconductor (chemically and work function value). In addition, the mechanical behaviour of the conductor in a cell that can flex or twist will depend on its composition, thickness and adhesion.

There are several options to be considered, using both physical and chemical coating techniques, to obtain the correct composition and conformality with a non-planar surface. These will deposit metals or TCOs at the limited temperatures allowed by the preferred textiles. Physical coating methods include sputtering from a solid target in an argon atmosphere and simple evaporation from a solid source (either thermally or with an electron beam), and both are widely used in the optics, electronics, and engineering industries. Chemical methods are based either on the decomposition of a volatile compound of the conductor by heat or electrical plasma, or on the deposition of a coating from a solution, usually driven by an electrical current. These methods are also in widespread use within industries. There are a few additional methods that are applicable for certain materials, such as dip- and spray- and spin-coating, that would be appropriate for the conducting polymers that are being developed now. These do not conduct well enough for photovoltaic cells, because they generate low voltage electrons that cannot overcome the resistance of a long path through a poor conductor.

The compromise solution to retaining substrate flexibility with high electrical conductivity may be to use a conducting polymer layer on the textile, or incorporated within its structure, superimposed by a thin, more-conducting metallic layer. If this layer breaks during flexure then the minute gaps will be bridged by the underlying organic conductor, whose limited conductivity will not be a problem over such short paths. This also allows the semiconductor to be in contact with its preferred metal (say aluminium for silicon), avoiding the inclusion of resistive barriers that can exist at other semiconductor/conductor interfaces.

12.9 Future trends

The successful integration, on a commercial basis, of solar cells into textiles opens up a whole range of fresh applications. Many of these applications are likely to exploit the flexibility and light weight of textile fabrics. They can, for example, be placed over the curved surfaces of buildings. They can also be installed in spaces which may otherwise be inaccessible, as in automotive, marine and aerospace equipment. In addition, textile fabrics can be rolled up, transported to a desired location and then unrolled at that location. Thus, the technology would then be beneficial to those living in remote areas, where there is no supply of electricity from the grid, fuel may be scarce and expensive, and maintenance of equipment is uncertain. Moreover, it could be used to get power quickly to disaster areas, hit by earthquakes, hurricanes, floods or fire.

However, as is already becoming apparent, many applications will also exploit existing applications of textiles. As already noted, there is interest in incorporating solar cells into awnings and canopies, and even into sails. It may also be that the presence of solar cells in apparel will be particularly beneficial, for example, to maintain the temperatures of young children, the elderly and hospital patients, or to power sensing elements for temperature and movement.

As polymer, and other organic, semiconductors become better developed, following a similar but slower path to that of organic light-emitting diodes, they will become increasingly attractive alternatives to inorganic semiconductors in solar cells. Their use would enable lower cost deposition technology to be used, replacing chemical vapour deposition, even in its low-temperature, plasma-driven form, by coating methods based on liquids rather than vapours. Dye-sensitisation for solar spectrum matching may eventually become a practical reality, thus leading to 'high tech' solar cells being produced by means similar to conventional printing and dyeing of cloth, but this is now a fantasy.

12.10 Sources of further information

Basic Research for Solar Energy Utilisation provides a survey of the current status of solar technology and discusses future developments, from a US Department of Energy workshop in Bethesda, Maryland, USA, in April 2005 (published by Argonne National Laboratory).

The International Solar Energy Society is a major professional institution that encourages the research, development, and use of solar energy through a variety of channels including conferences, publications, and its website.

Electrotextiles and Giant Area Flexible Circuits, editors Shur M.S., Wilson P.M. and Urban D., MRS Symposium Proceedings Volume 736, Materials Processing Society, Warrendale, Pennsylvania, USA, 2003. The volume contains papers presented at a symposium in Boston, Massachusetts, USA in 2002. It includes a wide range of papers on electrotextiles, i.e., textiles containing electronic components and circuits.

The website, http://www.solarbuzz.com/ provides up-to-date news and information about photovoltaic technology, prices and markets.

12.11 References

1. Grätzel M., 'Photoelectrochemical cells', *Nature*, 414, 2001, 338–344.
2. Goetzberger A., Hebling C., Schock H.-W., 'Photovoltaic materials, history, status and outlook', *Materials Science and Engineering: R: Reports*, 40, 2003, 1–46.
3. Konarka Technologies, Inc., *Konarka Builds Power Plastic*, http://www.konarka.com/technology/photovoltaics.php
4. Lee Y.-C., Chen S. M.-S., Lin Y.-L., Mason B.G., Novakovskaia E.A., Connell, V.R., *Integrated Flexible Solar Cell Material and Method of Production* (Sky Station International Inc.), US Patent 6,224,016, May 2001.
5. Iowa Thin Film Technologies, Inc., *Revolutionary Package of Proprietary Solar/Semiconductor Technologies*, http://www.iowathinfilm.com/technology/index.html
6. Mowbray J.L., 'Wider choice for narrow fabrics', *Knitting International*, 2002, **109**, 36–38.
7. Ellis J.G., 'Embroidery for engineering and surgery' in *Proceedings of the Textile Institute World Conference*, Manchester, 2000.
8. Karamuk E., Mayer J., Düring M., Wagner B., Bischoff B., Ferrario R., Billia M., Seidl R., Panizzon R., Wintermantel E., 'Embroidery technology for medical textiles' in *Medical Textiles*, editor Anand S., Woodhead Publishing Limited, Cambridge, 2001.
9. Goldsmith N., 'Solar tensile pavilion', http://ndm.si.edu/EXHIBITIONS/sun/2/obj_tent.mtm
10. Zabetakis A., Stamelaki A., Teloniati T., 'Solar textiles: Production and distribution of electricity coming from the solar radiation. Applications.' in *Proceedings of Conference on Fibrous Assemblies at the Design and Engineering Interface*, INTEDEC 2003, Edinburgh, 2003.
11. Muller, H.-F., 'Sailcloth arrangement for sails of water-going vessels', US Patent Office, 6237521, May 2001.
12. Martin T., 'Sun shade canopy has solar energy cells within woven textile material that is wound onto a roller' (Warema Renkhoff GmbH), German Patent Office, DE10134314, January 2003.
13. Koch, C., Ito M., Schubert M., 'Low temperature deposition of amorphous silicon solar cells', *Solar Energy Materials and Solar Cells*, 68, 2001, 227–236.
14. Ji J.-Y., Shen T.-C., 'Low-temperature silicon epitaxy on hydrogen-terminated Si (001) surfaces', *Physical Review B* 70, 2004, 115–309.
15. Moncrieff R.W., *Man-made Fibres*, Newnes-Butterworths, 6th edn, London and Boston, 1975.
16. Jones F.R., 'Glass fibre' in *High-performance Fibres*, editor Hearle J.W.S., Woodhead Publishing Limited, Cambridge, 2001, pp 191–238.
17. Hearle J.W.S., 'Introduction' in *High-performance Fibres*, editor Hearle J.W.S., Woodhead Publishing Limited, Cambridge, 2001, pp 1–22.
18. Abdelfattah M.S., 'Formation of textile structures for giant-area applications' in *Electronics on Unconventional Substrates – Electrotextiles and Giant-Area Flexible Circuits*, editors Shur M.S., Wilson P.M. and Urban D., MRS Symposium Proceedings Volume 736, Materials Processing Society, Warrendale, Pennsylvania, USA, 2003, pp 25–36.
19. Bonderover E., Wagner S., Suo Z., 'Amorphous silicon thin transistors on kapton fibers', in *Electronics on Unconventional Substrates – Electrotextiles and Giant-Area Flexible Circuits*, editors Shur M.S., Wilson P.M. and Urban D., MRS Symposium Proceedings Volume 736, Materials Processing Society, Warrendale, Pennsylvania, USA, 2003, pp 109–114.

13
Introduction to conductive materials

A HARLIN, Technical Research Centre of Finland and
M FERENETS, Tampere University of Technology, Finland

13.1 Electric conductivity

When a voltage, U (in volts V), is applied between two points on a material an *electric field*, E, is set up. Depending upon the electrode geometry and other factors the electric field may or may not be uniform. Electrically charged particles in the material will experience an electrostatic force in the direction of the electric field equal to the product of their charge and the electric field. If the charges are free to move, the force causes them to *drift* through the material. The rate at which the charge is transported is known as electric *current*, I. This is a system property whose material equivalent is the *current density*, J (amps per square metre A/m^2).

Ohm's law, the result of empirical observations, states that current through a conductor *due to drift* is proportional to the voltage across it:

$$I = (1/R)\ U, \qquad 13.1$$

where R (in ohms Ω) is the resistance of the conductor to current flow. Ohm's law can also be expressed in terms of the material properties, field E and current density J as:

$$J = (1/\rho)E = \sigma E, \qquad 13.2$$

where ρ is the *resistivity* of the material ($[\Omega\ m]$) and $\sigma = (1/\rho)$ is the *conductivity* ($[\Omega\ m]^{-1}$). This implies that the mean velocity of the charged particles in the direction of the field, the so called *drift velocity*, is proportional to the applied field.

Materials display an extremely wide range of resistivity, possibly wider than any other physical parameter. Superconductors have resistivity $<10^{-25}$ Ω m, typical metal conductors $\sim 10^{-8}$ Ω m, semiconductors $10^{-4} \ldots 10^{10}$ Ω m, and insulators $10^{10}\ \Omega \ldots 10^{20}\ \Omega$ m. Conductors exhibit a linear Ohmic current-density dependence upon electric field up to high current densities, even up to the point where they fuse. In semiconductors Ohmic behaviour is also observed at low fields. However, above a critical field the drift velocity, and hence the current, reaches a limiting value in semiconductors, resulting in an apparent increase of resistivity with the field. The basic unit of resistance measurement is the Ohm, and the basic unit of conductivity measurement is the siemens. Conductivity is the conductance as measured between opposite faces of a 1 cm cube of a given material. This measurement is in units of siemens/cm. The corresponding terms for specific resistivity are ohm-cm.

The conductivity of a material due to charge carriers is dependent on the electric charge on each carrier, its mobility, and the concentration of charges. Charge carrier species are many, e.g., electrons, holes, protons, and ions, but normally one charge carrier is likely to dominate. Mobility is dependent on the charge carrier's size and material structure, as in polymers, insulators, covalent bonding and crystal structures. The concentration of charge carriers may vary from near zero to more than one per atom ($<10^{29}$ m^{-3}).

13.1.1 Conductivity in metals

Early attempts to explain the electronic configurations of the metals with ionized atoms in which the free electrons form a homogeneous free moving sea of negative charge failed. The principal objection to this theory was that the metals should then have higher specific heats than they do. The *Pauli exclusion principle* states that no two electrons occupying the same space can have exactly the same energy, although two electrons can occupy the same energy level if they have opposite spin.

According to the band theory, any given metal atom has only a limited number of valence electrons with which to bond to all of its nearest neighbours. Extensive sharing of electrons among individual atoms is therefore required. This sharing of electrons is accomplished through overlap of equivalent-energy atomic orbitals on the metal atoms that are immediately adjacent to one another. This overlap is delocalised throughout the entire metal sample to form extensive orbitals that span the entire solid rather than being part of individual atoms. Each of these orbitals lies at different energies. The orbitals, equal in number to the number of individual atomic orbitals that have been combined, hold two electrons, and are filled in order from the lowest to the highest energy until the number of available electrons has been used up. Groups of electrons are then said to reside in *bands*, which are collections of orbitals. Each band has a range of energy values that the electrons must possess to be part of that band.

The highest energy band in a metal is not filled with electrons because metals characteristically possess too few electrons to fill it. The high thermal electrical conductivities of metals is then explained by the notion that electrons may be promoted by absorption of thermal energy into these unfilled energy levels of the band.[1]

13.1.2 Conductivity in plastics

Typical polymers are a type of molecular crystals, albeit with extremely large molecules, and therefore have very strong intra-molecular bonding but only very weak, van der Waals, inter-molecular binding forces. The relatively simple band theory model must therefore be modified for polymers. The energy bands in perfect crystalline structures, like metals, have infinite periodicity in all directions and the nuclei interact with the three-dimensional electron wave in the same way in any unit cell. On the other hand, the long polymer chains are essentially one-dimensional.

Intrinsically conductive polymers contain long conjugated chains formed by double bonds and heteroatoms. The polymers can be rendered conductive by modifying the π- and π-p-electron systems in their double bonds and heteroatoms. Charge carriers are electrons or holes originated by adding to the polymer certain blending or doping agents, which will serve as electron receptors or electron donors in the polymer. Thereby electron holes or extra electrons are formed in the polymer chain, enabling electric current to travel along the conjugated chain.[1] The conductive polymers were based on conjugated electron structures, so-called resonant structures. Typically the conductive polymer materials are semiconductors, say four decades less conductive than metal substances. Their light, moisture, and pH reactions were considered major disadvantages for the materials but, e.g., in sensors the features are more than welcome.[2]

Electrically conductive plastics are usually manufactured by mixing together electrically insulating polymers and electrically conductive fillers such as carbon black and metal particles or, for example, electrically conductive polymers. Charge carrier is electron or hole. The electrical conductivity of filled conductive polymers is dependent on mutual contacts between the conductive filler particles. The distance between the particles should be less than 100 Å when tunnelling of electrons is possible. Close to this value the conductivity may become nonlinearly temperature dependent.

Usually a well-dispersed conductive filler like carbon is needed in amounts of approximately 15–35 vol-% to reach percolation and produce composites having good conductance. Filler structure, especially aspect ratio, has a marked influence on the properties. In comparison acetylene black, structured carbon, and single walled carbon nano-tubes are needed 25, 15, and 2.5 wt-% respectively to reach percolation and good electrical conductivity. However, such conductive composites involve problems. Their mechanical and certain of their chemical properties are crucially impaired as the filler content increases and the polymer content decreases; their conductivity in dispersing of the filler into the matrix plastic is difficult.

13.2 Metal conductors

Metals are useful in practical applications when high conductivity is required, e.g., microelectromechanical systems (MEMS), electromagnetic shielding EMI, resistive heating, or signal transmission. Benefits are low cost and high electric performance. Carbon fibres are in many ways comparable with the metal fibre and discussed under the same title. Although relatively high conductivities (10^5 $[\Omega\ cm]^{-1}$) can be obtained by the drawn blending of metal slivers and slivers composed of textile fibres, inherent problems persist.[3] Not only are metal fibres expensive, they can also be as much as five times heavier than some textile fibres – making a homogeneous blend difficult to produce. In addition, metal fibres are brittle and can abrade the spinning equipment. A final concern is that metal-containing fibres can produce a fabric with undesired touch – a 'metallised hand'.

Regardless of these limitations, successful attempts at blending metals and stabile slivers have been reported. Yarns, fabric, and non-woven materials of metal fibres are available. Combined materials are processable with normal textile machinery, but comfort properties, wear resistance, and maintenance robustness may be limited especially when metal content is increased. Especially steel fibres have found many applications such as antistatic textiles.

13.2.1 Ferric materials as conductors

Conductivity of steel is moderate, and the material is mainly suitable for conductive backing of carpeting and antistatic protective clothing. Another example of use is resistor applications, like electrically heating cloths. In practical applications stainless steel fibres should be used, and made from low-temperature ductile stainless steel AISI 316L (Ni 10-14% and Cr 16-18%) that has proved to have excellent use properties. Stainless fibres are markedly resistive. For continuous fibre, and 20% mixed to PES the electrical conductivities are $\sim 10^2$ $[\Omega\ cm]^{-1}$ and $\sim 10^0$ $[\Omega\ cm]^{-1}$ respectively for Tex 200 g/1000 m yarns. Conductivity is also related to the yarn size.

Bending hardening of the fibre is evident, and limits the wrinkling of the steel fabrics. Also the nickel in the stainless steel may be a risk for allergy, especially with stainless steel like NY 601 or NY 845 containing 60 and 40 wt% of nickel. Combination of all these features limits the typical use of steel in technical textiles and certain sensors, e.g., bordered indicators.

13.2.2 Noble and colour metals

Copper, silver, and with certain limitations also gold, are markedly more conductive than any steel. Vacuum spraying is a relatively inexpensive method that produces metallic coated fibres with conductive properties as high as

Introduction to conductive materials

10^4 $[\Omega\ \text{cm}]^{-1}$. Unfortunately, this process has several limitations. There is adhesion between the metal and fibre creating an unstable construction, the process is difficult, and the fibres produced have low resistance to corrosion and wear.[3] Galvanic coating is comparable with vacuum spraying, but possible only for conductive fibre. Lately plasma sputtering has improved the quality, especially the coating thickness control, but adhesion between the metal and fabric is a problem.

The cost of colour metals is high, but due to their high electrical performance and ability to be drawn down to a very fine fibre they may introduce themselves as an attractive choice in various applications. Different mimic cable structures are developed for signal transportations and power. They are textile compatible, solid or stranded wires of copper, and silver with especially high conductivities (10^6 $[\Omega\ \text{cm}]^{-1}$) coated with Silicon rubber, PTFE, PFA, FEP, or PE layer.

Recently silver has become more readily available due to new digital photography technology. Even if silver is reasonably reactive, its electric performance is attractive. Also the antibacterial performance of silver is emphasised.

13.2.3 Carbon as a conductor

Electrostatic discharge (ESD) protective clothing benefits from carbon fibre added textiles structures. Carbon fibre intraconductivity is close to that of metal. The conductivities of carbon fibre is ($10^5 \ldots 10^0$ $[\Omega\ \text{cm}]^{-1}$) and especially if pure carbon is used conductivities as high as (10^2 $[\Omega\ \text{cm}]^{-1}$) can be reached. Unfortunately carbon can be difficult to process. Combined structures in fibres, like Nylon, Carbon and Polyester-Carbon sandwich fibre result in markedly less conductivity ($10^{-10} \ldots 10^{-6}$ $[\Omega\ \text{cm}]^{-1}$). Any amount of carbon in fibres will impart a black colour to the end product. Carbon is chemically but not very wear resistant. The thermal insulation properties are markedly better than those of metal. Even if the carbon fibre is black, high conductive material enables coloured cloth structures. The combination of features has made carbon fibre a desired conductive fibre for ESD application.

13.2.4 Carbon nano-tubes and their conductivity

Carbon nano-tubes and especially single walled CNTs are developing materials with extreme properties like super fibre mechanical strength, high thermal insulation, combined with metal conductivity. The conductivity properties are highly dependent on the molecular structural orientation in the CNT and number of the walls. Typical commercially available materials are multi-walled 10–12 nm fibre of 10–15 µm length, and 20–30 layered structure with intermediate properties. Grades are developing rapidly. High aspect ratios up to L/D ~1000 of the CNT and good electric performance makes it possible

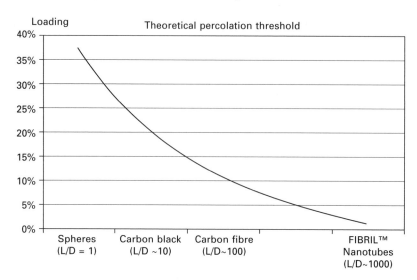

13.1 Theoretical percolation threshold for different carbon fillers [4].

to produce conductive polymer-based fibre markedly with textile properties in the fibre. EMI fibre requires acetylene black 25%, which is hardly processable as a fibre, while single walled carbon nano-tubes require some 2.5 wt-% for respective properties, see Fig. 13.1.

Loading of 3 wt-% of CNT in plastic leads to conductivities $\sim 10^{-6}$ [Ω cm]$^{-1}$. Lately polymer composite CNT yarns and even nano-fibre having a CNT core structure has been manufactured.[5] This development suggests that in future mechanically and electrically high-performing fibres can be manufactured from nano-tubes having properties similar to metal or better.

13.3 Ionic conductors

13.3.1 Metal salts

Certain metal salts like copper sulphide and copper iodide are predominantly utilised for electrically conductive coatings on fibres, as they are easily processable using ordinary textile technology. However, only low conductivities can be obtained. A variety of fibres have been used including nylon, polyester, wool and acrylic for antistatic applications such as carpets. Depending on the chemical and the process used, conductivities range from ($10^{-6} \ldots 10^{-1}$ [Ω cm]$^{-1}$).

White metal compounds are low reactive and thus useful for bi-component fibre manufacturing. A typical structure is cell-core or side-by-side bi-component structure. A typical benefit of the solution is markedly textile-fibre-like properties with reasonable conductivity. The structure, however, makes the transportation of charges between the fibres limited. Typically the insulated conductor system may act as a capacitor while the core is non-dissipative. Conductivities range from $(10^{-10}...10^{-8}$ $[\Omega$ $cm]^{-1})$ with nylon and polyester.

13.3.2 Ionic polymer conductors

Ionomers and ionic polyelectrolytes are organic polymers which are conductive modified by forming ionic salts through their functional groups, for example, the neutralisation of carboxylic acid functional groups on the polymer with metal hydroxide. The carboxylic acid could be, e.g., from acrylic acids. Metal is preferably M^{1+} with oxidation number one like lithium Li or even more preferably potassium K providing a decade better conductivities.

Contrary to the inherently conducting polymers, the charge carrier is ionic, including proton. Ions are more bulky in size and thus their mobility is low. The differentiation makes the conductive ionomers less conductive than most inherently conducting polymers. However, the ion systems are chemically markedly more sustainable and less coloured than conjugated polymer systems, which make the ionomers more robust against pH and oxygen, as well as easy to colour.

Polyelectrolyte containing polyester fibre materials have been developed for antistatic products. However, bi-component structured dissipative antistatic fibres are desired in a wide range for various fibre materials with different constructions and polymers with conductivity ranging from $(10^{-10}...10^{-6}$ $[\Omega$ $cm]^{-1})$.

13.4 Inherently conducting polymers

In 1977 Hideki Shirakawa, Alan MacDiarmid, and Alan Heeger published their discovery that led to the 2000 Nobel Prize in chemistry. Since the early 1980s development has been great and today the inherently conductive polymers, such as polyaniline (PANI), polypyrrole (PPy), polytiophene (PT), and poly (perinaphtalene) (Pna) have achieved a level where many industrial applications have become reality. The materials have become more practical because of improved resistance to oxygen and better processability. The advantage of the intrinsically conductive plastics is the ease of varying their conductivity as a function of the amount of the doping agent, i.e., the degree of doping, especially within low conductivity ranges. Achieving the low conductivities with, e.g., filled conductive plastics is difficult.

The level of conductivity achieved in inherently conductive polymers (ICP) depends on the molecular structure of the polymer backbone, the degree of doping, and the nature of the counter ion species incorporated. Controlled doping of polymers enables engineering their conductivity values

13.2 Chemical structure of various electrically conducting polymers.

in a large range. Orientation is one of the most important factors to enhance the conductivity of π-conjugated polyenes.[6] Today a wide variety of conducting polymers is available (see Fig. 13.2).

A common feature of all ICPs is poor processability. This derives from the fact that ICPs form a rigid, tightly packed chain structure that is essential for interchain charge hopping, but prevents the polymer intermixing with solvent molecules. To overcome this problem the chemical structure of ICPs has been modified by substituted derivatives, copolymers, polyblends, colloidal dispersions, coated latexes and composites.

The electric properties of polymers are essential for their successful application in the electronics industry. There is an ever-increasing demand for polymers which show antistatic behaviour. These polymers are employed as packing materials for microelectronic devices that are sensitive to electric discharges. The use of modern organic conductors (as opposed to the more traditional carbon black or salts) as fillers for the production of antistatic polymers is advantageous, especially for the formation of films.

13.4.1 Highly conductive materials

Polyacetylene (PA) was the first polymer exhibiting high conductivity comparable with metals if exposed to oxidising agents like iodine vapour. Oxidation forms charge carriers on the conjugated polyene structure, especially

13.3 Isomeric forms of polyacetylene.

of p-type conductivity. PA exists in two isomeric forms: *trans* and *cis*, see Fig. 13.3. The *trans* form is the thermodynamically stable form at room temperature.

In 1977 A. J. Heeger *et al.* reported on AsF_5-doped PA films showing electrical conductivity that could be systematically and continuously varied over a range of 11 orders of magnitude. The highest conductivity achieved was 2.2 $[\Omega\ m]^{-1}$ in the case of *trans*-rich PA with AsF_5-doping. Halogen-doped (iodine) *trans*-PA led to conductivity of about 1.6 $[\Omega\ m]^{-1}$. The highest room temperature results were obtained in the case of *cis*-PA doped with AsF_5 showing an electrical conductivity value of 5.6 $[\Omega\ m]^{-1}$.[7] Those films were described as flexible, crystalline and silvery.[8]

Gaseous acetylene was interface polymerised on a uniform solution of polymerisation catalyst, or on a metal or glass surface coated with the catalyst solution. Linear high molecular weight PA was formed into films and fibres. PA fibres were obtained by polymerising PA on the surface of a glass fibre or steel wire that had been dipped in the catalyst solution.[9]

Polyacetylene is one of the most promising materials for applications in optoelectronics. The conductivity of the polymer after doping is equal to that of copper, and some forms of polyacetylene have record values of non-linear third-order optical susceptibility.

Low stability by polyene chain to defects of polyacetylene has been the major obstacle in the way of practical applications of this polymer. Conformational defects in acetylene polymerisation relate to the conditions of solid state formation, especially in *cis-trans* isomerisation. Acetylene polymerisation with rhenium catalysts provides a route to highly ordered polyacetylene blends to oxidation and doping.

When acetylene is polymerised in solutions of certain saturated polymers, nanoparticles of polyacetylene are formed. These nanopolyacetylene blends appear to be stable, have low defect content, and exhibit a set of unique optical properties, which are not characteristics of standard polyacetylene modifications. The electronic structure with low Raman scattering, thermochromism and the transparency band in the optical spectrum in the near infra-red field of the highly ordered nanopolyacetylene outlines new potential applications.[10]

Polyacetylene films exhibit conductivity dependence on film orientation. Films are oriented by stretching. Stretched films show about twenty times increased conductivity compared with as-grown films.[11] Films of I_2-doped polyacetylene and polyisoprene diblock copolymer have been made. The microstructure of those films turned out to be a thermoplastic pseudo interpenetrating polymer network. At nano-scale the rod-like structure of I_2-doped polyacetylene was formed.[12]

In order to overcome problems of practically non-processibility of PA, different kind of copolymers incorporating PA has successfully been synthesised. It has been demonstrated that PA can be made soluble by making copolymers with polyisoprene, polystyrene or polybutadiene.[13] The stability of electrical conductivity and antioxidation ability of PA films have been improved by air or argon (Ar) plasma treatment. The *cis*-configuration content decreased while the *trans*-configuration content of PA increased with increasing Ar plasma power.[14]

13.4.2 Polyaniline and its derivates

Polyaniline (PANI), see Fig. 13.4, is one of the most widely studied ICPs. There are several reasons for that including simple preparation, relatively good environmental stability and good electrical conductivity. As a common feature to all ICPs PANI has a poor solubility in most common organic solvents that leads straight to the difficulty of processability. But it can be overcome by substituting nitrogen atom or benzene ring with a suitable atom or group.[15] Aniline monomer has toxicological limitations.

The general structure of PANI is depicted in Fig. 13.4. This is called emeraldine base (EB) PANI. It is partly oxidised. The other possible configurations are luecoemeraldine base (fully reduced) and pernigraniline (fully oxidised) base PANI that are unstable in air and not electrically conductive. The emeraldine salt form of PANI can be synthesised by the chemical or electrochemical oxidation of aniline in an aqueous acid solution. It can also be synthesised by doping EB with aqueous protonic acids, which yields a nine to ten orders of magnitude increase in conductivity. This is accompanied by colour changes of the polymer varying from blue to green.[16]

Conductivity of PANI pellet reduces while temperature decreases, which is contrary to metals typically (see Fig. 13.5). Neither is the voltage-current dependency of PANI at various temperatures does resemble that of metals as can be seen in Fig. 13.6. However, at temperatures above 300 K polymer PANI starts to act according to Ohm's law.[17]

In Fig. 13.7 is expressed current density against the electric field strength for bulk PANI-EB plaque, with a thickness of 2.2 mm. In the electric field above 4 kV/mm current becomes unstable, and at approximately 4.5 kV/mm

13.4 Molecular structure of polyaniline.

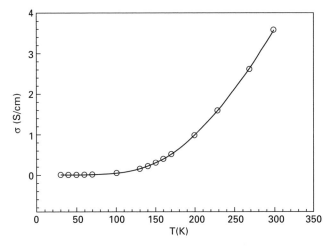

13.5 Conductivity of polyaniline as a function of absolute temperature [17].

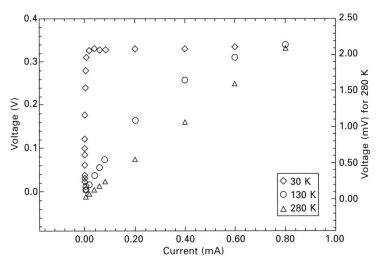

13.6 Polyaniline voltage-current behaviour at various temperatures [17].

electric breakthrough takes place. Breakthrough tests indicate that the PANI-EB has less probability of failure at impulse voltages of 300 kV/mm than carbon black mixed compounds.[18]

13.4.3 Semiconductive polymers

Polypyrrole (PPy) is a widely researched π-electron conjugated conducting polymer, because of its good electrical conductivity, good environmental

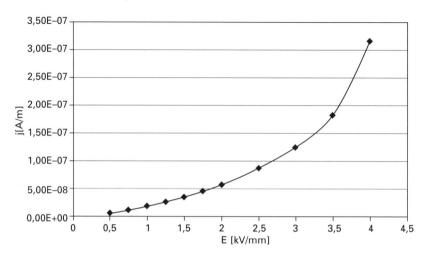

13.7 Current density (A/mm^2) against the electric field strength (kV/mm) for 2.2 mm thick PANI-EB Plaque [18].

stability in ambient conditions, and fewer toxicological problems. However, as a conjugated conducting polymer, the brittleness of PPy limits its practical uses. The processability and mechanical properties of this material can be improved either by blending PPy with some polymers or by forming copolymers of PPy. Thus PPy-based composites may provide the fibres or fabrics with electrical properties similar to metals or semiconductors.[19]

PPy (see Fig. 13.8) is prepared by an oxidative polymerisation process. It can be made either chemically through solution processing or electrochemically through polymer deposition at an electrode. Both processes involve electron transfer. The polymerisation proceeds via the radical cation of the monomer that reacts with a second radical cation to give a dimer by elimination of two protons. Dimers and higher oligomers are also oxidised and react further with the radical cations to build up the PPy chain.

PPy is readily polymerised in aqueous solution by a wide variety of oxidising agents like $FeCl_3$[20, 21], $(NH_4)_2S_2O_8$[22], $CuCl_2$[21]. The reaction speed is higher in the case of $(NH_4)_2S_2O_8$ (a few minutes) and is slower for $FeCl_3$ (e.g. 6 h[21]). Also different kinds of dopants are used. Dopant concentration has the same effect on the reaction speed as oxidants; increase in dopant concentration slows down the polymerisation reaction.

13.8 Molecular structure of polypyrrole.

It has been demonstrated that additives like anion surfactants can increase the conductivity of chemically synthesised PPy. The addition of sodium bis(2-ethyl) sulfosuccinate (AOT) into the polymerisation solution including $Fe_2(SO_4)_3$ as an oxidant, improved the conductivity of PPy several orders of magnitude depending on the PPy/AOT ratio[23] (see Table 13.1). The chemical oxidative polymerisation method for PPy synthesis is recommended if a large amount of polymer is needed. By applying this method PPy can be deposited both onto conducting and non-conducting substrates like metals, glass, plastic, textile, etc. In the case of an electrochemical preparation typically PPy films are galvanostatically deposited on a platinum electrode surface using a one-compartment cell containing an aqueous solution of pyrrole and oxidising agent.

Another method of preparing pyrrole and also thiophene polymers is plasma polymerisation. The usual procedure in this case is that the monomer is first plasma polymerised in the form of a thin film on top of the substrate. Then the specimens are exposed to the vapours of the doping agent (e.g. iodine) in order to introduce charge carriers into the plasma-polymerised structures. Resulting polymer structures have proven to have a higher degree of cross-linking and branching. The morphology of those films is smoother and more uniform than chemically polymerised analogues, but electrical conductivity (in the order of $10^7 \ldots 10^9$ Ω/sq mm) and environmental stability are poor.[25] Although PPy is prepared in its oxidised conducting state, the resulting polymer can be subsequently reduced to give the insulating form. Reversible electrochemical switching between the conducting and insulated state causes a colour change from blue-black to yellow-green.

Polythiophene (PT, see Fig. 13.9) is a material researched especially for field effect transistors (FET). An important requirement for the successful

Table 13.1 Electrical conductivity of chemically polymerised polypyrrole using different oxidisers/dopants

Oxidiser/dopant	Conductivity, $[\Omega \text{ m}]^{-1}$
$FeCl_3$	0.02 [21]
$CuCl_2$	$6 \cdot 10^{-7}$ [21]
DBSA (dodecylbenzenesulfonic acid)	0.01 [24]
$Fe_2(SO_4)_3$	$3.9 \cdot 10^{-4}$ [23]
$Fe_2(SO_4)_3$: AOT (20:1)	0.045 [23]

13.9 Molecular structure of polythiophene.

implementation of organic semiconductors in microelectronic switching and memory devices is the availability of both p- and n-type organic semiconductor materials. Organic n-type conducting polymers remain less developed than p-dopable materials partly due to the lack of electron-poor polymers. One promising attempt to create n-type conducting polymer is poly(3-perfluorooctylthiophene)[26] that was prepared by organometallic coupling. This material has shown reversible reduction, large Stokes shift and is soluble in supercritical CO_2. Another claim of synthesis of the n-type semiconductor has been made using fluoroarene thiophene.[27]

It has been found that unstretched PT films have a highly planar structure that is converted to a uniaxial structure by means of one-way stretching. The same study on PANI showed random orientation for unstretched PANI film. One-way stretching led to the uniaxial orientation that was incorporated with an increase in electrical conductivity.[28]

13.4.4 Optically active inherently conducting polymers

Poly(p-phenylene vinylene) (PPV, see Fig. 13.10) is an electrically conductive polymer that has good physical, electrical and optical properties essential for applications. The properties of PPV can be conveniently engineered over a wide range by the inclusion of functional side groups. The optical band gap of ~ 2,6 eV and bright yellow fluorescence make PPV suitable in applications like light-emitting diodes and photovoltaic devices. This is the polymer that can be both n- or p-doped by chemical or electrochemical means.[29]

Poly-3,4-Ethylenedioxythiophene, (PEDOT, see Fig. 13.11) is a conducting polymer based on a heterocyclic thiophene ring bridged by a diether. This means it has the same conjugated backbone as polythiophene. Its properties make it very easily processable. It can be spin-coated onto a huge variety of conductive and non-conductive substrates including glass, silicon, chromium, gold, etc.

13.10 Molecular structure of poly(p-phenylene vinylene).

13.11 Molecular structure of poly-3,4-ethylenedioxythiophene.

The combination of properties obtained by coating polymer films with polypyrrole or 3,4-polyalkylendioxythiophenes was heretofore unavailable in other systems. The coated films are permanently antistatic (independent of relative humidity), have a surface resistance of 10^2–10^5 Ω, are transparent, heat sealable and vacuum mouldable. Films coated with 3,4-polyalkylenedioxythiophene show a higher conductivity and possess greater environmental stability than polypyrrole.[30]

Nano-scale films of PEDOT and PPy have been prepared in a continuous roll-to-roll process using the vapour-phase polymerisation method. Thin films of ferric chloride doped conductive PPy and PEDOT on plastic substrates, in which neither matrix polymers nor binders are used for the film forming process, are obtained at nano-level thickness. The thickness of the resulting films varied depending on reaction time and temperature. The surface resistance changed with the time of exposure to monomer vapours, and the value was in the range of 10^3 ... 10^5 Ω/sq mm for 20 ... 60 nm thick films and 600 Ω/sq mm for >600 nm thick films. These films had a very highly ordered surface morphology. PEDOT films especially demonstrated a highly ordered crystal microstructure, with high anisotropy, parallel to the substrate film.[31]

A novel method for the preparation of transparent conducting polymer patterns on flexible substrates has been presented. This method, line patterning, employs mostly standard office equipment, such as drawing software, a laser printer, and commercial overhead transparencies, together with a solution or dispersion of a conducting polymer. A seven-segment polymer-dispersed liquid-crystal display using electrodes of the conducting polymer poly(3,4-ethylenedioxythiphene) doped with poly(4-styrene sulfonate) (PEDOT/PSS) was prepared. Also a method to fabricate an eleven-key push-button array for keypad application was presented. Electrode films were characterised using microscopy images, AFM, conductivity measurements, and tests of film stability.[32] A p-type thin film transistor (TFT) was prepared using PEDOT and PPy for gate electrode and active layer, which were made by photolithographic micro patterning. Polyvinyle cinnamate and epoxy were used for the insulating layer through spin coating.[33]

13.5 Application technologies for conducting fibre materials

Generally unmodified ICPs have very poor mechanical properties, they are insoluble, non-melting, and thus not processable. Some progress in mechanical processing has been achieved by mixing ICPs with conventional processable polymers. It has been recognised for some time that the electrical properties of ICPs can best be exploited by their incorporation into host structures that provide the required mechanical and physical properties for a given application.

Textiles produced both from naturally occurring and synthetic fibres are suited to this purpose.

ICPs immobilised by a textile substrate could be used for a number of applications. These electroconductive textiles can be used in the production of clothing which functions as wearable strain gauges for use in biomechanical monitoring, or direct biofeedback devices for sports training and rehabilitation. In this clothing physical changes in the textile causes changes to electrical resistance or electrical conductivity that can be monitored. Other applications include the production of clothing articles that change their thermal insulation or moisture transport characteristics in response to changing climate conditions. Electrically conductive textiles can also be used in applications where antistatic or EMI shielding properties are required. A further application is for use in heating devices such as car seats, car seat covers and gloves.

Requirements for ideal electrically conductive textiles are the possibility to integrate electronic components seamlessly into the conventional textile structure, stable electrical properties and withstanding normal wear and be washable. From the technological point of view it would also be desirable for conventional textile dyeing or printing techniques to be used in the production of the electrically conductive textiles. Limitations in this case are poor solubility of the ICPs and some monomer precursors in water.

In the case of using *in-situ* polymerisation of the ICPs onto a substantially non-conductive textile substrate there is no apparent bonding between substrate and ICP layer. That normally leads to the abrasion or displacement of the polymer layer from the substrate. During laundering the textile may suffer from rapid loss of conductivity. Also ICPs can suffer from oxidation or dedoping that causes changes (usually for the worse) in electrical conductivity. In the following sub-sections the most common techniques for applying ICPs on textiles and other surfaces are discussed.

13.5.1 Extrusion and compounds

Doped PANI sustains high temperature for a short time like dozen of seconds up to 350...380 °C without irreversible loss of conductivity. However, the doped PANI salt is not melt processable but is more like organic filler. For processing the doped PANI salt it has to be plasticised. This is possible through coordinating to the PANI a doping able surfactant like sulfonic acid, and then complex the system with certain metal soaps like zinc stearate. The plasticised system is then processable in an extruder and can form under controlled conditions an interpenetrating network (IPN) with a selected matrix plastic.

The IPN structure is necessary for conductivity, but it is sensitive to heat >230 °C and shear rates >300 s^{-1} for several minutes. The feature limits the processability of the PANI blends. Typical polyamide and polyester fibre materials blended with PANI are most difficult to melt spin, even if that is possible with polypropylene. The three-component complex system also has limited long-term stability.[34]

13.5.2 Solution coating

Solution coating of various materials with a layer of ICP includes spreading the solution on the surface of the substrate material, followed by the evaporation of the solvent. Besides the difficulties concerned with the desired uniform coating the main problem is the insolubility or even poor solubility of most ICPs in nearly all solvents of practical interest. During recent years some progress has been made in the chemical modification of the parent ICP structures in order to get more soluble polymer derivatives. Some examples are the introduction of an alkyl chain into the 3-position of pyrrole monomer, or an alkoxy group into the ortho-position of the aniline monomer. But such a distortion of the molecular structure leads to a drastic decrease in electrical conductivity of the resulting polymers.[35]

Doped PANI can form solution with toluene and in certain conditions with water. The solution can be blended in polymer fixative solution. The combined solution is useful for coating cloth structures with controlled amounts of PANI reaching the ESD properties and reasonable sustainability on the textile fibres.

13.5.3 *In-situ* polymerisation

Chemical oxidative deposition of ICPs onto different kinds of fibres and textile materials makes new composite materials useful for several applications. *In-situ* chemical polymerisation can be made by two distinct methods. The first one is principally bulk polymerisation taking place in solution, where part of the resulting polymer deposits onto the surface of the substrate material immersed into the polymerisation solution, but part of it precipitates as a bulk polymer. In this case it is clearly recommended to maximise the amount of polymer deposited on the surface of the substrate and this can be achieved by choosing a suitable concentration of reaction components, oxidant to monomer ratio, pre-treatment of substrate surfaces.[35]

The other approach is a chemical polymerisation provided directly onto the surface. In this case, the surface to be coated is enriched either with monomer or an oxidising agent, and after that is treated with a solution or vapour of either oxidiser or monomer respectively. This ensures that polymerisation takes place almost exclusively on the surface. The disadvantage of this method is that in order to get a uniform layer of ICP onto the surface, the preliminary layer of either monomer or oxidiser should also be uniform, but achieving this is usually not easy. The reason for that can be the poor or non-porosity of the substrate that prevents the proper absorption of the initial solution. In the case of non-porous substrates the surface needs pre-treatment (corona treatment, etc.) in order to reduce the surface energy of the substrate.

One of the first attempts to cover textile fabric with an electrically conductive coating was made by Gregory *et al*. Polyamide 6 and 6,6 as well as polyethylene terephthalate and quartz fabrics were covered by *in-situ* chemical polymerisation with PANI or PPy coatings. The resulting films had smooth and coherent structures. Each single fibre of the composite structure was uniformly coated by ICP. Resistivity varied from 120 to 12,000 Ω/sq mm depending on the polymer.[20]

In-situ PANI-coated cotton fabrics have been demonstrated. Besides improved electrical conductivity such a fabric showed improved flame retardancy compared to pure cotton. The reduced electrical conductivity resulting from washing was compensated by redoping the material in 1 M HCl for two hours. That led to even lower resistivity (~10^3 Ω) values than initially after polymerisation (~10^4 Ω). The possible applications proposed are in industry, stealth technology and special anti-flammable clothing.[36]

A PPy-carbon fibre composite with a carbon fibre content of 15–20% was prepared by the chemical polymerisation of pyrrole. The resulting composite has high tensile strength and good thermal stability. Such a material could be used as a cathode material for a rechargeable battery, where carbon fibres act as an electrically conductive skeletal electrode for current collection with a specific capacity of 91 mA h/g.[37]

Wet spinning of PANI fibres has been demonstrated from a solution of Polyaniline doped with 2-acrylamido-2-methyl-1-propanesulfonic acid (AMPSA) in dichloroacetic acid (DCA) into various coagulation solvents, including acetone and butyl acetate. The resulting fibres are inherently electrically conductive and may be cold drawn. The conductivity of an 'as-spun' fibre is typically between 0.7 and 1.3 $[\Omega\ m]^{-1}$, rising to 12 … 19 $[\Omega\ m]^{-1}$ after drawing at room temperature to a draw ratio of ~500%. The Young's modulus of the fibre spun into butyl acetate is 40 ± 10 MPa, and the tensile strength at break is 50 ± 10 MPa. The diameter of 'as-spun' fibres is about 200 μm.[38]

13.5.4 Polymerisation in supercritical fluid

Non-conductive fibre materials can successfully be covered by thin films of ICPs. The use of such coated fibres in textile products that require washing is still problematic because the coating is worn out quite soon. Improved washing resistivity has been achieved using *in-situ* polymerisation of PANI and PPy in supercritical CO_2 onto polyester and polyamide fibres.

The main advance of using this technology during coating by *in-situ* polymerisation is that the environment of super critical fluid, the structure of matrix polymr expands so that the monomer is able to penetrate into the matrix. That way the interpenetrating network (IPN) of covering ICP and matrix polymer is created. The IPN is conductive and well protected, and thus is formed into a markedly more sustainable conductive polymer coating layer.[34]

13.5.5 Electrochemical polymerisation

Another possible method for *in-situ* polymerisation is an electrochemical polymerisation of monomers at electrodes, performed either in aqueous or in organic solutions. Although much knowledge on the polymerisation mechanism and redox transformations taking place during reversible oxidation and reduction of ICPs was obtained using electrochemical polymerisation, its use is strongly restricted by the use of conducting substrates.[35] Another aspect is the limited size of substrates; no large area surfaces can be covered by ICPs by electrochemical methods. Several studies have shown that films of ICPs obtained by this method are also very porous and of poor quality.

13.5.6 Electrostatic spinning

Electrostatic spinning has been used as a coating method. This is the only industrially viable method to produce nanofibres in scales of sub-micron (especially 20–500 nm) from solutions of different polymers and polymer blends. Polymer melt can be used in the electro-spinning process, but fibres thinner than 1 µm are more easily made from polymer solutions in an acceptable solvent.

In the electrospinning process, a high-voltage electric field is generated between an oppositely charged polymer fluid contained in a glass syringe with a capillary tip and a metallic collection screen. As the voltage is increased, the charged polymer solution is attracted to the screen. Once the voltage reaches a critical value, the charge overcomes the surface tension of the suspended polymer cone formed on the capillary tip of the syringe of the glass pipette and a jet of ultra-fine fibres is produced. However, the field 1...5 kV/cm is markedly lower than that required for partial discharge in the material. As the charged fibres are spread out, the solvent quickly evaporates and the fibres are accumulated randomly on the surface of the collection screen. Typically the operation voltages are 20 ... 40 kV from device to earth that makes the operation of the process still possible in a converting line without special arrangements.[39]

If the process and material parameters are correctly set, electro spinning of polymer materials from ultra light non-woven-like layers with density of 0.001 ... 0.05 g/m^2 and fibre diameter 0.1 ... 0.8 µm. This means that the consumption of the inherently conductive material is very low. In the case of blended polymers it is even lower. The nanofibres appear light or light green while normally PANI fibres are dark green or black.

It has been demonstrated that conducting polymer blends can be electro-spun using low hazard solvents like water-alcohol and organic acids (formic acid). Various polymers like polyvinyl alcohol, polyethylene oxide and polyamide blends with about 20% conducting PANI doped with 2-acrylamido-2-methyl-1-propanesulfonic acid have been electro-spun utilising fields in magnitude of 2.5 kV/cm providing fibre size of 100 ... 400 nm. Fibre thickness can be controlled by means of molecular weight of the base polymer, and the electric field strength. Very thin coating (< 0.15 g/m^2) can reduce surface resistance down to ESD requirements.[39]

It has been found that no fibres of pure PANI can be electrospun. The fabrication of nanofibres of PANI doped with camphorsulfonic acid (HCSA) blended with polyethylene oxide (PEO) is discussed by Ko.[40] It was found that at least 2 wt% PEO is essential in the blend as no fibre formation occurs in PANI/HCSA dissolved in chloroform alone since the viscosity and surface tension of the solution is not high enough to maintain a stable drop at the end of the capillary tip. Further, addition of more doped PANI does not increase the viscosity of the polymer solution as PANI/HCSA has a very low solubility in chloroform.

13.6 Future trends in conductive materials

Electrically conductive fibres usually lack typical textile fibres' properties such as comfort, wear resistance, and dyeability. These problems can be solved by modifying the fibre structure, finishing technologies and using new materials. Wear resistance problems are increasingly solved by means of coating of traditional textile fibres with conductive materials. Alternatively, the bi-component and other advanced fibre structures are utilised.

Heat and especially washing sensitivity requires limitation in cloth handling such as special washing processes. New, promising and more attractive solutions are the more tolerant materials, especially the ionomer and carbon nano-tube blends. The same development trend also enables improved colour and appearance for the cloths. Further progress is towards the creation of entirely new, superconductive and performing fibre materials based on the new nanofibres and carbon nano-tubes.

Printable electronics is based on conductive polymers. Printing is made on smooth substrates that are typically plastic films. Application to textiles is then most suitable through lamination technology. MEMS based on, e.g., low-temperature co-fired ceramics can be integrated as well.

13.7 References

1. Dissado L A, Fothergill J C, *Electrical Degradation and Breakdown in Polymer*, IEE Materials and Devices Series 9., 1992, Peter Peregrinus Ltd., UK.
2. Harlin A, Nousiainen P, Electrically and Optically Conductive Synthetic Polymer Fibre, *41. International Man-Made-Fibres Congress*, 18–20 September 2002, Dornbirn/Austria.
3. Anderson K, Seyam A M, 'The road to true wearable electronics', The Textile Institute's 82nd World Conference, March 23–27, 2002, Cairo, Egypt.
4. http://www.fibrils.com, web page visited on 13 June 2005.
5. Sun Z, Zussman E, Yarin A, Wendorff J, Greiner A, 'Compound core-shell polymer nanofibers by co-electrospinning', *Adv Mater*, 2003 **15**, 1929–1932.
6. Ulrich H, *Introduction to Industrial Polymers*, Munich 1993.
7. Chiang C K, Fincher jr C R, Park Y W, Heeger A J, *et al.*, 'Electrical conductivity in doped polyacetylene', *Phys Lett*, 1977 **39**, 1098–1101.
8. Chiang C K, Druy M A, Gau S C, Heeger A J *et al.*, 'Synthesis of highly conducting films of derivatives of polyacetylene, (CH)x', *J Am Chem Soc*, 1978 **100**, 1013–1015.
9. Shirakawa H, Ikeda S, *Patent No JP 48032581*, 1973.
10. Kobryanskii V M, *Developments in Polyacetylene – Nanopolyacetylene*, Rapra Review Reports, **10**, Rapra Technology Ltd., UK, 2000.
11. Soga M, 'Orientation and conductivity of polyacetylene', *Proc of the Pacific Polymer Conference*, 1992, 245–256.
12. Dai L, White J W, 'Morphology and electrical properties of polyacetylene-polyisoprene conducting copolymers', *Polymer*, 1997 **38**, 775–783.
13. Ferraro J R, Williams J M, *Introduction to Synthetic Electrical Conductors*, Academic Press, California, 1987.
14. Wang D, Zhao X, Wang F, 'Enhancement of stability of polyacetylene by plasma surface treatment', *Gaofenzi Xuebao*, 1989 **5**, 612–615.
15. Kondratowicz B, Narayanaswamy R, Persaud K C, 'An investigation into the use of electrochromic polymers in optical fibre gas sensors', *Sensors and Actuators B*, 2001 **B74**, 138–144.
16. Shalaby W *et al.*, 'Molecularly bonded inherently conductive polymers on substrates and shaped articles thereof', *US Patent 5691062*, 1997.
17. Mzenda V M, Goodman S A, Auret F D, 'Conduction models in polyaniline. The effect of temperature on the current-voltage properties of polyaniline over the temperature range 30<T(K)<300', *Synth Met*, 2002 **127**, 285–289.
18. Pelto J, 'Intrinsically Conductive Polymers (ICPs) in high voltage applications', *Nordic Insulation Symposium*, Tampere, 2003, 7.
19. Xue P, Tao X M, Keith W Y, Leung M Y, 'Electromechanical behavior of fibers coated with an electrically conductive polymer', *Textile Res J*, 2004 **74**, 929–936.
20. Gregory R V, Kimbrell W C, Kuhn H H, 'Conductive textiles', *Synth Met*, 1989 **28**, C823-C835.
21. Sak-Bosnar M, Budimir M V, Kovac S, Kukulj D, Duic L, 'Chemical and electrochemical characterization of chemically synthesized conducting polypyrrole' *J Polym Sci Pol Chem* 1992 **30**, 1609–1614.
22. Kudoh Y, 'Properties of polypyrrole prepared by chemical polymerization using aqueous solution containing $Fe_2(SO_4)_3$ and anionic surfactant' *Synth Met*, 1996 **79**, 17–22.

23. Omastova M, Trchova M, Piontec J, Prokes J, Stejskal J, 'Effect of polymerization conditions on the properties of polypyrrole prepared in the presence of sodium bis(2-ethylhexyl) sulfosuccinate' *Synth Met*, 2004 **143**, 153–161.
24. Lee J Y, Kim D Y, Kim C Y, 'Synthesis of soluble polypyrrole of the doped state in organic solvents', *Synth Met*, 1995 **74**, 103.
25. Wang J, Neoh K G, Kang E T, 'Comparative study of chemically synthesized and plasma polymerized pyrrole and thiophene thin films', *Thin Solid Films*, 2004 **446**, 205–217.
26. Li L, Counts K E, Kurosawa S, Teja A S, Collard D M, 'Tuning the electronic structure and solubility of conjugated polymers with perfluoroalkyl substituents: Poly(3-perfluorooctylthiophene), the first supercritical-CO_2-soluble conjugated polymer', *Adv Mater*, 2004 **16**, 180–183.
27. Facchetti A, Yoon M-H, Katz H E, Marks T J, 'Materials for n-type organic electronics: synthesis and properties of fluoro-thiophene semiconductors', *Proc of SPIE – The International Society for Optical Engineering*, **5217** 2003, 52–56.
28. Ou R, Samuels R J, Gerhardt R A, 'Three dimensional structure-electrical properties of oriented PPV and PANI films', *International Packaging Technical Conference, USA*, 2003 **1**, 87–92.
29. Damlin P, *Electrochemical Polymerization of Poly(paraphenylene vinylene)*, Åbo Akademi University, 2001.
30. Jonas F, Scharader L, 'Conductive modifications of polymers with polypyrroles and polythiophenes', *Synth Met*, 1991 **41**, 831–836.
31. Kwon S, Han S, Ihm D, Kim E, Kim J, 'Preparation and characterization of conductive polymer nano-films', *Mol Cryst Liq Cryst*, 2004 **425**, 77–83.
32. Hohnholz D, Okuzaki H, MacDiarmid A G, 'Plastic electronic devices through line patterning of conducting polymers', *Adv Funct Mat*, 2005 **15**, 51–56.
33. Kang S, Kang H S, Joo J, Lee M S, Kim M S, Lee J Y, Epstein A J, 'Electrical characteristics of all polymer based thin film transistors using poly (3,4-ethylenedioxythiophene) and polypyrrole', *Mol Cryst Liq Cryst*, 2004 **424**, 203–208.
34. Harlin A, Nousiainen P, Puolakka A, Pelto J, Sarlin J, 'Development of polyester and polyamide conductive fibre', *Proc of the 4th AUTEX Conference*, Roubaix, France, 22–24. June 2004.
35. Malinauskas A, 'Chemical deposition of conducting polymers', *Polymer*, 2001 **42**, 3957–3972.
36. Bhat N V, Seshadri D T, Radhakrishnan S, 'Preparation, characterization, and performance of conductive fabrics: Cotton + PANi', *Textile Res J*, 2004 74, 155–166.
37. Li H H, Shi C Q, Ye W, Li C, Liang Y Q, 'Polypyrrole-carbon fiber composite film prepared by chemical oxidative polymerization of pyrrole', *J Appl Polym Sci*, 1997 **64**, 2149–54.
38. Pomfret S J, Adams P N, Comfort N G, Monkman A P, 'Inherently electrically conductive fibers wet spun from a sulfonic acid-doped polyaniline solution', *Adv Mat*, 10 1998, 1351–1353.
39. Heikkilä P, Pääkkö M, Harlin A, 'Electrostatic coating with conducting polymer blend fibre', *Proc of the 4th AUTEX Conference*, Roubaix, France, 22–24 June 2004.
40. Ko F K, 'Electrospinning ultrafine conductive polymeric fibers', *US Patent Application 20030137083*, 2003.

14
Formation of electrical circuits in textile structures

T K GHOSH, A DHAWAN and J F MUTH,
North Carolina State University, USA

14.1 Introduction

In its simplest definition, an electrical circuit is meant to interconnect electrical elements to form a functional system. The 'electrical elements' could be sensors, actuators, transistors, power sources, etc. Electrotextiles represent enormous potential in creating a new generation of flexible/conformable and multifunctional textile structures for many electrical and electronic systems. In their current state, in many cases, sensors and other electronic devices have been built on or incorporated into textile-based structures using the available technologies; whereas in other cases new technologies, materials, and systems are being developed to form fabric-based electrical devices and systems. A fabric-based electrical circuit is an essential infrastructure necessity for these ongoing efforts.

This chapter provides an overview of recent developments in the area of textile-based electrical circuits, describing processes used to fabricate these circuits and highlighting issues and problems associated with these. Some of the issues that are involved in the development of fabric-based electrical circuits include formation of interconnects and disconnects between orthogonally and otherwise intersecting conductive threads at certain points in the electrical circuit. Evaluation of mechanical and electrical properties of fabric-based electrical circuits and important issues related to these properties are also discussed. This chapter also covers methodologies for attaching electrical devices to the conductive elements incorporated into or on the surface of the fabrics (similar to surface mount technology in printed circuit boards). The research carried out in this area has already been applied to develop commercial products for civilian and military use. The discussion includes applications of the electrical circuits including flexible multi-chip modules, fabric-based antennas, and fabric-based sensors and sensor systems.

Achieving electrical switching is an important part of developing integrated circuits. The chapter also covers ongoing efforts by researchers to develop transistors on fibers or thin films that could eventually be integrated into fabrics to form integrated circuits.

The chapter concludes with a discussion on the opportunities and challenges in the development of fabric-based electrical circuits, a critical component for the evolution of fully integrated electronic textiles with transistors and integrated circuits, sensors and other electronic devices built into the textile structures.

14.2 Development of textile-based circuits

Conventional printed circuit boards are multi-layered structures that have a conductive wiring pattern inscribed on insulating substrates.[1-4] Printed circuit boards are usually made from copper-clad organic laminated materials like epoxy or phenolic resins embedded with reinforcements like glass. The interconnect patterns are inscribed on each copper clad laminate using photolithography or electron beam lithography. These conventional printed circuit boards are not flexible beyond a certain point. In many applications like hand-held electronic devices and laptops, it is desirable to have circuit boards that can be bent or flexed easily. In order to form flexible circuit boards, printing of circuit patterns is carried out on polymeric substrates such as films. Fabric based circuits potentially offer additional benefits of higher flexibility in bending and shear, higher tear resistance, as well as better fatigue resistance in case of repeated deformation.

Needless to mention that textiles used in everyday life already offer platforms to functionalize many useful applications. Extensive work is being carried out in the area of electronic textiles to develop electrical circuits on fabric-based substrates so as to achieve a higher level of circuit flexibility. In section 14.4 we discuss the different materials and processes employed for the development of fabric circuits. We also discuss certain steps that are required to form fabric-based circuits like the formation of cross-over point interconnects and disconnects the next part of section 14.4 discusses the different methodologies that have been used by researchers to attach integrated circuits (chips) and rigid multi-chip modules to flexible electrical circuits. This is followed by a discussion on the properties of thin film and fiber-based transistors.

14.3 Fabrication processes

Different processes that have been described in literature for the fabrication of fabric based circuits include embroidery of conductive threads on fabric substrates, weaving and knitting of conductive threads along with non-conductive threads, printing or deposition and chemical patterning of conductive elements on textile substrates.

14.3.1 Embroidery

Embroidery is a traditional variety of usually decorative needlework in which designs are created by stitching strands of some material onto an appropriate substrate. Circuits can be formed by employing manual or numerically controlled embroidery of conductive threads on fabrics made of insulating materials. The insulating fabric could be woven, non-woven, or knitted. One of the advantages of this process of fabric circuit formation is that the conductive threads can be embroidered in any shape on the fabric irrespective of the constituent yarn path in a fabric. Moreover, one does not need to perform a lot of machine preparation before fabric circuits can be formed. In contrast, in woven circuit formation, conductive threads in the warp direction have to be placed in the appropriate positions before weaving. Orth[5] has described machine sewing and embroidery as one of the most stressful textile manufacturing processes. In sewing conductive threads on fabric substrates, the threads encounter various levels of stresses and friction. This requires an embroidery thread to have relatively high strength and flexibility because of the stresses due to bending and shear and the tortuous path it has to traverse to form a secure stitch.

Formation of electrical circuits on textiles, by numerically controlled sewing processes has been described, by Orth et al.[6] and Post et al.[7,8] Their primary method of circuit patterning was numerically controlled embroidery using conductive threads. Different circuit elements that were embroidered on woven fabric substrates included pads for attachment of devices to the embroidered circuit, conductive bus lines interconnecting different devices on fabric-based multi-chip modules, and elements of a tactile sensor. Orth et al.[6] developed embroidered circuitry to form a conformable fabric keyboard, pressure sensors, a transmitter system, and a musical jacket having embroidered keypad and bus lines. Orth et al.[6] also developed a polymeric thin film-based module which could be incorporated into textile products to form a sensor readout network.

One of the primary disadvantages of embroidery as a means of circuit formation is that it does not allow formation of multi-layered circuits involving conductive threads traversing through different layers as is possible in the case of woven circuits. Moreover, the stresses mentioned earlier can lead to yarn breaks. Although discontinuities in normal sewn fabrics may not be a problem, discontinuities are not desirable in fabric circuits. Joins or splices to fix the discontinuities in sewn conductor lines may lead to undesirable additional impedance. This problem of breakage of conductive threads may also be encountered in other processes of circuit development like weaving and knitting but is slightly less pronounced as compared to sewing.

14.3.2 Weaving and knitting

Conductive threads can be woven into a fabric structure along with non-conductive threads to form an electrical circuit. In the weaving process, two sets of orthogonal threads, called warp and filling, are interlaced with each other. In the process of weaving, the warp yarns lying in the long direction of the fabric are separated in two layers to form an orthogonal interlacing path for the other set of yarns called weft.[9-10] Weaving of electrical circuits can be carried out on conventional weaving machines by making appropriate modifications required to weave conductive threads. One can have a high degree of control in the placement of conductive elements in these circuits by using a jacquard shedding system that allows each warp thread to be addressed individually in order to control the order of interlacement of the threads most precisely. One can form a weave design based on the electrical circuit design and feed it into the jacquard weaving machine controller. This enables one to form complicated circuit patterns on fabrics in an automated manner. One of the limitations of using weaving for making electrical circuits is that the conductive threads have to be placed at predetermined locations in the warp direction while forming the warp beam or from a creel during set up of the machine. Different kinds of conductive threads can be supplied in the weft or filling direction and inserted using the weft selectors provided on a weaving machine. Some modifications to the yarn supply system of the machine may be needed in order to process the conductive threads that are more rigid.

Formation of woven circuits, according to a given fabric design, has been reported by Dhawan *et al.*[11-16] In their work they used a rapier weaving machine and placed the conducting threads on a separate creel. They reported using different kinds of conductive yarns like steel, copper, and silver-coated polyester yarns. Signal integrity issues, like crosstalk noise in certain conductive lines due to switching of signals in neighboring lines, were reported. They also investigated multi-layered fabric circuits and woven fabric-based transmission lines. To improve signal quality, Dhawan *et al.*[11-16] developed coaxial transmission-line-like thread structures by wrapping an insulated conductive thread with a conductive thread and twisted-pair thread structures by twisting two insulated copper threads. They reported significant reduction in crosstalk noise with twisted-pair (see Fig. 14.1) and coaxial yarns. A woven transmission cable, in which a number of conductive warp elements are interwoven with weft elements, has been described by Piper.[17] In this transmission cable, the relative arrangement of ground and signal lines determines the characteristic impedance of the cable. This transmission cable showed a low value of impedance even at very high signal frequencies. Placement of ground lines in between the signal carrying lines leads to a lower level of crosstalk noise in neighboring signal lines of the cable.

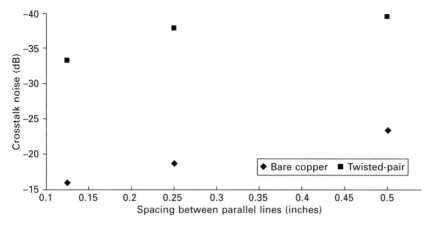

14.1 Crosstalk noise between woven interconnect lines for a 1 Volt signal (1 MHz frequency) sent into an aggressor line: crosstalk noise vs. spacing between the twisted pair and bare copper lines.[12]

Investigation of woven fabric-based signal transmission lines has been described by Kirstein et al.,[18] Cottet et al.,[19] and Locher et al.[20] Polyester yarns twisted with an insulated copper filament were woven to develop structures similar to coplanar waveguides.[18, 19] The conductive threads were separated from each other by any number of non-conductive threads. In coplanar waveguides, the signal lines are surrounded by ground lines in order to prevent coupling between neighboring signal lines at high frequencies.[18,19] In the case of fabric-based coplanar waveguides, one, two, and three signal (S) lines were woven with one or two ground (G) lines on each side of the signal lines to form GS, GSG, GSSG, and GSSSG configurations of signal and ground transmission lines. Electrical parameters like line impedance, insertion loss, and crosstalk noise were determined using time and frequency domain analysis.[18,19, 20]

Jayaraman et al.[21] have described the development of a woven garment (called the sensate liner) with intelligence capabilities. The garment consists of woven optical fibers running in a helical path along the length of a tubular woven fabric. The optical fibers lie between a signal transmitter and a receiver. An optical signal sent by a transmitter travels along the optical fiber, is reflected back from the receiver, and goes back to the transmitter. The time taken for the full signal traverse is known based on the length of the optical fiber. In case of penetration of the fabric by a ballistic object, the incident signal is reflected back from the point of the penetration. Hence, the signal reaches the transmitter before the normal time, implying that it was disrupted by a penetration and rebounded back. In this garment, electrically conducting fibers were also woven into the fabric in the longitudinal direction. These fibers formed an interconnect between a computer and sensors attached to the fibers.

Lebby et al.[22] described the development of a conductive fabric-based communicative wristwatch. The wristband of this watch was developed by weaving conductive and non-conducting yarns. An electronic unit and a power source were integrated into the wristband. The conductive yarns provided interconnection between the different electronic elements attached to the conductive threads of the wristband fabric. Development of a conductive fabric, with electrical anisotropy, has also been reported by Yoshida et al.[23] In this fabric, conducting and non-conducting yarns were arranged in orthogonal directions. This allowed current to flow only in the direction of the conductive threads, thereby leading to the anisotropic electrical properties.

Formation of knitted conductive fabrics was carried out by Farringdon et al.[24] to develop fabric-based stretch sensors. These sensors measure strain from changes in resistance of the knitted conductive strips. Although circuit boards or multi-chip modules have not been developed on knitted fabrics, development of a conductive knitted fabric to form stretch sensors[24] can provide some insight into the feasibility of using knitting[25] as a process for developing electrical circuits. In most conventional weft knitting machines, like a flatbed machine, the conductive threads can be knitted in the fabric only in one direction, i.e., the course (or cross) direction. Hence one can form transmission line cables made up of knitted conductive threads interspaced by non-conductive threads. As knitted fabrics are generally stretchable, the conductive transmission lines developed from knitted fabrics would be expected to exhibit variability in impedance characteristics due to opening and closing of the conductive loops. In order to keep the conductive element in a knit structure straight, one can also insert a conductive thread in the course direction such that the conductive thread is embedded into the fabric between two courses formed from non-conductive threads.

14.3.3 Patterning conductive elements on fabrics

Patterning conductive materials on fabric substrates to form electrical circuits and to achieve conductivity gradients have been reported in the literature.[26-27] Processes that have been employed to form a patterned conductive path on fabric surfaces include deposition of polymeric or non-polymeric conducting materials and subsequent etching, reducing, or physical removal of the conductive materials from certain regions. Thus, the conductive material that is not removed forms a patterned electrical circuit or a region of higher conductivity. Other processes include *in-situ* formation of conductive polymeric films on fabric substrates and subsequent patterning of the films using one of the techniques described above.[26-28] While some have investigated printing of conductive lines on fabric substrates, others[18-28] have looked at chemical plating and subsequent patterning of fabrics. The biggest problem associated with patterning of circuits from thin conductive films (polymeric or metallic) deposited on fabric substrates is that use of an etching agent for forming a circuit pattern leads to non-uniform etching, as some of the etching liquid is absorbed by the threads of the underlying substrate fabric.[20,26-27] Another problem with deposition of conductive films on fabric substrates is that bending the fabric may lead to discontinuities in conductivity at certain points. This problem also exists in printed electrical circuits as conductive materials such as inks or pastes with metallic particles as fillers[29] printed on fabric substrates are stiff and are prone to cracking on bending. Electrical properties, mainly resistivity, of conductive inks and conductive silver loaded epoxies, were evaluated by Cadogan et al.[29] They employed the conductive inks and epoxies to form interconnects with conductive threads of a sensate liner.[21]

DeAngelis et al.[26] described the development of conductive fabrics by selectively patterning a conductive polymeric film deposited on fabrics. Firstly, a conductive polymer film was coated all over a woven or knitted fabric substrate. After coating the whole fabric, a mask was applied in certain areas of the fabric. In the unmasked areas, the conductive polymer degrades on application of the etching material. Thus one gets a design or pattern of conductive polymer on a flexible textile substrate. The area of the fabric having the conductive polymer has high conductivity and the uncovered area has low conductivity. Instead of depositing a layer of conducting polymer on top of the substrate, conductive polymers can be synthesized on the surface of the fabric itself.[26] Kuhn et al.[27] described the formation of a conductive polymer film on the surface of textile materials. A yarn or a fabric was placed in an aqueous solution of a polymerizable material and a conductive polymer layer was formed on the surface of the fabric in the presence of an oxidizing agent. This process can be used to form electrical circuit patterns on a fabric.

Adams et al.[30] described the development of an electrically conductive fabric having non-uniform conductivity. Development of this fabric involved coating a substrate fabric with a thin layer of conductive polymer and subsequent removal of the polymeric coating from certain regions of the fabric using high-velocity water jets. These regions have a lower value of electrical conductivity as compared to the regions where the conductive coating has not been removed. Gregory et al.[31] also described the development of a conductive textile fabric having a conductivity gradient. A conductive polymeric film (doped polypyrrole) was coated on a fabric substrate and a chemical reducing agent was applied to certain regions of the fabric. Application of a reducing agent to the doped conductive polymer on top of the fabric substrate caused conductivity of that region to be changed. The reducing agent was applied to certain parts of the conductive fabric for varying conditions of time, concentration and temperature thereby changing the conductivity of the conductive polymeric film in varying degrees. The process described by

Gregory et al.[31] could also be potentially employed for forming patterned electrical circuits on fabric substrates.

14.4 Materials used

In order to form electrical circuits in textile structures different kinds of non-conductive, and conductive materials are assembled into fabrics. These materials include conductive fibers and yarns, conductive materials deposited on thin films, or organic and inorganic conducting materials deposited on top of fabric substrates.

14.4.1 Conductive fibers and films

In general, the materials employed for the formation of fabric-based electrical circuits include metallic yarns, yarns made from conductive polymers, polymeric threads containing high levels of conducting particles (carbon, silver, etc.), and conducting thin inorganic films. Resistivity values of some of the commercially available conductive fabrics are given in Table 14.1.

Intrinsically conductive polymers like polyaniline, polypyrrole, polyacetylene, polythiophene, poly(p-phenylene), poly(p-phenylene vinylene) and poly(para-phenylene) are some of the materials that have been investigated for forming conductive fibers employing solution spinning processes.[32] The structure of some of these polymers is shown in Fig. 14.2. Conductivity in most of these polymers is based on a π-conjugated system which is formed by the overlap of carbon p_z orbitals and the presence of alternating single and double bonds in the polymer chains.[33-34] Intrinsically conductive polymers are doped to achieve a higher value of conductivity. A comparison of conductivities of conducting polymers as compared to those of metals, inorganic semi-conductors, and insulators is shown in Fig. 14.3.[35] Electrical conductivities of the intrinsically conducting polymers range from that typical of insulators $<10^{-10}$ S/cm to that typical of semiconductors such as silicon ($\sim 10^{-5}$ S/cm). Conductivity levels for doped conductive polymers are much higher. Iodine doped acetylene has a conductivity of $\sim 10^5$ S/cm, a value very close to that of metals. Conductivities of electronic polymers are transformed through the process of doping such that increasing the doping level increases the conductivity of these polymers. Both n-type (electron donating) and p-type (electron accepting) dopants are employed to introduce conducting properties in electronic polymers. The doping process involves exposing intrinsically conductive polymer fibers or films to dopant vapors or solutions.

Conductive polymeric fibers were developed by Mattes et al.[36-37] and these fibers could be integrated into fabrics to form electrical circuits. A 20.1 wt%. solution for spinning emeraldine base (polyaniline) fibers was prepared by Mattes et al.[37] in the solvent N-methyl-2-pyrrolidinone (NMP). The polymer

Table 14.1 Resistivity values of commercially available conductive fabrics[28]

Name	Composition	Production process	Resistivity	Manufacturer
FlecTron	Copper coated polyester taffeta	Electroless copper plating process	<0.1 ohms/sq	Less EMF
Phantom Fabric	Copper coated polyester mesh (90 threads/inch)	Copper plating process	~10 ohms/sq	Less EMF
Zell Conductive Fabric	Tin/copper coated plain weave nylon taffeta ripstop	Copper and tin plating process	<0.09 ohms/sq	Less EMF
FlecTron N-Conductive	Nickel/copper coated polyester taffeta	Electroless copper and nickel plating process	<0.1 ohms/sq	Less EMF
Softmail Fabric	Microfine stainless steel fibers and nylon plain weave	Plain weave process	300 ohms/sq	Less EMF
Shieldex Supra	Tin/copper coated plain weave nylon overcoated with conductive acrylic and polyethylene	Copper and tin plating process	~0.1 ohms/sq	Less EMF
Conductive Felt	Silver coated nylon non-woven	Silver plating process	~0.3 ohms/sq	Saquoir Industries
SeeThru Conductive	Silver coated nylon knit	Silver plating process	5.0 ohms/sq	Less EMF
Snowtex 001.43	Textured 56 AWG copper wire and polyester plain weave	Texturing (kinking) process for yarn making to strain relieve Cu wire	<0.1 ohms/sq	Snowtex
Snowtex 001.44	Textured 56 AWG copper wire with thin insulating coating and polyester plain weave	Texturing (kinking) process for yarn making to strain relieve CU wire	Insulated	Snowtex
Snowtex 001.45	Textured 56 AWG copper wire with thin insulating coating and polyester plain weave	Texturing (kinking) process for yarn making to strain relieve Cu wire	Insulated	Snowtex

14.2 Repeat units of some intrinsically conductive polymers.[32-33]

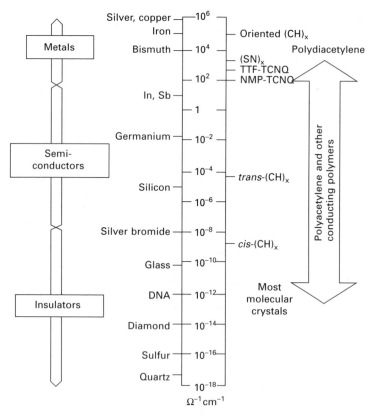

14.3 Comparison chart of conductivities in S-cm⁻¹ for conducting polymers when compared with metals, semi-conductors, and insulators.[35]

solution was extruded through a one-inch air gap directly into a water coagulation bath at 5 °C where the solvent was removed from the nascent polyaniline. The take-up speed was varied between three to ten feet per minute. Conductive fibers can also be developed by slitting a Kapton® polymeric film and depositing gold on top of the slit film pieces as described by Bonderover et al.[38] They took a 50-micron thick sheet of ®Kapton E and deposited silicon nitride on it at 150 °C. The silicon nitride layer was patterned using conventional photolithography and wet etching. Using this patterned silicon nitride layer as an etch mask, the Kapton sheet was etched right through by employing a high-power-density oxygen plasma. A gold film of 100 nm thickness was then deposited on top of the Kapton slit films using e-beam evaporation.

Conducting polymeric films have also been deposited on woven, knitted, and nonwoven fabric surfaces and patterned using photolithography and subsequent chemical etching. Some of the polymeric materials employed are polyaniline, polypyrrole, polyacetylene, and polythiophene.[32–34,39] Instead of depositing polymer on top of a fabric substrate as described earlier[26,31] some researchers have described the development of polymers on the fabrics by first impregnating the fabrics with monomers and subsequently carrying out polymerization at the appropriate conditions.[27]

14.4.2 Non-conductive elements

Some of the materials that have been employed for the formation of fabric-based electrical circuits are nylon, polyester, or acrylic fibers, yarns, or fabrics.[11–16] These non-conductive threads serve to separate the conductive threads from shorting each other. Employing a high-temperature-resistant material like Nomex for forming the insulating material is useful as signal-carrying conductive threads can heat up while carrying a high current. Gorlick et al.[40] employed Nomex threads to form conductive fabric webbing by weaving the Nomex yarns along with steel threads. The other non-conductive materials that could be used to form a substrate, on which conducting (polymeric or metallic particles) material could be deposited, include polyester spun-bonded non-woven fabrics, polyurethane electrospun non-woven fabrics, and polyester electrospun fabric.[41] The main considerations in choosing the non-conducting fibers are ease of processing, temperature requirements and, maybe most importantly, requirements for the primary function of the substrate. Depending on the application, the primary function may be comfort, protection, etc.

®Kapton is a polyimide film by DuPont.

14.4.3 Formation of crossover point interconnects

As described in earlier sections, electrical circuits can be developed by weaving or knitting of conductive threads into a fabric structure in order to interconnect electronic devices placed at different points on the fabric. In order to route the signals in these circuits, interconnections need to be developed between orthogonal conductive threads and these interconnections can be termed as crossover point interconnections. Different techniques can be employed to form crossover point interconnections like resistance welding, adhesive bonding, air splicing, and soldering.

Dhawan et al.[11–16] described resistance welding as one of the best methods for forming crossover point interconnects[11,13,16] for its simplicity and resulting minimal damage to neighboring non-conductive fibers. The process led to the formation of less rigid fabric-based circuits as it relied on controlled heating and melting of the orthogonal conductive threads at the crossover point. Two kinds of resistance welding processes, i.e., top-bottom probe welding and parallel gap welding were used to form crossover point interconnects in fabrics containing steel and copper threads. DC resistance values of the fabric samples having welds at the crossover points were found to be less than those for woven copper samples without any weld.

Another method to form interconnections between orthogonal threads is soldering. Generally, this leads to a rigid fabric as the solder blobs are larger than the weld points formed using resistance welding. Moreover, one may need to use a controlled amount of solder for this process to be reliable for multiple crossover point interconnects.

Locher et al.[42] employed a conductive adhesive to form interconnection between two orthogonally woven conductive wires. Firstly, insulation was removed (at the crossover point of these conductive wires) from the two orthogonally woven conducting wires using a laser ablation process. Subsequently, the crossover point interconnect was formed using a conductive adhesive.[42]

In order to develop integrated circuits using transistors developed on fibers, one may need to form interconnections between drain, gate or source metal pads on the fibers containing the transistors and orthogonal conductive threads. Lee et al.[44] have reported the formation of transistors directly on fibers.[44] They employed conductive aluminum and stainless steel wires as the gate of the transistor. For the operation of this back-gated field effect transistor, interconnections were made to the metallic wires forming the transistor gate. These interconnections could be made anywhere along the fiber, where the coatings are not present or are removed. Contacts to the source and drain regions of the transistor could be made by conductive lines woven into the fabric, orthogonal to the direction of the fiber containing the transistor. They proposed physical contact as being the only means of interconnection between

Formation of electrical circuits in textile structures 251

the conductive threads and the source and drain contact pads and did not employ thermal, electrical or light energies to form crossover point interconnections as suggested in the patent by Dhawan et al.[43]

14.4.4 Formation of disconnects of conductors

Woven electrical circuits require formation of disconnects at certain locations along a conducting yarn so as to control the signal path in the circuitry. These disconnects may be cuts or switches that could be switched on and off. In woven fabrics, the disconnects can be formed by creating a fabric with 'floats' at certain points. These floats could be selectively cut by appropriate cutters to form the disconnects. Interestingly, Dhawan et al.[14] reported formation of disconnects using a properly set up parallel gap resistance welding technique. They formed disconnects by using a very high value of weld current such that a break in the conductive fibers could be created.

A patent by Dhawan et al.[43] describes the formation of disconnects employing one of thermal, electrical or light energies. This patent also describes a process by which a programmable grid array could be developed in textile-based electrical circuits by developing switching elements like transistors on woven threads such that there is a transistor on every thread around the crossover point of the orthogonal threads, as shown in Fig. 14.4. Thus current

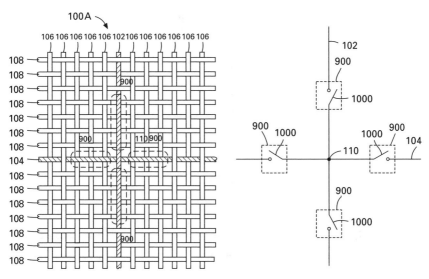

14.4 Figures demonstrating means of developing programmable circuitry in woven fabrics by developing switching elements like transistors on woven threads such that there is a transistor on every thread around the crossover point of the orthogonal threads.[43]

flowing into and from the crossover point can be switched according to the switching of the transistors in a programmable manner.

Lee et al.[44] have proposed fabrication of transistors at the crossover points of orthogonally woven fibers. This would allow formation of gate array switching grids such as the one shown in Fig. 14.5. Lee et al.[44] developed one of the most important elements in the formation of such a gate array switching grid, which is the formation of transistors directly on fibers.[44]

A conventional printed wiring board layout has 45° bendings when the trace direction changes from horizontal to vertical.[42] On the other hand, in a woven fabric structure, only perpendicular interconnection layouts can be implemented. This is illustrated in Fig. 14.6,[42] where the black lines indicate the path used for routing electrical signals in a textile fabric. Recently, Locher et al.[42] have also described the process of making disconnects in order to avoid short-circuits with the rest of the routing. They formed these disconnects or cuts by using laser light with a fluence greater than 4 J/cm^2.

14.4.5 Attachment of devices

Attachment of electronic devices to conducting elements integrated in the fabric poses numerous challenges in the field of electronic textiles. Devices

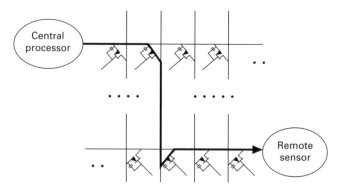

14.5 A schematic demonstrating the concept of a gate array switching network[44] © 2005 IEEE.

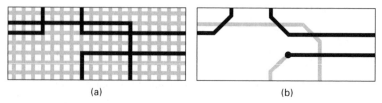

14.6 (a) Perpendicular routing layout in a woven fabric and (b) 45° bend in the trace when going from horizontal to vertical direction.[42]

that may be attached to the conductive elements of a textile fabric include rigid and flexible circuit boards, multi-chip modules and individual integrated circuit packages. It is desirable that individual chip packages be directly connected or bonded to conductive threads or conductive elements of a fabric, so as to form a more flexible and conformable electronic textile. The attachment of these devices could be carried out either by using physical contact between snap-connectors, conducting yarns, and electronic devices or by employing soldering or thermal bonding techniques.

Gorlick[40] described the development of a fabric-based power bus and data network for wearable electronic devices. To form a conductive fabric-based power bus and data network, a webbing was woven from Nomex and steel yarns with two pairs of multi-stranded stainless steel conductors woven directly into the webbing as shown in Fig. 14.7. A means of connecting conventional electronics with this fabric webbing was also proposed. The electronic devices can be attached to these suspenders (the conductors of the fabric webbing) by using electrical surface connectors like sew-on ball and socket snap connectors as shown in Figs. 7(c) and (d). As shown in Fig. 14.7(d), the ball of the snap connector is connected to a stinger (an L-shaped wire soldered to the ball of the connector), which is inserted into the conductive fiber bundle of the fabric webbing, thereby making an electrical contact with the conductive fiber of the webbing. A socket connector, shown in Fig. 14.7(c), could also be employed to interconnect electronic devices with the connectors. Electric

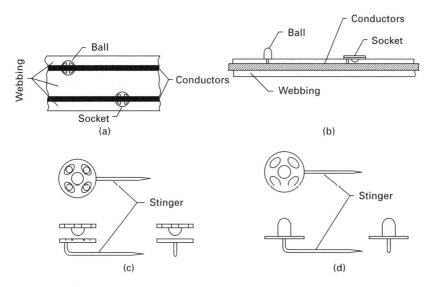

14.7 (a) and (b) Woven fabric webbing used to interconnect electronic devices to conductive fabrics using snap connectors, (c) a socket connector, and (d) a ball connector.[40]

suspenders made from the fabric webbing were transformed into a shared low-voltage, DC power bus for digital wearable devices.

Jung et al.[45] carried out attachment of electronic devices to conductive elements of a woven fabric. In one of the device attachment methods proposed by them, a flexible circuit board acts as an interface between the woven conductive threads and the integrated circuit. The integrated circuit is attached to the flexible circuit board as shown in Fig. 14.8. Conducting threads are connected to the electrodes and pads on the flexible circuit board by soldering. In another interconnect method, bonding wires are used to connect integrated circuits and conductive (having an insulating coating) threads woven into fabric strips. Insulation of the woven conductive threads is first removed by laser treatment and then tiny metal contact plates are soldered to these threads. Then, insulated bonding wires are used to connect the metal plates to the integrated circuit bond pads (Fig. 14.9).

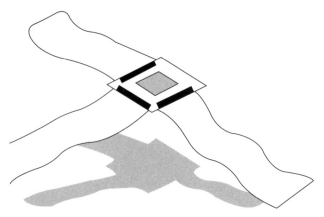

14.8 Flexible circuit board used to form interconnects between an integrated circuit and the conducting threads woven into narrow fabric strips.[45]

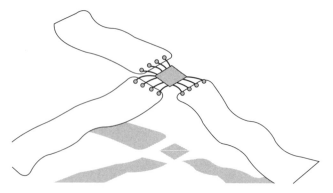

14.9 Wire bonding of narrow fabric strips containing conducting threads to an integrated circuit.[45]

Post et al.[7] have described several methods for producing device-fabric interconnects. They found that soldering devices to the conducting threads in the fabric, bonding the electronic components using conductive adhesives, or stapling components into conductive stitched circuits were not very effective processes in producing these interconnects. Packages called plastic threaded chip carrier (PTCC) packages, having long conducting threads, were designed by them.[7] The leads of the plastic thread chip carrier packages were woven or stitched into the fabric circuitry. In order to manufacture threaded packages, a bare die or integrated circuit was taken and fine gold wires were bonded onto gold plated bond pads of the integrated circuit chip using thermo-compression. The other end of the fine gold wire was connected to a copper lead frame. Flexible and conformable conductive threads were micro-spotwelded to the lead frame stubs and the structure was sealed in a plastic carrier.

Lehn et al.[46] have suggested some designs for attaching electronic modules and devices to textile fabrics such that these devices are easily attachable and interchangeable.[46] A comparison between the different device attachment methods like raised wire connectors, solders, snap connectors, and ribbon cable connectors, is provided in Table 14.2. According to them, soldering produces reliable electrical connections to conductive threads of an electronic textile fabric but has the disadvantage of not being compatible with several conductive threads or materials like stainless steel. Moreover, soldering of electronic devices to threads that are insulated is a more complex process

Table 14.2 Advantages and disadvantages of different device attachment methods[46]

Type	Pros	Cons
Solder	solid electrical connection, strong physical connection, small, light, comfortable, not, noticeable	slow connection process, wire compatibility issues, wire breaks, alignment issues, mid-wire stripping, exposed wire protection expensive
Snaps	connection/reconnection ease, common use	slow connection process, soldering or welding issues, connection size, weak physical connection exposed leads
Ribbon Cable Connector	insulation displacement, common part, insulated connection, alighment tolerance, reliable connection	size, installation difficulty wire breaks
Raised Wire Connector	single textile side, no threads, low profile	weak fabric connection

involving an initial step of removal of insulation from the conductive threads in the regions where the device attachments are desired and insulation of the soldered region after completion of the soldering process. Another device connection method described by Lehn et al.[46] is to attach snap connectors to conductive fibers or wires integrated into a textile fabric and subsequently attaching electronic devices or circuit board elements to these snap connectors. According to Lehn et al.[46] different ways of attaching snap connectors to conductive threads integrated into fabrics include soldering or welding of these connectors to the conductive threads or sewing of the snap connectors to the base fabric employing conductive or non-conductive sewing threads. The main advantage of employing snap connectors is the ease of attachment or removal of electronic devices from these connectors, whereas the main disadvantages are the large size of the device and the weak physical connection formed between the snap connectors and the devices. Ribbon cable connectors employ insulation displacement in order to form an interconnection with insulated conductor elements integrated into the textiles. A v-shaped contact cuts through the insulation to form a connection to the conductor. Firstly, the ribbon cable connector is attached to the conductive threads in an e-textile fabric and subsequent electronic devices and printed circuit boards are attached to the ribbon cable connector. One of the advantages of employing ribbon cable connectors for device attachment is the ease of attachment and removal of the electronic devices to form the electronic textiles.

14.4.6 Development of film or fiberbased transistors and integrated circuits

As mentioned before, electrical switching is an important part of developing integrated circuits. The ability to form transistors on fibers, yarns, or thin films would enable fabrication of integrated circuits on flexbile textile-based substrates. Initial studies of thin film transistors and transistors developed on fibers are starting to be carried out.[47-53]

The main problems associated with transistors developed on flexible polymeric substrates is that the channel layer is based on either amorphous hydrogenated silicon or on nano-crystalline silicon layers deposited on the flexible substrates. The fact that these depositions are carried out on non-crystalline substrates and at a lower temperature (due to a lower glass transition temperature of the polymeric substrates), leads to the non-crystalline nature of the deposited films. Thus, the carrier mobilities are generally lower in these thin film transistors as compared to transistors developed on rigid planar silicon wafers. Another problem encountered in thin film transistors is a large gate leakage current due to porosity of the films, leading to a non-ideal turn-on in the transfer characteristics. Silicon dioxide gate dielectrics deposited at lower temperature may also have high levels of hydrogen in them and can be porous. Moreover, the interface between the gate dielectric and the amorphous or nano-crystalline channel of the thin film transistors has a higher defect density as compared to transistors developed on silicon wafers or other rigid substrates. A brief overview of the different kinds of thin film transistors is provided below, which is followed by a section that describes the development of these transistors on thread-like structures. However, using such transistors on flexible planar substrates has been useful for flexible display applications where the switching rate of pixels is low, and the size of the transistor can be large. For computing applications thin film transistors are typically avoided since the speed and performance of conventional transistors is significantly superior. In terms of consumer applications of thin film transistors for flexible displays the main problem remains packaging to handle the mechanical stresses, and in the case where organic materials are used protection from moisture penetration and protection from ultraviolet light.

Amorphous and nanocrystalline thin film transistors

As one typical example, Gleskova et al.[50–51] described the development of hydrogenated amorphous silicon thin film transistors on flexible substrates. They employed RF-excited plasma-enhanced chemical vapor deposition to deposit a hydrogenated amorphous silicon at a low temperature (150 °C) on a polyimide foil substrate. A silicon nitride (SiN_x) layer was used as a gate dielectric in this transistor, which had a bottom gate structure as shown in Fig. 14.10. The dielectric constant of the silicon nitride deposited at 150°C, and measured at 1 MHz, was found to be 7.46. Chromium was used as the gate electrode material and un-doped hydrogenated amorphous silicon was used as the transistor channel. The transfer characteristics of this transistor are shown in Fig. 14.11. In these transistors, there is substantial gate leakage as shown by the existence of a substantial source-drain current for gate voltage less than zero volts. The value of linear mobility for these transistors was found to be 0.5 $cm^2V^{-1}s^{-1}$. Gleskova et al.[54–55] also studied the effect of

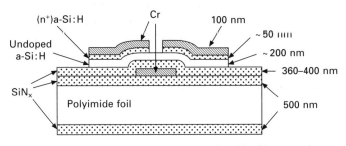

14.10 Schematic cross-sectional view of a thin film transistor fabricated on a flexible polyimide substrate[51] © 1999 IEEE.

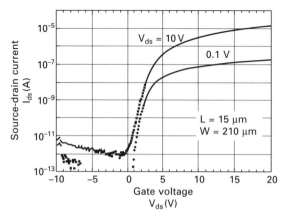

14.11 Source-drain current as a function of gate voltage for $V_{ds} = 0.1$ and 10 V. The substrate thickness is 2 mils[51] © 1999 IEEE.

mechanical strain on the properties of thin film transistors and reported that application of a compressive strain leads to an instantaneous increase in the mobility of carriers in the transistor.

As another example, Cheng et al.[49] developed thin film transistors based on nanocrystalline hydrogenated silicon deposited on Kapton substrates employing a staggered top gate, bottom source-drain geometry. The thin film transistors were developed on a nanocrystalline silicon seed layer on top of the Kapton substrate. Development of the crystalline structure of the channel layer takes place on top of this seed layer as shown in Fig. 14.12. The depositions and processing were carried out at 150 °C. This process for fabrication of thin film transistors ensures that the nanocrystalline channel layer is formed towards the end of the fabrication process sequence ensuring no plasma etch damage to it. Hole mobility of approximately 0.17 $cm^2V^{-1}s^{-1}$ was observed for p-channel thin film transistors and an electron mobility of approximately 23 $cm^2V^{-1}s^{-1}$ was observed for the n-channel thin film transistors developed on Kapton substrates. Despite showing a much better hole mobility as compared to hydrogenated amorphous silicon thin film transistors,[50–51] these transistors were not found to be suitable for the development of CMOS transistors as they had high gate leakage characteristics, as shown in Fig. 14.13. The silicon dioxide gate layer deposited at 150 °C contains excessive hydrogen, is electrically unstable and porous. The large leakage current is due to the porosity of the films and leads to a non-ideal turn-on in the transfer characteristics.

Organic thin film transistors

Organic thin film transistors have a channel layer composed of conjugated polymers, oligomers, or small organic molecules. These transistors generally

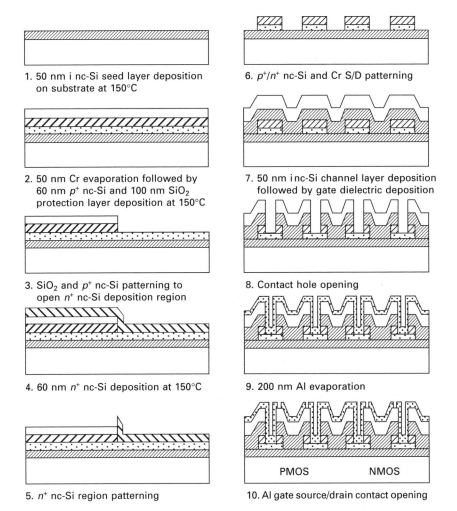

14.12 Process sequence for fabricating a CMOS device on a flexible Kapton substrate. A nanocrystalline (nc) silicon seed layer is first deposited at 150 °C followed by the steps described above[49] © 2003 IEEE.

have lower mobilities as compared to those developed from inorganic single crystalline semiconductors and therefore cannot be employed for high-speed switching applications. Although these transistors are inferior to transistors developed on rigid single crystalline substrates, they offer a few advantages over hydrogenated amorphous and polycrystalline silicon-based thin film transistors. Potentially, organic thin film transistors could be developed at lower costs as compared to amorphous and polycrystalline silicon-based thin film transistors.[56] Moreover, the ability to develop organic transistors at room temperature can enable their fabrication on thin polymeric films, which

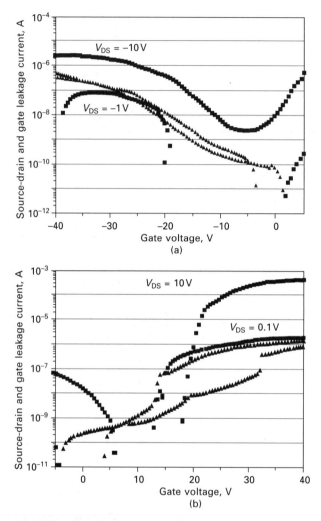

14.13 Transfer characteristics of (a) n-channel and (b) p-channel thin film transistors fabricated using a nanocrystalline hydrogenated silicon channel layer. The width (W) and length (L) of the channel were 200 and 30 microns respectively[49] © 2003 IEEE.

have a much lower glass transition temperature as compared to silicon or glass. This is in contrast to amorphous and polycrystalline thin film transistors which require processing at elevated temperatures of above 350 °C.

Some of the materials from which organic thin film transistors have been developed are polymers like Polyacetylene, Polythiophene, Poly(3-hexylthiophene), and Polythienylenevinylene or organic molecules like Pentacene and α-ω-dihexyl-quaterthiophene. Pentacene-based thin film transistors with a mobility as high as 1.5 $cm^2V^{-1}s^{-1}$ have been reported, whereas mobilities of polymeric thin film transistors are in the range of 0.045–0.22.[56] Pentacene based organic thin film transistors also demonstrate higher levels of on to off current ratios (I_{on}/I_{off}) as compared to polymeric thin film transistors. Two organic thin film device configurations are shown in Fig. 14.14. Transfer characteristics of a Pentacene-based organic transistor developed by Dimitrakopoulos et al.[57] are shown in Fig. 14.15. The mobility of this transistor was calculated from this plot as 0.16 $cm^2V^{-1}s^{-1}$. Inoue et al.[58] developed organic thin film transistors on flexible polycarbonate substrates by using Tantalum oxide as the gate insulator. Pentacene was used as the organic semiconductor channel layer material and the thickness of its layer was 50 nm. A layered structure of aluminum (50 nm) and tantalum (200 nm) was used for forming the gate electrode and the gate oxide was developed by anodizing the tantalum metal deposited as the gate electrode, as shown in Fig. 14.16. These organic thin film transistors demonstrated good transistor characteristics like low threshold voltage (approximately 1.1 V) and a low leakage current. Development of organic transistors on flexible substrates was also carried out by Yoshida et al.[59] and Bonfiglio et al.[60] Park et al.[61] also report developing organic thin film transistors using poly-3-hexylthiophene as the channel layer. However, much work remains to be done to ensure the environmental stability of organic transistors outside of laboratory conditions.

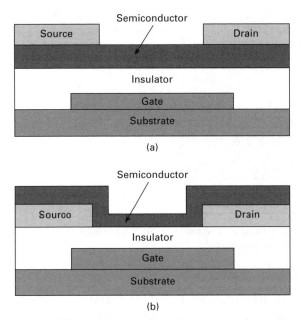

14.14 (a) Top-contact device, with an organic semi-conducting layer onto which source and drain electrodes are evaporated, and (b) bottom-contact device, with the organic semiconductor deposited on source and drain electrodes.[56]

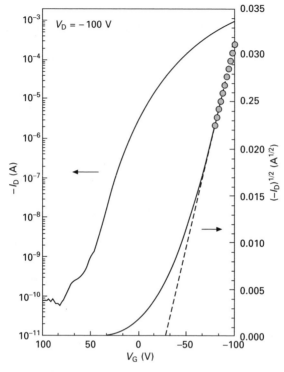

14.15 Drain current I_D versus gate voltage V_G and $|I_D|^{1/2}$ versus V_G for a pentacene OTFT having a heavily doped n-type Si gate electrode, with L.4.4 m and W.1.5 mm.[56–57]

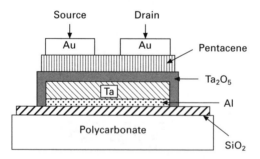

14.16 Structure of an organic thin film transistor developed on a flexible plastic substrate.[58]

Thin film transistors on threads or polymeric slit films

The development of transistors on threads is potentially exciting since it implies the possibility of the formation of integrated circuit structures in fabrics by appropriately forming interconnects using textile processes. However, this excitement should be tempered by the realization that forming the correct

Formation of electrical circuits in textile structures 263

interconnections is a formidable task. One also would need to design the circuitry such that it is fault tolerant and able to withstand a much higher number of defects than circuits on a silicon chip. At this stage of technology development, transistor fabrication on fibers is a major step in electronic textile development and may provide a methodology for signal routing, signal amplification, sensing, or low-level computation within a textile.

Lee et al.[44,52] fabricated organic pentacene-based back-gated thin film transistors directly on fibers employing a weave-masking process as shown in Fig. 14.17. Conductive aluminum and stainless steel wires (125–500 µm) formed the gate of the transistor and these wires were coated with a thin layer of gate dielectric. Silicon-dioxide, deposited using low temperature chemical vapor deposition, and cross-linked poly-4-vinylphenol were used as the gate dielectric materials. A 150–250 nm silicon-dioxide layer was deposited at 450 °C employing the low-temperature chemical vapor deposition process and using a high ratio of oxygen to SiH_4 in the gas mixture. The fiber was coated with a poly-4-vinylphenol dielectric layer by preparing a poly-4-vinylphenol casting solution, adding poly(melamine-co-formaldehyde) cross-linking agent into the solution, and finally annealing and cross-linking the polymer at 100 °C and 200 °C respectively. The fiber was subsequently coated, via the evaporation process, with a 90 nm layer of pentacene as the channel material. Interconnections to the metallic core of the fiber, which forms the gate of the transistors, could be made anywhere along this coated fiber. The active areas of the transistor are defined by the over-woven masking fibers having a diameter of 50 µm. Channel length of these transistors is thus equal to diameter of the masking fiber (50 µm). Subsequently, 100 nm of gold is evaporated in the gaps between the masking fibers to form the source and drain pad regions. Contacts to the source and drain regions of the transistor

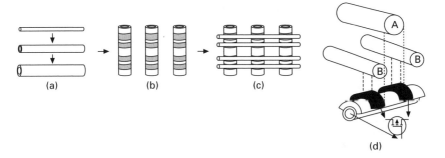

14.17 The process flow for developing transistors on fibers via the weave masking technique: (a) deposition of gate oxide and channel layers, (b) pattering of the source and drain pads employing overwoven fibers, (c) interconnections with orthogonal yarns, and (d) a schematic demonstrating the masking fibers A and the interconnect fibers B[44] © 2005 IEEE.

could be made by conductive lines woven into the fabric, orthogonal to the direction of the fiber containing the transistor. Transfer characteristic of these transistors are shown in Figs. 14.18(a) and (b). The fiber transistors, having silicon dioxide as the gate dielectric, had a significant gate leakage and a high defect density. High defect density of the silicon dioxide layer was primarily due to the high surface roughness of the substrate aluminum wires. This resulted in low dielectric strength and high low-field leakage current through the gate oxide layer. These fiber-based organic transistors exhibited mobility of approximately 0.5 $cm^2V^{-1}s^{-1}$.

Gnade et al.[62] fabricated thin film transistors by depositing layers of n-doped and intrinsic amorphous silicon on silicon nitride films on flexible Kapton films employing plasma enhanced chemical vapor deposition. Gate, source and the drain gold contact pads were deposited by using the thermal deposition process and all the patterning was carried out using photolithography. These transistors exhibited a high threshold voltage of 4 V and a relatively high gate leakage current of 4 pA. Gnade et al.[62] and Bonderover et al.[63] described the development of yarn-like transistors by slitting films, with multiple transistors formed on it, into thin strips using the plasma etching process. These films were integrated into woven fabric structures as yarns. Gnade et al.[62] demonstrated formation of digital logic functionality on these woven structures by making appropriate interconnections, using conductive strips, between the different yarn-like thin-film transistors.

Bonderover et al.[38] have developed an integrated circuit prototype using amorphous silicon-based transistors developed on silicon nitride-coated Kapton films. A schematic of a thin film transistor, comprising of an intrinsic amorphous

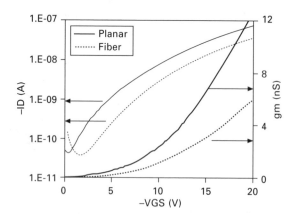

14.18 Transfer characteristics for fiber-based and planar field effect transistors using the silicon-di-oxide gate dielectric. For the planar transistor, W/L = 250 μm/25 μm and gate oxide thickness = 220 nm. For the fiber transistor, W/L = 250 μm/25 μm and gate oxide thickness = 250 nm[44] © 2005 IEEE.

silicon-based channel layer, is shown in Fig. 14.19. Fabrication of these transistors is carried out at approximately 150 °C employing conventional processing methods. Multiple transistors are arranged in the form of columns on the wide film and individual transistor fibers are detached from this wide film by employing the slit-film technique. These thin film transistors have a threshold voltage of 7.5 V and carrier mobility of 0.13 cm^{2}V^{-1}s^{-1}. These slit transistor fibers are interwoven along with spacer fibers and conductive fibers to form an active load N-MOS inverter circuit having appropriate interconnections to the input, output, ground, and power lines. The inverter is reported to operate at frequencies of up to 1 kHz, the operation frequency being dependent on the time taken to turn the transistor on, which is in turn dependent on the overlap and gate capacitances of the circuit.[38]

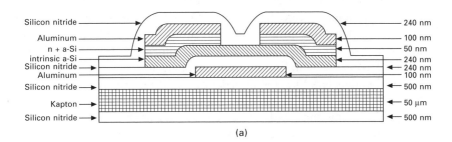

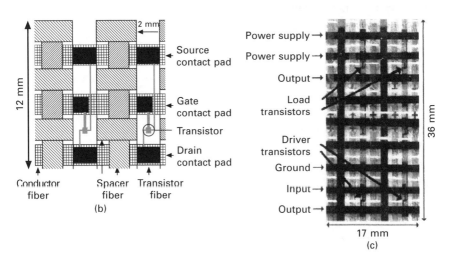

14.19 (a) Schematic of a thin film transistor developed on a Kapton thin film, (b) schematic of arrangement of the conductor fibers, the spacer fibers, and the transistor fibers to form the transistors, and (c) circuit woven from active component fibers, a logic inverter[38] © 2004 IEEE.

14.5 Characterization

Characterization of electronic textile circuits obviously involves evaluation of their electrical behavior. However, these being primarily textile structures, properties that are of value in a textile product may be of significant importance. These properties are flexibility, breathability, comfort, etc. These and other such properties may be more or less important depending on the application.

14.5.1 Evaluation of electrical properties

Resistance and impedance evaluations

Kirstein et al.[18–20] applied methods of microwave technology to evaluate the frequency response of woven transmission lines and to develop a theoretical model to describe signal transmission in textiles.[19] Configurations similar to coplanar waveguides were developed in the woven circuits by weaving the copper-polyester yarns in the warp direction. In the coplanar waveguide structures, the signal line is surrounded on either side by ground lines to minimize parasitic coupling between neighboring signal-carrying lines. Electrical parameters like line impedance, insertion loss, far end crosstalk, etc., were determined for the different configurations of woven coplanar waveguides using time and frequency domain analysis. The impedance measurements were made by measuring the signal reflections along the transmission line with time domain reflectometry (TDR). The block diagram of the measurement setup is depicted in Fig. 14.20.

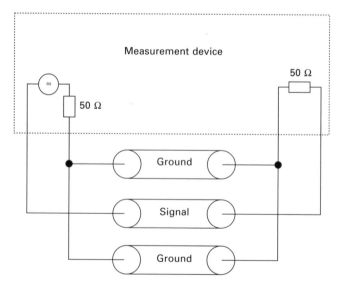

14.20 Setup to evaluate woven fabric-based transmission lines using time domain reflectometry (TDR)[19] © 2003 IEEE.

Reliable connection of the textile samples with the measurement equipment was made by connecting the fabric transmission lines with patterned solder pads on one side of a FR4 laminate on SMA connector on the other side. It was reported that when the GSSSG configuration was used, the value of coplanar waveguide impedance was lower as compared to the GSSG or GSG configurations as shown in Fig. 14.21. Moreover, it was observed that the impedance of the woven transmission lines showed variation with time and this was attributed to irregularities in the fabrics. Based on the measured values of impedance, inductance, capacitance, and resistance for the different transmission lines, a theoretical transmission line model was also developed through simulation using the SPICE software.[19] Kirstein et al.[18] evaluated the impedances, capacitances, signal rise times, and output signal profiles of three configurations of textile-based co-planar waveguides – the first configuration consisting of two ground (G) lines enclosing one signal (S) line, the second consisting of two ground (G) lines enclosing three signal (S) lines, and the third consisting of two ground (G) lines enclosing six signal (S) lines. Variation of these parameters as a function of number of signal lines in the textile-based coplanar waveguides is shown in Fig. 14.22 and is attributed to dependence of transmission line properties of coplanar waveguides on the width or number of signal lines and spacing between signal and ground lines.[18]

Circuits were developed by Dhawan et al.[11–16] on a rapier weaving machine by weaving conductive threads into a fabric in order to interconnect electronic

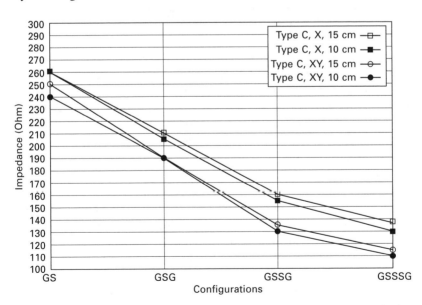

14.21 Impedance measurements for different woven transmission lines.[20]

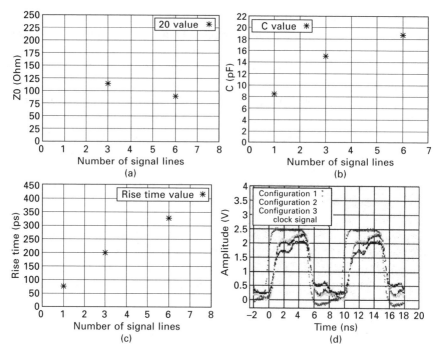

14.22 Variation of the following properties of a textile-based coplanar waveguide transmission line as a function of number of conductive signal threads in the coplanar waveguide – (a) characteristic impedance, (b) line capacitance, (c) signal rise time, and (d) output signal for a 100 MHz clock input signal.[18]

devices placed at different points on the fabric. In order to route the signals in these circuits interconnections were developed between orthogonal conductive threads (crossover point interconnections) employing different techniques like resistance welding, adhesive bonding, air splicing, and soldering. In order to evaluate the resistance of the crossover point interconnects, Dhawan *et al.*[11–16] employed two-point probe resistance measurements using a multimeter. Four-point probe measurements of these fabric circuits were also investigated. DC resistance values of the fabric samples having welds at the crossover points, were found to be less than those for woven copper samples without any weld. Impedance values of the different crossover point interconnects were also determined by measuring AC signal (square wave) reflections from the crossover point interconnects. Slade *et al.*[28] also reported DC resistance evaluations of conductive fabrics using the four-point probe measurement system. They measured the resistance of one inch by one inch conductive fabric samples employing four-point probe measurements on an HP LCR meter. The conductive fabrics evaluated by them included metal-coated fabrics and fabrics woven from metal-plated and metal-foil wrapped yarns. The four-point probe technique is conventionally employed for characterization of semiconductor devices and materials and has been described by Runyan et al.[64] and Schroeder.[65]

Signal integrity issues

Researchers working in the area of electronic textiles realize that there are numerous challenging issues to be addressed like achieving reliable and robust interconnect formation, improving signal integrity (crosstalk noise), maintaining textile characteristics (lightweight, flexibility, strength, conformability, etc.), providing efficient means of power generation/harvesting, and addressing washability and weatherability for wearable electrotextiles.[66–68] In order to study and evaluate AC signal crosstalk noise in woven circuits, Dhawan et al. used AC signals varying between 10 KHz–15 MHz on woven transmission lines (called aggressor lines) and evaluated the effect of these signals on the neighboring quiet lines (carrying no signal).[12,66–67] They observed that the magnitude of crosstalk noise increased as the spacing between the conducting lines is decreased, see Fig. 14.1. Moreover, as the frequency of the signals transmitted through the aggressor line is increased, the magnitude of the crosstalk noise also increased. Coaxial transmission-line-like thread structures were developed by wrapping an insulated conductive thread with a conductive thread and twisted-pair thread structures were developed by twisting two insulated copper threads. Dhawan et al. also observed that there was a significant reduction in crosstalk noise when twisted-pair (see Fig. 14.1) and coaxial yarns were used to form these circuits.

Crosstalk noise, i.e., noise induced on a quiet line or a signal line due to change in signal on neighboring signal-carrying lines, between two neighboring signal lines was evaluated by Kirstein et al.[18] and it was reported that the value of crosstalk reduced substantially if the neighboring signal lines were separated by a ground line.[18]

14.5.2 Mechanical properties

Several researchers have investigated the physical properties of fabric-based circuits for their mechanical and endurance characteristics.[5,28,69–70] Some researchers have evaluated the mechanical properties to determine the lifetime of these fabric circuits, when used in everyday life. The effect of flexing, bending, or abrasion on electrical properties of these circuits have also been evaluated.

The effects of washing and dry-cleaning on the electrical properties of conductive fabrics were evaluated by Slade et al.[28] The fabrics evaluated by them included narrow fabrics woven from Ni-Ag-Cu coated fibers and Ag-Cu metal foil wrapped yarns and metal-coated fabrics shown in Table 14.1.

One of the metal-coated fabrics evaluated by them was FlecTron, a polyester fabric on which copper and nickel were deposited by electroless plating. The narrow-woven conductive fabrics and the metal-coated fabrics were washed in a commercial washing machine ten times and then dried in a dryer for 40 minutes. The washing cycle included a hot wash and a cold rinse cycle using bleach powder and detergent. After the wash and drying cycles, fabric samples were evaluated for electrical resistance using a four-point probe measurement system on an LCR meter. This procedure was carried out 50 times. Similarly, the conductive fabric samples were dry-cleaned ten times and the electrical resistance was measured. This procedure was also carried out 50 times. The conductive material was removed from the fabric surface when the metal-coated fabrics were washed or dry-cleaned. SEM images of the fabric samples also confirm that metal-coated fabrics are more abraded as compared to the fabrics made from foil-based yarns. The resistance values, as seen in Fig. 14.23, for conductive fabrics made from metal-coated yarns increased as more washing and dry-cleaning cycles were performed while those made from foil-based yarns had very little increase in resistance. The main reason for increase in resistance of conductive fabrics made from metal-coated yarns is that some of the coating material was abraded and removed on washing and dry cleaning.

Conductive threads used for making fabric-based circuits need to have certain characteristics depending on the process employed for making the fabric circuits. As mentioned earlier, for sewing and embroidery of conductive threads high levels of flexibility and tenacity are desired. The conductive threads have to withstand high levels of stresses that are involved in the

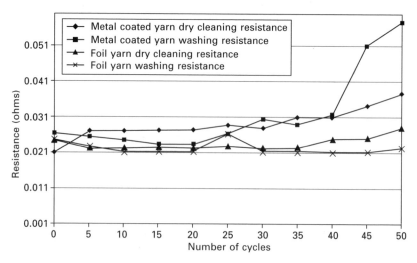

14.23 Increase in resistance as a function of wash cycles for woven conductive fabrics made from different conductive yarns.[28]

process of sewing or embroidery.[5] They also need to be flexible as they have to change shape a number of times to unwind from the bobbin and to form loops. Orth[5] evaluated mechanical properties of conductive threads that are used for producing embroidered circuit fabrics. The objective of her research was to observe if the conductive threads are permanently deformed if they are bent and loaded in a manner shown in Fig. 14.24. This testing procedure was labeled as the 'Curl test' by Orth[5] and involved running a yarn loop over a top metal point fixture and under two rotatable steel rods to provide tension, looping, and rotation of the thread by a certain distance. The yarn loops were loaded to around 85% of the breaking load of the yarn, reloaded to 50–90% of the original loading, and then removed and monitored for machine sewability. The yarns were considered machine sewable only if the loop diameter was less than 1/16 inch and the ratio of the loop diameter to the loop length (loop aspect ratio) was less than 1/3. It was determined that a 100% continuous stainless steel yarn was not suitable for sewing due to the large diameter of its loops and a loop aspect ratio close to 1. Similarly, a steel-polyester composite yarn containing 70% non-continuous steel fibers and 30% Kevlar was found to be unsuitable for sewing. A steel-polyester composite yarn, containing 20% non-continuous stainless steel fibers and 80% polyester fibers, and a rayon embroidery thread were considered machine sewable according to the criteria described above.

Slade *et al.*[69] carried out tests for tensile fatigue and tensile strength of fabric-based USB cables. The USB cables were developed by them by integrating 28 AWG twisted pair copper wires into narrow woven fabrics and

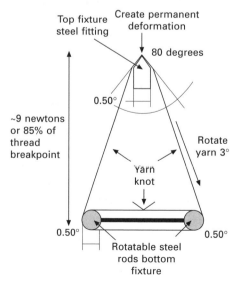

14.24 The 'Curl test' used for measuring the permanent deformation of conductive threads upon bending and loading.

passed group 6 signal integrity test requirements. The tensile fatigue and tensile strength of these USB cables were carried out on a strength tester using a load profile that varied from 0 to 250 pounds for 5,000 to 40,000 cycles. The test results showed that these USB cables passed signal integrity tests even after 40,000 cycles of loading and unloading and that the first version of the USB cable developed by Slade et al.[69] had an ultimate strength of nearly 2,000 lb. These USB cables were also tested for bending fatigue resistance under a load of 250 pounds and passed signal integrity testing after 35,000 cycles. Moreover, Slade et al.[69] carried out abrasion testing by draping a length of cable over a hexagonal steel bar such that one end of the cable was attached to an oscillating drum while the other was attached to a two-pound weight. The abrasion test indicated that the power conductor wires protruded from the USB cable fabric after 1,000 abrasion cycles and the shielded twisted pair wires after 4,000 cycles.

The mechanical properties of fabric-based electrical circuits are determined not only by the structure of the fabric but also by types of fibers, films or yarns forming these circuits. As the mechanical properties of the conducting and non-conducting elements of fabric-based circuits are very different, one has to ensure that the best electrical characteristics of fabric circuits are achieved without sacrificing the often desirable physical characteristics like conformability and flexibility. Slade et al.[70] developed a number of different kinds of textile-based USB cables and evaluated them for their bending stiffness. They developed a flexible USB cable with shieldings of braided conductive threads. In the USB cable called USB v3.0, the inner conductors were made up of insulated metal-clad textile-based yarns shielded by a flexible braided shield made from metal-clad textile-based yarns. The different USB cables were tested for their bending stiffness using a heart loop test (FED STD 191A-5200). This standard textile test procedure was slightly modified to measure bending stiffness of electronic circuit fabrics. The procedure involved making deflection measurements as weights were applied. In order to ensure accuracy, an optical comparator was employed to make the deflection measurements instead of a ruler. Load-deflection curves were obtained from these measurements. Their results indicated that the USB cable called USB v3.0 was less stiff as compared to the cable called USB v2.0, which had copper inner conductors with a foil shielding. The USB v3.0 cable was not only flexible but also exhibited a shielding effectiveness that complied with USB specifications for signal frequencies lying in the 30 MHz–1 GHz range.

14.6   Applications

14.6.1 Fabric-based antennas

Loop antennas, helical antennas, whip antennas, patch antennas, thin wire dipole antennas, cylindrical dipole antennas, and biconical antennas are some of the different kinds of antennas developed on rigid substrates.[71–78] Antennas are characterized on the basis of properties like radiation pattern, beam width, radiation intensity, directivity, gain, efficiency, impedance, range, quality factor, and loop inductance.[76] Currently, research is being carried out to develop antennas on flexible and conformable fabric substrates so that easily deployable large-sized antennas or arrays of antennas could be fabricated.[74–78]

Slade et al.[74] and Massey et al.[75] have described the development of fabric-based dual loop antennas. Antennas described by Slade et al.[74] were based on the Merenda antenna design developed by BAE systems. The design consists of two loops that are orthogonal to each other. These antennas produce a nearly spherical radiation pattern and are operational in the 30–88 MHz frequency range. The biggest advantage of employing these antennas is the decrease in consumed power as compared to conventional antenna technology. This dual loop antenna was formed on a fabric attachment of conductive fabric strips incorporated in a vest so that two orthogonally intersecting loops are formed.

Slade et al.[74] have also developed loop antennas that could be integrated into pockets or flaps of battledress uniforms. These antennas were employed for signal transmission or reception in the 225–400 MHz frequency range. Conventional loop antennas that are developed on chips and rigid circuit boards are small and have a very high electrical loss resistance. In order to have a lower value of loss resistance as compared to radiation resistance, large sized loops of the antenna are desired. Fabricating antennas on fabric substrates provides the ability to develop very large sized antennas as fabric circuits (with the antenna pattern) several meters long and can be easily developed on woven or non-woven substrates.

Massey[71] reported the development of patch and planar inverted F antennas on textiles.[71–73] These patch antennas could be integrated into several parts of a textile garment such as the shoulders or the back.[72] Planar inverted F antennas consist of a conducting patch, which is connected to the center conductor of a coaxial feed cable as shown in Fig. 14.25. The conducting patch has an inverted L section that is connected to the ground plane. The conducting patch and the ground planes were developed by electroless plating of a rip-stop nylon fabric with copper. It was reported that the patch could also be developed by coating of knitted fabrics. These planes were separated from each other by a thin foam layer. The ground plane in the patch antenna isolates the electromagnetic fields of the human body (wearing the textile-based antenna fabric) from those of the antenna. A coaxial cable feeds transmitter or receiver signals to the patch antenna by connecting its center conductor to the conducting fabric patch and its outer conductor to ground. Massey[71] has reported that efficiency of the fabric-based patch antennas can be improved by having better impedance matching between the feed coaxial

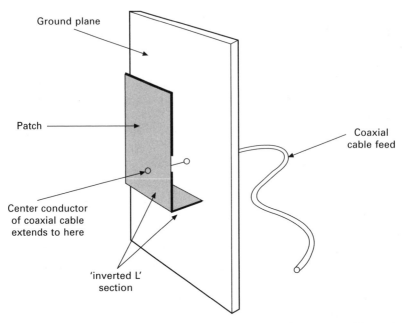

14.25 Planar inverted F antenna consisting of an inverted L section of a conducting patch, a ground plane, and a coaxial cable feed[71] © 2001 IEEE.

cable and the patch antenna. The distance between the point, where there is a short between the patch and the ground plane and where the coaxial cable joins the patch, was optimized to obtain the best value of impedance matching in these antennas.

14.6.2 Sensors

Patent number 6341504[79] describes the development of an inductive sensor that is based on integrating conductive threads into a stretch fabric in the form of loops. These loops change dimension when the fabric is stretched, thereby leading to a change in inductance of the loops. This inductive sensor fabric is used as a device to monitor the physiological characteristics of patients. A commercial product called life-shirt, based on an inductive sensor, has been developed by Vivometrics[80] and this shirt simultaneously monitors several parameters without affecting the patient.

De Rossi and his coworkers[81–85] at the University of Pisa have done some of the pioneering work in the area of strain sensing using textile substrates. Their work is relevant for many applications including motion sensing and physiological monitoring. De Rossi *et al.*[81] demonstrated remarkable strain and temperature sensing capability of conventional elastomeric fabrics coated with a thin layer of conducting polymer polypyrrole. They demonstrated the piezoresistive properties of these fabrics by fabricating a sensing glove. The change in resistance caused by fabric strain due to finger movement was measured. Lorussi et al.[83] applied the same concept to instrument a sensing sleeve that is capable of sensing body posture. They used printed carbon-filled rubber layers on fabric and created an interconnected array of sensors to determine the surface strain fields induced by movement. The composite material used in the sensor fabrication has an appropriate level of conductive particles embedded in an elastomeric binder. Resistivity of the composite is altered due to application of pressure.

Various sensing devices and systems compatible with textiles (or sometimes textile based) for continuous monitoring of physiological and behavioral parameters for health care have been proposed.[84–87] In all cases use of functionalized materials enabled design and fabrication of sensors and electrodes. Scilingo[84] reported using knitted fabrics made of stainless-steel-wrapped viscose yarn as electrodes to monitor electrocardiogram and electromyogram signals. Additionally they also reported using elastic fabrics coated with carbon-loaded rubber to monitor respiratory changes. Similar efforts to design electrodes and strain sensors for health monitoring have also been reported by Paradiso et al.[86]

Jones[88] has also developed flexible tactile sensor fabrics that show substantial resistance change when compressed. A composite material, made up of metal particles embedded in an elastomeric binder, is employed to form this sensor. On application of force to the composite material, metal particles move closer to each other, leading to greater quantum tunneling between the conducting particles. This leads to a substantial reduction in the measured resistance of the sensor fabrics.

Inaba et al.[89] have developed fabric-based tactile sensors that are distributed on the body of the robot. This sensor suit has 192 sensing regions and each of these works as a switch, i.e., it allows current to flow wherever pressure is applied on the garment in that region. The tactile sensor fabric consists of six layers of fabric as shown in Fig. 14.26. On applying pressure to the sensor fabric, an electrical contact is made between the switch pattern of one of the layers of the sensor and the ground layer, thereby completing the circuit. The electrical signal produced upon application of pressure to the sensor fabric is multiplexed on a video signal to produce sensor images.

Capacitive and resistive fabric-based weight sensors for automotive seats have been described by Gilbert et al.[90] Such weight sensors could be employed for controlled deployment of automotive airbags in accordance with weight of the occupant sitting on an automotive seat. The capacitive weight sensor comprises a foam layer between two conductive fabric layers. When pressure or weight is applied to the top fabric of this sensor assembly, the foam layer is compressed and the capacitance between the conductive fabrics is changed.

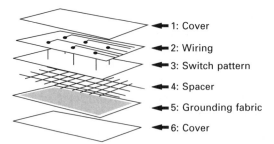

14.26 Structure of a tactile sensor made of an electrically conductive fabric[89] © 1996 IEEE.

Hence, change in the value of the capacitance between the two conductive fabrics is employed to detect the change in thickness of the foam layer between the conductive fabrics. Gilbert *et al.*[90] also describe the development of a resistive weight sensor. In this application, a compressible foam layer having a series of holes is placed between two conductive fabric layers. On application of force on the fabric, contact between the top and bottom conductive fabrics is made. This leads to a decrease in the measured resistance between the two conductive fabric layers, which can be measured.

Textile-based acoustic sensor arrays have been developed by Luthy *et al.*[91] and Grant *et al.*[92] for accurate tracking and location of targets via beam-forming and triangulation. These acoustic arrays act as a passive listening system where the sensitivity can be directed in a particular direction similar to the passive sonar system present on submarines. The acoustic arrays can be employed for detection of vehicles, aeroplanes, or troop movements. When using the time delay triangulation method the location of the origin of gunshots can be detected. Fabric-based acoustic arrays can be easily and quickly deployed, be camouflaged to suit the surroundings, and have the ability to flex and conform to the surroundings. They can also be integrated into tents and fabric vehicle covers. They fabric-based acoustic arrays (5 × 4 arrays) were developed by weaving (using a jacquard weaving machine) conductive elements, like twisted-pair yarns and copper wires, into large area textile substrates and subsequent attachment of microphones at certain locations.

14.7 Potential for the future

The field of electronic textiles has advanced to the point where increasingly more sophisticated products are being marketed. At the last Techtextil Exhibition (2005) in Frankfurt there were at least eight different companies showing their electronic textile products. However, the products are limited to a few areas of flexible input devices (keyboards, etc.), limited vital sign monitoring devices, and others. The vast potential of electrotextiles in areas of military, biomedical, homeland security/public safety, transportation, consumer products, and other applications are not yet explored.

One of the key elements in the integration of electronics in textiles is textile-based circuits and relevant interconnect technologies. Some of the challenges obviously are materials related. Development of better polymer-based conductive fibers is essential to this effort. Additionally, development of robust and reliable interconnect technologies compatible with polymer/ textile materials are key in the realization of the full potential of electrotextiles. There are obviously other challenges, conductive fibers and yarns in fabric-based circuits do not maintain fixed positions compared to rigid circuit boards or an integrated circuit. There are problems encountered in forming continuous ground planes between different layers of the multi-layered fabric circuits. Signal integrity problems such as variability of signal noise due to crosstalk or simultaneous switching associated with these must be addressed. Better fabric circuit design including computer-aided design tools to lay out the fabric circuits, as well as development of appropriate fault-tolerant yarn and fabric structures are needed.

Potential application of electrotextiles is numerous. It spans through all the segments where textile materials or products are used today and beyond. Most of the applications proposed in the literature seek to multi-functionalize existing textile products, e.g., heated undergarments. Other proposed applications, such as a flexible keyboard is entirely new. Textile-based circuits are particularly attractive for very large area applications, e.g., deployment of sensors, for its light weight and ease of deployment and removal.

Much of the ongoing research in the area of textile-based electrical circuits is likely to pave the way toward development of flexible and conformable integrated circuits entirely from textile-based materials. Development of fiber/ yarn based transistors would potentially allow switching in textile circuits. These transistors then may be configured to form inverters, gates and eventually, large-scale integrated circuits on fabrics. The eventual goal is to turn existing trends in hanging mostly silicon-based hard devices on textiles into truly conformable, rollable, washable, and sometimes wearable form.

14.8 Bibliography

1. H. K. Bakoglu, *Interconnections and Packaging for VLSI*, Addison Wesley, 1987.
2. R. K. Poon, *Computer Circuits Electrical Design*, Prentice Hall, 1994.
3. R. R. Tummala, and E. J. Rymaszewksi, *Microelectronics packaging Handbook*, Van Nostrand Reinhold, 1989.
4. J. D. Plummer, P. B. Griffin, and M. D. Deal, *Silicon VLSI Technology: Fundamentals, Practice, and Modeling*, Prentice Hall, 2000.
5. M. Orth, 'Defining Flexibility and Sewability in Conductive Yarns', *Proceedings of the Materials Research Society Conference Symposium D*, Boston, 736, D1.4, 2002.
6. M. Orth, and E. R. Post, 'Smart fabric, or 'wearable clothing', *Proc. of the First International Symposium on Wearable Computers*, 13–14, Oct 1997.

7. E. R. Post, M. Orth, P. R. Russo, and N. Gershenfeld, 'E-broidery: Design and fabrication of textile-based computing', *IBM Systems Journal*, 39(3 & 4), 840–860, 2000.
8. E. R. Post, and M. Orth, 'Smart Fabric, or Washable Computing', *Proc. First International Symposium on Wearable Computers*, Cambridge, MA, IEEE Computer Society, 167–168, 1997.
9. P. R. Lord, and M. H. Mohamed, *Weaving: Conversion of yarn to fabric*, Merrow Publishing Company Ltd, 1973.
10. H. W. Kipp, *Narrow fabric weaving*, Sauerlander, 1989.
11. A. Dhawan, T. K. Ghosh, A. M. Seyam, and J. F. Muth, 'Woven Fabric-based Electrical Circuits Part I: Evaluation of Interconnect Methods', *Textile Research Journal*, October, 2004.
12. A. Dhawan, T. K. Ghosh, A. M. Seyam, and J. F. Muth, 'Woven Fabric-based Electrical Circuits Part II: Yarn and fabric structures to reduce crosstalk noise in woven fabric-based circuits', *Textile Research Journal*, November, 2004.
13. A. Dhawan, T. K. Ghosh, A. M. Seyam, and J. F. Muth, 'Woven Fabric-based Electrical Circuits,' *Proc. of Textile Technology Forum, IFAI and the Textile Institute*, Charlotte, NC (2002).
14. A. Dhawan, 'Woven fabric-based electrical circuits', Masters Thesis, North Carolina State University, 2001.
15. A. Dhawan, T. K. Ghosh, A. M. Seyam, and J. F. Muth, 'Development of Woven fabric-based electrical circuits', *Proc. of the Materials Research Society Conference Symposium D*, Boston (2002).
16. A. Dhawan, T. Ghosh, A. Seyam, and J. Muth, 'Formation of Electrical Circuits in Woven Structures', *Proc. of the International Interactive Textiles for the Warrior Conference organized by the Soldier and Biological Chemical Command, NASA, & DARPA*, Cambridge, MA, July 9–11, 2002.
17. D. E. Piper, 'Woven low impedance electrical transmission cable and method', US Patent 4463323.
18. T. Kirstein, D. Cottet, J. Grzyb, G. Tröster, 'Textiles for Signal Transmission in Wearables', *Proc. of the Workshop on Modeling, Analysis and Middleware Support for Electronic Textiles (MAMSET)*, in conjunction with ASPLOS-X (Tenth International Conference on Architectural Support for Programming Languages and Operating Systems), San Jose, California, October 6, 2002.
19. D. Cottet, J. Grzyb, T. Kirstein, and G. Tröster, 'Electrical Characterization of Textile Transmission Lines', *IEEE Transactions on Advanced Packaging*, Vol. 26, No. 2, May 2003, pages 182–190.
20. I. Locher, T. Kirstein, and G. Tröster, 'Electronic Textiles', *Proc. of ICEWES Conference*, Cottbus, Germany, December 2002.
21. S. Jayaraman, S. Park, and R. Rajamanickam, 'Full-fashioned weaving process for production of a woven garment with intelligence capability', US Patent 6145551, 1998.
22. M. S. Lebby, K. E. Jachimowicz, and D. H. Hartman, 'Integrated communicative watch', US Patent 6158884, 1998.
23. M. Yoshida, M. Yoshikawa, Y. Yoshimoto, H. Wada, T. Suzuki, and S. Nakajima, 'Planar conductive piece with electrical anisotropy', US Patent 4929803, 1989.
24. J. Farringdon, A. J. Moore, N. Tilbury, J. Church, and P. D. Biemond, 'Wearable Sensor Badge & Sensor Jacket for Context Awareness', *Proc. 3rd International Symposium on Wearable Computers*, San Francisco, 107–113, 1999.

25. J. E. McIntyre, and P. N. Daniels, *Textile terms and definitions*, tenth edn, The Textile Institute, 1995.
26. A. R. DeAngelis, A. D. Child, and D. E. Green, 'Patterned conductive textiles', US Patent 5624736, 1995.
27. H. H. Kuhn, and W. C. Kimbrell Jr., 'Electrically conductive textile materials and method for making same', US Patent 4803096, 1987.
28. J. Slade, M. Agpaoa-Kraus, J. Bowman, A. Riecker, T. Tiano, C. Carey, and P. Wilson, 'Washing of Electrotextiles', *Proceedings of the Materials Research Society Conference Symposium D*, Boston, 736, D3.1, 2002.
29. D. P. Cadogan, and L. S. Shook, 'Manufacturing and Performance Assessments of Several Applications of Electrotextiles and Large-Area Flexible Circuits', *Proceedings of the Materials Research Society Conference Symposium D*, Boston (2002).
30. L. W. Adams Jr., M. W. Gilpatrick, and R. V. Gregory, 'Fabric having non-uniform electrical conductivity', US Patent 5316830, 1992.
31. R. V. Gregory, W. C. Kimbrell Jr., and M. E. Cuddihee, 'Electrically conductive polymer material having conductivity gradient', US Patent 5162135, 1990.
32. L. Rupprecht (ed.), *Conductive Polymers and Plastics in Industrial Applications*, 1999 William Andrew Publishing/Plastics Design Library.
33. R. S. Kohlman, and A. J. Epstein, in *Handbook of Conducting Polymers*, edited by T. Skothern, R. L. Elsenbaumer, and J. Reynolds (Marcel Dekker, Inc., New York) p. 85 (1997).
34. D. Baeriswyl, D. K. Campbell, and S. Mazumdar, in *Conjugated Conducting Polymers*, edited by H. G. Keiss, p. 7., Springer-Verlag, Berlin, 1992.
35. M. Aldissi, *Inherently Conducting Polymers: Processing, Fabrication, Applications, Limitations*. Noyes Data Corporation: Park Ridge, N. J., 1989.
36. B. Mattes, 'Electro-mechanical actuators based upon conducting polymer fibre and ionic liquid electrolytes', *Proc. International Interactive Textiles for the Warrior Conference*, Cambridge, MA, July 9–11, 2002.
37. B. R. Mattes, Hsing-LinWang, and Dali Yang, 'Electrically Conductive Polyaniline Fibers Prepared by Dry-Wet Spinning Techniques' in *Conductive Polymers and Plastics in Industrial Applications*, edited by Rupprecht, L. © 1999 William Andrew Publishing/Plastics Design Library.
38. E. Bonderover, and S. Wagner, 'A Woven Inverter Circuit For E-Textile Applications', *IEEE Electron Device Letters*, 25(5), May 2004, 295–297.
39. A. J. Epstein, J. Joo, C.-Y. Wu, A. Benatar, C. F. Faisst, Jr., J. Zegarski, and A. G. MacDiarmid, in *Intrinsically Conducting Polymers: An Emerging Technology*, M. Aldissi, ed. (Kluwer Academic Pubs., Netherlands, 1993) p. 165.
40. M. Gorlick, 'Electric suspenders: A fabric power bus and data network for wearable digital devices', in *Dig. Papers Int. Symp. Wearable Computers*, 1999, pp. 114–121.
41. A. Dhawan, T. K. Ghosh, and J. F. Muth, 'Incorporating Optical Fiber Sensors into Fabrics', *Proceedings of the Materials Research Society Conference Symposium D*, Boston (2002).
42. I. Locher, T. Kirstein, and G. Tröster, 'Routing Methods Adapted to e-Textiles,' *Proc. 37th International Symposium on Microelectronics* (IMAPS 2004), Long Beach CA, Nov. 14–18, 2004.
43. A. Dhawan, T. K. Ghosh, J. Muth, and A. Seyam, 'Methods and systems for selectively connecting and disconnecting conductors in a fabric', US Patent 6852395, 2002.
44. J. B. Lee, and V. Subramanian 'Weave Patterned Organic Transistors on Fiber for E-Textiles', *IEEE Transactions on Electron Devices*, 52(2) 2005, 269–275.

45. S. Jung, C. Lauterbach, and W. Weber, 'Integrated Microelectronics for Smart Textiles', *Proc. of Workshop on Modeling, Analysis and Middleware Support for Electronic Textiles* (MAMSET), in conjunction with ASPLOS-X (Tenth International Conference on Architectural Support for Programming Languages and Operating Systems), San Jose, October 6, 2002.
46. D. I. Lehn, C. W. Neely, K. Schoonover, T. L. Martin, and M. T. Jones, 'e-TAGs: e-Textile Attached Gadgets,' *Proceedings of the Communication Networks and Distributed Systems Modeling and Simulation Conference*, January 2004.
47. H. Gleskova, D. Shen, and R. Wagner R., 'Electrographic patterning of thin film circuits', US Patent No. 6080606, 2000.
48. H. Gleskova, S. Wagner, V. Gasparik, and P. Kovac, '150 degrees centigrade amorphous silicon thin-film transistor technology for polyimide substrates', *J. Electrochem. Soc.*, 148, G370–G374, 2001.
49. I-C. Cheng, and S. Wagner, 'Nanocrystalline silicon thin film transistors', *IEE Proc.- Circuits Devices Syst.*, Vol. 150, No. 4, August 2003, 339–344.
50. H. Gleskova, S. Wagner, and Z. Suo, 'a-Si:H TFTs made on polyimide foil by PECVD at 150 C', *Mater Res. Soc. Symp. Proc.*, 1998, 508, pp. 73–78.
51. H. Gleskova, and S. Wagner, 'Amorphous silicon thin-film transistors on compliant polyimide foil substrates', *IEEE Electron Device Lett.*, 1999, 20, (9), pp. 473–475.
52. J. B. Lee, and V. Subramanian, 'Organic transistors on fiber: a first step towards electronic textiles', *Electron Devices Meeting, 2003. IEDM '03 Technical Digest, 2003*, 8.3.1–8.3.4.
53. H. Gleskova, S. Wagner, V. Gasparik, and P. Kovac, 'Low-temperature silicon nitride for thin-film electronics on polyimide foil substrates', *Appl. Surf. Science*, 175–176, 12–16, 2001.
54. H. Gleskova, and S. Wagner, 'Electron mobility in amorphous-silicon thin-film transistors under compressive strain', *Appl. Phys. Lett.* 79, 3347–3349, 2001.
55. H. Gleskova, S. Wagner, W. Soboyejo, and Z. Suo, 'Electrical response of amorphous silicon thin-film transistors under compressive strain', *J. Appl. Phys.* 92, 6224–6229, 2002.
56. C. D. Dimitrakopoulos, and D. J. Mascaro, 'Organic thin-film transistors: a review of recent advances', *IBM J. Res. & Dev.*, 45(1), 2001, 11–27.
57. C. Dimitrakopoulos Purushothaman, S. J. Kymissis, A. Callegari, and J. M. Shaw, 'Low-Voltage Organic Transistors on Plastic Comprising High-Dielectric Constant Gate Insulators', *Science* 283, 822, 1999.
58. Y. Inoue, Y. Fujisaki, Y. Iino, H. Kikuchi, S. Tokito, and F. Sato, 'Low-Voltage Organic Thin Film Transistors on Flexible Plastic Substrates with Anodized Ta2O5 Gate Insulators', *Proceedings of the Materials Research Society Conference Symposium D*, Boston, 736, D4. 2, 2002.
59. M. Yoshida, S. Uemura, S. Hoshino, T. Kodzasa, S. Haraichi, and T. Kamata, 'High Performance Organic Field Effect Transistor With a Novel Top-and-Bottom Contact (TBC) Structure', *Proceedings of the Materials Research Society Conference Symposium D*, Boston (2002).
60. A. Bonfiglio, F. Mameli, O. Sanna, and L. Lutsen, 'An Organic FET Structure for Unconventional Substrates', *Proceedings of the Materials Research Society Conference Symposium D*, Boston (2002).
61. S. K. Park, J. I. Han, D. G. Moon, W. K. Kim, and Y. H. Kim Information Display Research Center, Korea Electronics Technology Institute, Pyungtaek, Kyunggi, Korea, 'High Performance Polymer Thin Film Transistors Array Printed on a Flexible

Polycarbonate Substrate', *Proceedings of the Materials Research Society Conference Symposium D*, Boston (2002).
62. B. Gnade, T. Akinwande, G. Parsons, S. Wagner, and R. Shashidhar, 'Active devices on fiber: the building blocks for electronic textiles', *Proc. International Interactive Textiles for the Warrior Conference*, Cambridge, MA (2002).
63. E. Bonderover, S. Wagner, and Z. Suo, 'Amorphous Silicon Thin Film Transistors on Kapton Fibers', *Proc. of the Materials Research Society Conference Symposium D*, Boston (2002).
64. W. R. Runyan and T. J. Shaffner, *Semiconductor measurements and instrumentation*, McGraw-Hill, New York, 1998.
65. D. K. Schroeder, *Semiconductor material and device characterization*, 2nd edn, 1998, John Wiley and Sons, New York.
66. K. Natarajan, A. Dhawan, A. M. Seyam, T. K. Ghosh, and J. F. Muth, 'Electrotextiles – Present and Future', *Proc. of Materials Research Society Conference Symposium D*, Boston, MA, Fall 2002.
67. J. F. Muth, E. Grant, A. Dhawan, A. Seyam, and T. Ghosh, 'Signal Propagation and Multiplexing challenges in Electronic Textiles', *Proc. of Materials Research Society Conference Symposium D*, Boston, MA, Fall 2002.
68. E. Ethridge, and D. Urban, 'ElectroTextiles – Technology to Applications', *Proc. Materials Research Society Conference Symposium D*, Boston, MA, Fall 2002.
69. J. Slade, P. Wilson, B. Farrell, J. Teverovsky, D. Thomson, J. Bowman, M. Agpaoa-Kraus, W. Horowitz, E. Tierney, and C. Winterhalter, 'Mechanical Testing of Electrotextile Cables and Connectors', *Proceedings of the Materials Research Society Conference Symposium D*, Boston, 736, D3.2, 2002.
70. J. Slade, P. Wilson, B. Farrell, J. Teverovsky, D. Thomson, J. Bowman, M. Agpaoa-Kraus, W. Horowitz, E. Tierney, and C. Winterhalter, 'Improving Electrotextile Wearability Using Stiffness Testing Methods', *Proceedings of the Materials Research Society Conference Symposium D*, Boston, 736, D3.3, 2002.
71. P. J. Massey, 'Mobile phone fabric antennas integrated within clothing', Eleventh International Conference on *Antennas and Propagation* (IEE Conf. Publ. No. 480), Volume 1, 2001, 344–347, 2001.
72. P. J. Massey, 'GSM fabric antenna for mobile phones integrated within clothing', *Antennas and Propagation Society, 2001 IEEE International Sym*, Volume 3, 452–455, 2001.
73. P. J. Massey, 'Fabric antennas for mobile telephony integrated within clothing', *London Communications Symposium*, London, 2000.
74. J. Slade, J. Teverovsky, B. Farrell, J. Bowman, M. Agpaoa-Kraus, P. Wilson, J. Pederson, J. Merenda, W. Horowitz, E. Tierney, and Carole Winterhalter, 'Textile Based Antennas', *Proceedings of the Materials Research Society Conference Symposium D*, Boston, 736, D3.9, 2002.
75. P. J. Massey, and P. Wilson, 'Fabric Antennas Integrated Within Clothing', *Proc. of the International Interactive Textiles for the Warrior Conference*, Cambridge, MA, July 9–11, 2002.
76. C. A. Balanis, '*Antenna Theory – Analysis and Design*', John Wiley & Sons, Inc., New York, 1996.
77. D.H. Werner, 'Exact integration procedure for vector potentials of thin circular loop antennas', *IEEE Transactions on Antennas and Propagation*, 44(2), 1996.
78. J. Slade, M. Agpaoa-Kraus, J. Bowman, A. Riecker, T. Tiano, C. Carey, and P. Wilson, 'Washing of Electrotextiles', *Proceedings of the Materials Research Society Conference Symposium D*, Boston (2002).

79. C. L. Istook, 'Composite elastic and wire fabric for physiological monitoring apparel', US Patent No. 6, 341, 504, (January 29, 2002).
80. website: http://www.vivometrics.com.
81. D. De Rossi, A. Della Santa, and A. Mazoldi, 'Dressware: wearable hardware', *Materials Science and Engineering* C 7–1999. 31–35.
82. E. P. Scilingo, F. Lorussi, A. Mazoldi, and D. De Rossi, 'Strain-Sensing Fabrics for Wearable Kinaesthetic-Like Systems', *IEEE Sensors Journal*, Vol. 3, No. 4, August 2003.
83. F. Lorussi, W. Rocchia, E. P. Scilingo, A. Tognetti, and D. De Rossi, 'Wearable, Redundant Fabric-based Sensor Arrays for Reconstruction of Body Segment Posture', *IEEE Sensors Journal*, Vol. 4, No. 6, December 2004.
84. E. P. Scilingo, A. Gemignani, R. Paradiso, N. Taccini, B. Ghelarducci, and D. De Rossi, 'Performance Evaluation of Sensing Fabrics for Monitoring Physiological and Biomechanical Variables', *IEEE Transactions on Information Technology in Biomedicine*, Vol. 9, No. 3, September 2005.
85. M. Pacelli, R. Paradiso, G. Anerdi, S. Ceccarini, M. Ghignoli, F. Lorussi, E. P. Scilingo, D. De Rossi, A. Gemignani, and B. Ghelarducci, 'Sensing threads and fabrics for monitoring body kinematic and vital signs', in *Fibers and Textiles for the Future*, M. Isosomppi and R. Salonen, eds, 2001, pp. 55–63.
86. R. Paradiso, 'Wearable health care system', in *Proc. 4th Int. IEEE EMBS Special Topic Conf. Inf. Technol. Applic. Biomed.*, Birmingham, UK, Apr. 2003, pp. 283–286.
87. L. Van Langenhove, C. Hertleer, M. Catrysse, R. Puers, H. Van Egmond, and D. Matthijs, 'Smart textiles', in *Wearable Health Systems for Personalised Health Management*, A. Lymberis and D. de Rossi, eds Amsterdam, The Netherlands: IOS Press, 2004, pp. 344–351.
88. D. Jones, 'Interactive fabrics using SOFTswitch technology', *Proc. of the International Interactive Textiles for the Warrior Conference* organized by the Soldier and Biological Chemical Command, NASA, & DARPA, Cambridge, MA, 2002.
89. M. Inaba, Y. Hoshino, K. Nagasaka, T. Ninomiya, S. Kagami, and H. Inoue, 'A Full-Body Tactile Sensor Suit Using Electrically Conductive Fabric and Strings', *Proc. of the IEEE/RSJ International Conference on Intelligent Robots and Systems*, 450–457, 1996.
90. S. C. Gilbert, and J. E. Boyce, 'Conductive fabric sensor for vehicle seats', US Patent 5878620, 1997.
91. K. Luthy, J. Braly, L. Mattos, E. Grant, J. F. Muth, A. Dhawan, T. K, Ghosh, A. M. Seyam, and K. Natarajan, 'An Acoustic Array as an example of a large-scale electronic fabric', *Proc. of Materials Research Society Conference Symposium D*, Boston, MA, Fall 2002.
92. E. Grant, K. A. Luthy, J. F. Muth, L. S. Mattos, J. C. Braly, A. Seyam, T. Ghosh, A. Dhawan, and K. Natarajan, 'Developing Portable Acoustic Arrays on a Large-Scale E-Textile Substrate', *Proc. of the INTEDEC conference – fibrous assemblies at the design and engineering interface*, Edinburgh, UK, 2003.

15
Stability enhancement of polypyrrole coated textiles

MYS LEUNG, J TSANG, X-M TAO, C-W M YUEN and Y LI, The Hong Kong Polytechnic University, Hong Kong

15.1 Introduction

Polypyrrole (Ppy) has been widely developed for actuators, chemical sensors, biomaterials, batteries, gas separation films, etc., owing to its easy preparation, good conductivity and non-toxicity.[1-7] When it forms composites with a number of fabrics, the obtained electrical conductivity can also be used in the field of electromagnetic insulation and microwave absorbance, etc. This is mainly due to the combination of the mechanical properties of the flexible fabrics, electrical and microwave properties, and bio-compatibility of the Ppy coatings.[8-12] In recent years, it has been found that the conductivity of the Ppy-coated fabrics is sensitive to strain deformation leading to the development of different sensing devices to measure and control the various movements of the human body for training, shape analysis or rehabilitation.[10,13-16] However, Ppy-coated textiles exhibit the problem of relatively low sensitivity and stability. The Ppy-coated Nylon-spandex shows a strain sensitivity $\left[\frac{\Delta R}{\varepsilon R_o}\right]$ of only about 2 for a deformation of 50%[13] and 3 for the Ppy-coated PET/Spandex.[10] They are also sensitive to external stimuli during applications. The stabilisation enhancement, without a great sacrifice of initial conductivity and, more importantly, the strain sensitivity, becomes a critical analysis.

Polypyrroles are formed by the oxidation of pyrrole or substituted pyrrole monomers as shown in Fig. 15.1. Electrical conductivity of Ppys coating or film involves the movement of positively charged carriers or electrons along their polymer chains or hopping of carriers between chains. The conductivity of Ppy can range up to 10^2 Scm^{-1}. With the inherently versatile molecular structures, Ppys are capable of undergoing many interactions. Oxidation of polypyrrole can be carried out either by (i) electro-polymerisation at a conductive substrate, i.e., electrode with an external potential applied, (ii) chemical polymerisation in solution or (iii) vapour deposition by means of a chemical oxidant.[17] In order to obtain a wide variety of chemical and electrical properties of polypyrrole, the control of fabrication condition and selection

15.1 Polymerisation of polypyrrole.

15.2 Polymerisation process of pyrrole and FeCl$_3$.[18]

after treatments is critical. The mechanical properties of Ppy vary widely from strong, tenacious materials to extremely brittle ones, depending on the chemical structure, fabrication condition and the application environment. The polymerisation process of polypyrrole based on the oxidative transition of metal ions, ferric chloride (FeCl$_3$), is summarised in Fig. 15.2.

15.1.1 Electrical properties – conductivity

The electrical properties of conducting polymer coated textiles are important as they can help determine the ability of polymers to transfer electrical signals along the conductive structure, store information, trigger responses, convert and store energy. Electrical conductivity in Ppy films involves movement of positively charged carriers or electrons along polymer chains and hopping of carriers between chains. The experimental parameters encountered during synthesis have an influence on the polymer conductivity. In electrochemical polymerisation, the solvent, counterion and monomers used during synthesis influence the electrical properties of the resulting polymer. By employing different metallic salts, such as FeCl$_3$, Fe(NO$_3$)$_3$, Fe(ClO$_4$)$_3$, Fe$_2$(SO$_4$)$_3$, etc., in polypyrrole polymerisation, the conductivity of polymers ranging from 10^{-5} to 200 Scm^{-1} can be obtained.[18] However, the quality of the Ppy conductive coating may be reduced by the presence of nucleophiles during polymerisation. Water itself is a nucleophile, attacking the pyrrole ring to form carboxyl groups that break down the polymer chain and lead to a decrease in conductivity and mechanical properties.[23] The presence of a small amount of water affects the rate of polymer growth and its resultant conductivity.

15.1.2 Chemical properties

The chemical properties of polymer structures determine the response of the conductive textiles to the external stimuli. Ppys are inherently versatile molecular structures, capable of undergoing many interactions such as Ppy/Cl, Ppy/dodecyl sulphate and Ppy/polyacrylic acid.[19] They are particularly strong anion exchangers that are capable of undergoing hydrophobic interactions. Other ion exchange groups such as carboxyl groups or self-doping sulfonate groups can be added to the pyrrole monomers prior to polymerisation. As polymerisation of polypyrrole can be carried out in aqueous solutions, it enables a wide range of counterions to be incorporated into the structure. Since the crystallinity of polypyrrole chains in all forms of fabrication are low,[20] polymerisation of polypyrrole mainly occurs at the α-carbons. This is the ideal Ppy chain formation with the spatial orientation of pyrrole moieties along the main chain, while the pyrrole units alternate in a planar arrangement to form, ideally α-α couple polypyrrole. However, magic angle spinning NMR and IR techniques indicate the involvement of β-carbons in the bonding of most polymers, which denotes the presence of branching.[18] Diffraction patterns of Ppy chains are not oriented parallel to each other as those of poly (β,β'-dimethylpyrrole) but aligned at random.[21] These variations in mechanical, chemical and electrical properties of the Ppy-coated fabrics as well as their strain sensitivity are governed by structural chain orientation but can be controlled by fabrication conditions. As reported by Oh et al.[13] the Ppy-coated Nylon-spandex showed a strain sensitivity $\left[\frac{\Delta R}{\varepsilon R_o}\right]$ of only 2 for strain deformation of 50% and Kim et al.[10] developed the Ppy-coated PET/Spandex strain sensors with strain sensitivity up to 3 for a large strain deformation of 50%. However, by modifying the fabrication conditions and a combination of sequences of after-treatments, we can develop the e-textile strain sensors with strain sensitivity up to ~120 or higher.[22]

15.1.3 Mechanical properties

The mechanical properties of Ppy coating on textiles vary widely from strong, tenacious materials to extremely brittle ones.[23] Mechanical properties including tensile strength, and elongation to break are affected by the polymerisation conditions. It was reported that there was a decrease in tensile strength and breaking elongation as the polymerisation temperature was decreased.[17] Apart from the solvent used during polymerisation and before chemical vapour deposition, different types of counterion used as well as the wet pick up of the solutions containing oxidising have an influence on the molecular weight of the Ppy films, degree of cross-linking, molecular order, conductivity and resistance to environmental effects.[64]

The aim of this chapter is to review the degradation of the Polypyrrole coating due to both intrinsic and extrinsic changes as mentioned in Section 15.2. The stabilisation methods developed by others through pre-ageing treatments for extrinsic instabilities and the control of fabrication for intrinsic instabilities are discussed in Section 15.3. Extrinsic pre-ageing treatment includes acid and alkaline treatment, as well as annealing processes[24,25] while intrinsic modifications of the polypyrrole films includes the applications of dopants or metal ions during the fabrication process or voltage applications.[26,46] In Section 15.4, a method of producing Ppy-coated textiles with high strain sensitivity and good stability is discussed. The newly developed method includes the combination of new fabrication processes such as the method of printing and chemical deposition, large anions dopants application and low-temperature after-treatments including purification and annealing processes.

15.2 Conductivity changes of polypyrrole films on textiles

The electrical conductivity of Ppy is ascribed to electrons hopping along and across polymer chains with conjugated double bonds.[26] The higher electrical conductivity of Ppy is due to the presence of a larger number of positive charge carriers. This is related to the doping or oxidation level of the polymer. In general, the presence of higher positively charged ions in the Ppy chains will result in more polarons being available, together with longer and parallel polymer chains which contribute to a higher level of conductivity.

As the mobility charges or number of the charges varies, the conductivity of polypyrrole also changes. The theoretical mechanism of the mobility of charges in Ppy is governed by the conjugation length of Ppy polymer chains and the distances between inter-chains.[27] Basically, the conductivity of the e-textile sensors and their stability are influenced by both or either intrinsic or extrinsic factors in nature.[28]

15.2.1 Intrinsic instability

Intrinsic instability is mainly dealt with by conformation changes in the polymer backbone activated during the thermal annealing process. The charged sites may be de-stabilised by oxidation which completely reverses the doping or undoping process. The oxidation of Ppy is more vigorous at high temperature, typically at temperatures higher than 230 °C.[29] This kind of effect is thermodynamic in origin and causes irreversible chemical reaction between the charged sites of the polymer chain and the dopant counterion or the π-system of an adjacent chain. Instability caused by thermal ageing may be due to the loss of structural order of the polymer backbone caused by oxygen attack on the counterion to the Ppy backbone. This irreversible chemical reaction introduces sp^3 defects along the polymer chain causing the irreversible breakage of the conjugation. As a result, the resultant polymer consists of a variety of conjugation lengths which will greatly reduce its conductivity.[30–33] Furthermore, other mechanical structures of the Ppy films including both thickness and surface morphology are also critical. In general, the Ppy films with higher conjugation lengths and doping level in the polymer backbone are characterised by higher molar mass and higher thickness. This type of film benefits by reducing the permeability of the Ppy films to external stimuli like oxygen and water. Experimental investigations have identified the proper fabrication methods including dopant applications, types of solvent, applied voltage, concentration of chemicals, and reaction temperature that can control the mechanical structures of the Ppy.[18] Longer deposition time, lower applied potential and reaction temperature, and higher concentrations of monomer and electrolyte are favourable for the conductivity and stability.[18]

15.2.2 Extrinsic instability

The instability of Ppy in ambient conditions is due to the reaction of the polymer backbone with oxygen, water or damaging radiation that lead to an irreversible loss of conjugation and conductivity.[24,26,29,34–38] This attack comes from the external environment, and is called 'extrinsic instability'.[30,32,33,39] Since oxygen and hydrogen will react with nitrogen and carbon in the Ppy polymer chain the aromatic ring will be opened and shortened. As a result, degradation of conductivity will occur.[26,40–41] The ideal process for polypyrrole polymer is the oxidation of pyrrole units, the α-α coupling during polymerisation. However, the reactions of Ppy with oxygen and water may attack the α and β positions of the pyrrole unit. There are two possibilities for attack. Firstly, the hydroxide ion (OH^-) attacking the β-position on the pyrrole unit may be considered as one of the most important solution degradation reactions. The oxidised polymer chains may undergo nucleophilic attack by the oxygen of the surrounding polymerisation media giving C-O bonds. Oxidation and phototropic processes may transform C-O bonds into C=O bonds.[42]

Figure 15.3 shows a possible degradation mechanism[36] indicating the formation of a carbonyl group and the shortening of the polymer chain. This chemical reaction between polymer chain and oxygen, water or damaging

15.3 Possible reaction ways for pyrrole-pyrrole linkages in Ppy generation.[18]

radiation causes the formation of carbonyl, ester and hydroxyl groups, leading to the reduction of positive charge carriers on the polymer chain.[39] Secondly, the hydroxide ion may attack the α-position of the pyrrole unit that disrupts conjugation.[43] This nucleophilic attack and pyrrole ring opening mechanism have been proposed by Otero *et al.*[44] to explain the degradation of Ppy at anodic potentials in aqueous solutions. The changes of chemical structure of Ppy in air[36] and the interaction in water[26] are shown in Fig. 15.4(b) and 15.4(c) respectively. The degradation due to the diffusion of oxygen into the conductive film and the reaction of oxygen with Ppy backbone leading to the loss of conductivity were verified by the XPS examination.[12] The amount of carbonyl functional groups, i.e., oxygen content of the Ppy-coated fabrics after ageing, indicated a significant increase. The rate of degradation is related to the oxygen concentration within the Ppy film and relative humidity. The difference between the initial conductivity ($\Delta\sigma$) versus time ($t^{1/2}$) was initially linear, and the rate of oxygen diffusion into the film would limit the rate of decay.[26,34,36,45]

Other than water or oxygen, polymer can be attacked by alkaline solutions during finishing treatments or laundering can promote polymer ring oxidation and ion exchange between Ppy and OH^- at the initial stage.[24,45] Prolonged base treatment can give rise to the formation of covalent bonds such as C-O and C=O that may create bonding between the hydroxyl ion and the pyrrole structure, and thus shorten the conjugation length of the polymer. This reaction causes the instability of the Ppy films at the initial stage. Although the Ppy films can be stabilised after reaching equilibrium the conductivity will be lost.

15.4 Mechanism for the attack of Ppy with (a) presence of oxygen, (b) air and (c) water.[26]

15.3 Stabilisation of the Ppy

In this section, the review of the stabilisation methods is based on the principles of intrinsic and extrinsic instabilities of polypyrrole films.

15.3.1 Dopant applications

There are two main ways to explain the stabilisation of Ppy films by dopant applications. Firstly, the carbonium chains doped with counter-anions balance the charge on the Ppy films and thus reduce the opportunities for oxygen to attack the C or N in the Ppy chain. Secondly, the surface structure of Ppy films is modified by dopants to produce a denser and smoother surface morphology so as to inhibit oxygen attacking the Ppy chains. Ppy conductive films with more porous morphology can be obtained through the use of smaller size dopants which contributes to the relative instability of the films. In contrast, the relative stability of the Ppy film can result from the formation of a smooth surface using large-sized dopants such as anthraquinone-2-sulfonic acid (AQSA)[47] or benzenesulfonate and perchlorate[31] and p-toluene sulphonate, p-chlorobenzene sulphonate.[61] These dopants with aromatic anions exhibit better stability with less easy detachment from the origin polymer structure as well as fewer attacks from external stimuli.[61] The stability enhancement induced by doping with large organic anions such as dodecylbenzenesulfonate (DBS)[48] is mainly due to the localisation of the large anions in the Ppy polymer chains as their mobility is greatly reduced once the Ppy films are doped. The large anions will not easily be affected by the external stimuli during the redox process and can prevent the doping and un-doping of Ppy films.

15.3.2 Acid and base treatments

Base treatment of the Ppy films renovates the (OH$^-$) blocking of the reactive sites and restrains the reactions of the Ppy film with oxygen and water. However, base treatment of the Ppy films may result in the de-protonation of the polypyrrole chains leading to the loss of initial conductivity (>40%).[24,37] A longer base treatment process enhances the stability of the polymer but at the expense of conductivity.[24] It is, however, important to note that the initial base treatment has already resulted in the significant loss of conductivity. On the other hand, doping Ppy films with various protonic acids such as HCl and sulfonic acid can provide stability without greatly reducing the initial conductivity.[25] The XPS characterisation for the surface chemical composition of Ppy film indicated that both acid and base treatments resulted in the elimination of reactive sites.[37] An increase in the C-Ox species was observed from the Ppy films after the acid and base treatments which result in the elimination of reactive sites in the Ppy structure. Munstedt[31] reported an increase in long-term stability for Ppy films by treating with sulfuric acid (H_2SO_4) and sodium hydroxide (NaOH) at the elevated temperature (80 °C and 140 °C). The enhanced stability was attributed to the decrease in permeability of the Ppy films to oxygen and water which could retain up to 70% of its initial conductivity after 100 days at 140 °C.

15.3.3 Thermal annealing

The annealing process causes the restructuring of the Ppy polymers which may be partly due to the evaporation of residual solvent. Solvent plays an important role in the specific salvation process by stabilising the π segments along the polymer chain. Reordering within the Ppy film based on thermal annealing improves the local ordering within the Ppy films and decreases the defect concentration.[49] Large numbers of voids and pores were observed in the micrographs with increasing annealing time, which hindered the path of conducting channels and thus decreased the conductivity.[38] Scanning Electron Microscopy (SEM) micrographs performed by Singh et al.[50] confined the cauliflower structure of the poly (N-methylpyrrole-pyrrole) film which had totally disappeared after annealing for 29 hours. After that no change in the conductivity was observed. Hence, reordering of the polymeric chain is a crucial factor in the determination of conductivity and stability.

15.3.4 Voltage applied

The degree of conductivity and the stability enhancement may be related to the nature of the counter-anions doped in the Ppy films. Onyang and Li investigated the effect of voltage on the conductivity of Ppy[46] and found that there was an increase of 65% and 47% for conductivity of the film doped with TsO^- and NO_3^- respectively after applying a higher voltage. The Ppy(TsO^-) film remained stable while the Ppy(NO_3^-) dropped back to its original value quickly after the treatments. The doped counter-anions are used for balancing the charges on the chains in the Ppy films. When a voltage is applied to the films, an electric field will be set up between the two electrodes. The counter-anions doped in the film will tend to move under the electric field. When reaching a certain voltage level, the counter-anions move to a quasi-stable, higher-energy position. At this time, the chain structure of the film changes and its conductivity increases. After the voltage treatment, the quasi-stable structure may relax. The relaxation of the TsO^- counter-anion is less easy when compared with NO_3^- and BF_4^-. After the voltage treatment for a longer period, the Ppy(TsO^-) film remains a higher conductivity, as the ion size for TsO^- is larger than NO_3^- and BF_4^-. In addition, the cyclic redox reaction of Ppy polymer structure by alternative potential applications may lead to the conformational changes of polymer structure from *trans* to gauche orientation and shorten the conjugation length.[51] However equilibrium can be reached at 1200 cycles of the redox reaction.[51]

15.3.5 Structural modification of the Ppy film

The Ppy film thickness and layer structure may contribute to the stability of conductivity.[31] As the outer layers of the Ppy films are oxidised, they create a barrier to oxygen diffusion from outer boundaries to the inside structure. The oxygen permeability of the material is decreased and the rate of conductivity decay is much reduced. Ultimately, the stage is reached where the oxidised outer skin protects the pristine core from further oxidation and this is called 'self-stabilisation'. It was found that when the Ppy films had greater thickness, up to 100 μm, the conductive film could withstand thermal stability up to 200 °C.[52] Thicker conductive films can be produced through longer deposition time, lower applied potential for electrochemical polymerisation, higher reaction temperature, and higher concentrations of monomer and electrolyte.

15.3.6 Addition of metal salts

In the early 20th century, Liu and Hwang[26] investigated the modification of Ppy in the electronic configuration by metal ions. Cu(I) was used to enhance the conductivity and stability of Ppy. The reason arises from the deposition of Cu(I) onto Ppy which is mainly used to stabilise the N^+ ions. Consequently, the original N^+ is stabilised into the Cu-Ppy complex. The Cu-Ppy complex is more stable to external stimuli and less porous morphology films will be produced. A more porous structure and larger specific surface area facilitate the attacking gas diffusing into the films. Therefore Cu-modified Ppy may contribute to the relative stability based on the viewpoint of the decrease in permeability of the Ppy film to water by adding large-sized metal ions into the film.

15.4 Experimental results of stability enhancement

The conductive fabrics prepared by the solution polymerisation of pyrrole on the surface of fabrics show a low strain sensitivity of not more than 3, although the deformation is as large as 50%.[6,11] This phenomenon is related to the thickness of the conductive coating on the textile. The conductive coating is composed of multi-layers of Ppy deposited on the fabric surface during the solution polymerisation. When the coated fabric is elongated, the multi-layers of Ppy may not change in the same way as the substrate. Unless it is under very large deformation the relative position and contact of the conductive Ppy coating may not change much with the elongation of the fabrics. This leads to a small change in conductivity resulting in low strain sensitivity. For this reason, the method of using printing and chemical vapour deposition (CVD) to obtain a relatively much thinner coating of Ppy on the surface of a Tactal (83%)/Lycra (17%) fabric was employed. It was probably due to the existence of elastic Lycra that the prepared composites showed good reversibility under large-strain deformation of up to 100%.

15.4.1 Printing and chemical vapour deposition

Polypyrrole-coated fabrics were prepared by combining the printing and chemical vapour deposition. Fabric composed of 83% Tactel and 17% lycra was first printed with the solution containing sodium dodecyl benzene sulfonate (DBS) and $FeCl_3$ in ethanol solvent, and the wet take-up after each immersion was controlled. The fabrics and pyrrole monomer were transferred to a desiccator for the chemical vapour deposition. The vapour phase polymerisation of polypyrrole proceeded on the surface of the fabrics under vacuum for 24 hours. The black fabrics so obtained were washed with de-ionized water and ethanol, respectively, and then purified at 40 °C in a vacuum. The strain-stress properties of the Ppy-coated fabrics were obtained using an Instron testing Instrument (Model 4466) under the standard testing conditions of 25 °C and 65% RH. The fabrics were repeatedly stretched and relaxed for ten cycles with the maximum extension up to 12.5 mm (50% deformation) in each cycle. The conductivity change of the sensing fabrics in both the stretched and relaxed states was recorded using a digital multi-meter (Keithley Model 2010) to investigate its strain sensitivity. SEM images of the Ppy-coated fabrics were obtained using a scanning electron microscope (Leica stereoscan 440). ATR-FTIR spectra were determined by means of a Bio-Rad FTS-6000 spectrometer.

The typical curves of conductivity change versus the strain of Ppy-coated fabrics prepared by printing and CVD as well as solution polymerisation during the ten cycles of elongation and relaxation measurements are illustrated in Figs 15.5(a) and 15.5(b), respectively. The resistance of the sensor prepared by the combination of printing and CVD increases sharply with extension up to a large deformation of 50%. The sensing curves are almost the same during the ten cycles of the test, indicating that the sensor has very good repeatability. The strain sensitivity is $\left[\frac{\Delta R}{\varepsilon R_o}\right]$, where ΔR is the resistance change of the fabric under extension and R_0 is the original resistance, i.e., the resistance of the fabric in the relaxed state. It can be calculated from Fig. 15.5 that the strain sensitivity of the sensor is as high as about 20. Based on the developed printing and chemical vapour deposition process as shown in Fig 15.6, the fabric sensors can be further enhanced through annealing treatment and dopant application. Higher strain sensitivity up to 120 can also be obtained

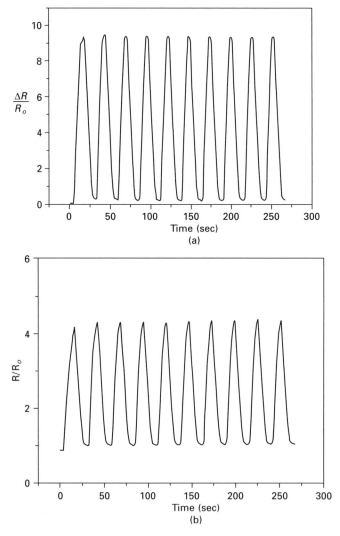

15.5 The conductivity change versus the elongation of Ppy-coated fabrics prepared by (a) printing and CVD and (b) solution polymerisation (deformation = 50%).

after further stabilisation processes including purification and low temperature annealing. In contrast, the sensor prepared by solution polymerisation also shows repeatable sensing curves during the cyclic tests, but its strain sensitivity is as low as about 6.

In addition to high sensitivity, good environmental stability is especially important for its possible application in smart garment, rehabilitation and biomedical fields. It was reported that a smooth and dense layer of Ppy film

Stability enhancement of polypyrrole coated textiles 295

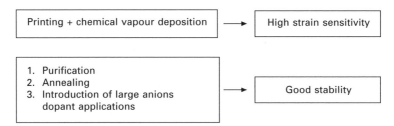

15.6 Strategy for preparing a flexible strain sensor from the Ppy-coated fabrics with both high sensitivity and good stability.

was more resistant to the attack of oxygen and water molecules, thus enhancing its stability.[7,15–16] Figures 15.7(a) and (b) show the scanning electronic micrograph of the film of Ppy-coated fabrics prepared by printing plus CVD and by solution polymerisation, respectively. It is revealed that the fabrics coated with Ppy formed by CVD exhibits a much smoother surface, while the coated fabric prepared by the solution polymerisation is featured with a rough surface. The smooth morphology further explains the advantages in

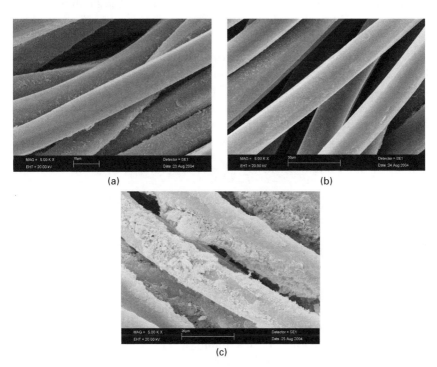

15.7 SEM images of Ppy-coated fabrics prepared by (a) printing and CVD without DBS doping, (b) by printing + CVD with DBS and (c) by solution polymerisation.

strain sensitivity. Strain sensitivity of the deposited conductive film can further be enhanced by the combination of purification and annealing processes. Annealing is a method used for improving the stability of the conductive fabrics. R. Singh *et al.* investigated the effect of heat annealing on the conductivity and surface structure of Ppy, and found that annealing would lead to the removal of residual solvent as well as the restructuring and reordering of the Ppy chain, thus stabilising the conductivity.[53] However, the most recommended annealing processes[50] involve a very high temperature, usually over 100 °C. This is not beneficial to textile substrates containing Lycra because it may degrade the quality of Ppy-coated textile materials. Therefore, low-temperature annealing processes for Ppy coated fabric containing Spandex materials are used.

15.4.2 Purification

Purification of samples under vacuum drying at around 40 °C shows a significant improvement in strain sensitivity. The vacuum drying process promotes the efficiency of removal of impurities or residual solvents, allowing the conductive coating to respond to strain deformation. The strain sensitivity $\left(\frac{\Delta R}{\varepsilon R_o}\right)$ approximately doubled the values from the samples without vacuum drying (~20) to those with vacuum drying about (~60). Figure 15.8(a) shows the diagram of the resistance change of the fabric sensors without any treatment in response to large strain deformations up to 50% extension. Under ten consecutive cyclic extensions, the change of resistance in the extension of Ppy-coated textiles indicates an excellent linearity and repeatability. Except for the first cycles of extension, the responses of all other cycles are nearly identical with the strain sensitivity $\left(\frac{\Delta R}{\varepsilon R_o}\right)$ approaching 20. Under the same dimension of resistance magnitude, it is very clear that vacuum purification treatment significantly improves the strain sensitivity of the Ppy-coated fabric As shown in Fig. 15.8(b). However, in all cases the linearity of the sensing curves remains excellent except for a slight variation in the sensitivity at peak strain (near 50%).

15.4.3 Annealing

In order to avoid destroying the structure and stability of the fabrics, annealing was carried out by heating at a maximum of 60 °C under vacuum for ~40 hours. Figure 15.8 shows the curves of conductivity change of a strain sensor under cyclic extension before and after annealing. It is obvious that the annealing process increases the strain sensitivity and improves the repeatability of sensing curves during the repeated cyclic extension, which indicates an improved stability. Although annealing at 60 °C will increase the initial conductivity, the relative effect is much lower when compared with high-temperature annealing processes.[22] More importantly, the strain sensitivity is also enhanced. The performance of the e-textile sensor can further enhance the strain sensitivity $\left(\dfrac{\Delta R}{\varepsilon R_o}\right)$ of the Ppy-coated e-sensor up to (~120) by the combination of low-temperature purification and vacuum heat treatment under

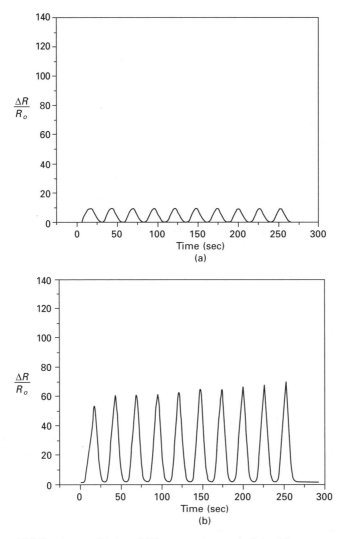

15.8 Strain sensitivity of PPy-coated sample (a) without vacuum drying and heat treatment, (b) with vacuum drying, (c) with vacuum drying and heat treatment under nitrogen atmosphere, (d) with vacuum drying and heat treatment under vacuum.

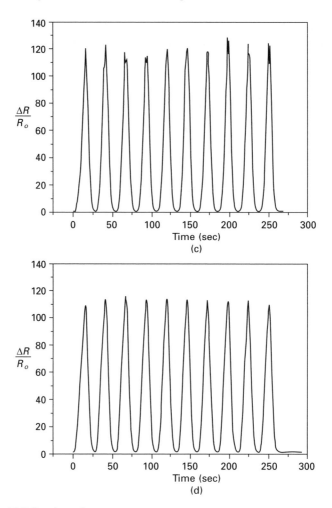

15.8 Continued

nitrogen medium at 60 °C for 40 hours. Figure 15.8(c) shows the great improvement of strain sensitivity of the Ppy-coated fabric sensor under heat treatment in nitrogen medium. The heat treatment process at a temperature of 60 °C significantly increases the strain sensitivity but without affecting the linearity of the textile sensor. Similarly, only light variations of resistance change at peak strain level are observed.

15.4.4 Introduction of large anions

As mentioned in Section 15.3, dopant can effectively affect the stability and conductivity of Ppy.[53–56] Ppy film doped with larger anionic species such as dodecylsulfate and dodecylbenzenesulfate displayed smoother morphology than the cauliflower-like appearance of Ppy/Cl⁻ or Ppy/ClO$_4^-$ film.[50] Ppy doped with dodecyl benzene sulfonate anion shows more compact morphology and higher conductivity than Ppy/Cl⁻. The introduction of large-sized anions in the Ppy film will result in enhancing the thermal and humidity stability. The large size dopants are difficult to be de-doped even at high temperature and humidity. Furthermore, the Ppy/Cl⁻ system was reported to have worse stability than those doped with organic acid. It was also found that the Ppy-coated film prepared by both the solution polymerisation or printing and chemical vapour deposition with only Cl⁻ as the counterion would increase the resistance by more than tenfold after storage for several months. The results agreed with those obtained from Scilingo *et al.* that the conductive fabrics prepared by solution polymerisation with inorganic ClO$_4^-$ as the counterion faced the problem of rapid decrease in conductivity.[57]

Considering the fact that the doping agent should be of low toxicity for practical application, as a result, sodium dodecyl benzene sulfonate, a bulky aromatic sulfonate, was consequently used as the additional doping agent for the preparation of conductive fabrics so as to provide good stability. This bulky anion has a long aliphatic chain and its large dopant ion is more stable than the small Cl⁻ ion, which has been extensively investigated for the preparation of Ppy film with potential applications in movable pixels or micro-actuators.[58-60,62,64] In addition, its inclusion in the Ppy film is expected to give a smoother morphology and more compact coatings. It may protect the Ppy from the penetration and attack of air and moisture, and improve the stability of the sensor of the Ppy-coated fabrics.

15.4.5 Temperature and humidity ageing

The conductivity degradation of Ppy is closely related to its reaction with oxygen, especially under elevated temperature and moisture environments. Therefore, the stability of the developed strain sensors based on the combination of printing and chemical vapour deposition with or without DBS anions was thoroughly investigated by the thermal ageing and humidity ageing tests. The thermal ageing test is composed of a few cycles of heating and cooling tests. Both thermal and humidity ageing tests were carried out by recording the conductivity change of the Ppy-coated fabrics placed in a climatic chamber (Hotpack Series 922) where both the temperature and humidity could be controlled. For thermal ageing, the humidity is always kept at 65% RH. The temperatures investigated include 20, 35, 50 and 60° C, and the temperature changed every two hours. The humidity ageing was carried out at 30 °C. The humidity investigated included 40, 55, 70 and 90% RH and the humidity changed every two hours.

Figure 15.9a shows that the difference in resistance between the heating and cooling cycle is significantly irregular during the ageing process. The

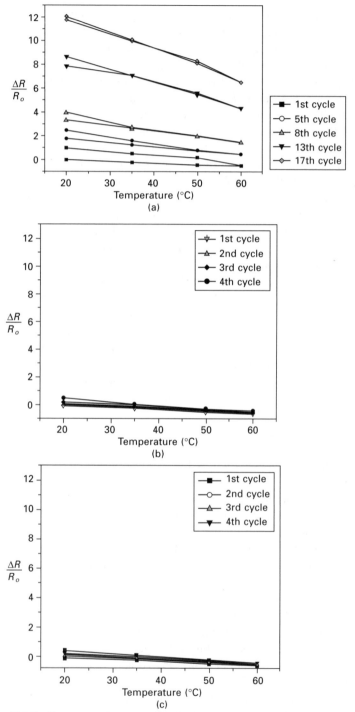

15.9 Cyclic temperature sensitivity of the samples fabricated (a) without any DBS doping and heat treatment, (b) without DBS doping but with vacuum heat treatment at 60 °C for 40 hours and (c) with both DBS doping and vacuum heat treatment.

cyclic temperature sensitivity of the sample showed a significant hysteresis at the starting point but was reduced to around 8% after eight cyclic ageings. They were further reduced to around 2% for a continuous heat ageing of 17 cycles. However, the thermal ageing process introduces the problem of drastically increasing the sensor resistance up to over 4500 K ohm after thermal ageing, which is too high for a practical strain sensor. The problem can be compensated by the vacuum heat treatment and DBS doping. It can be seen in Figs 15.9(b) and (c) that vacuum heat treatment reduces the hysteresis of the thermal changing curve effectively. The hysteresis of samples with vacuum heating at 60 °C successfully reduces the hysteresis from around 40% at the 4th cycle to around 10%. Application of DBS doping can further enhance the stability of the sensor by significantly reducing the hysteresis of heating and cooling curves to about 7.8%. In addition, the repeatability and linearity of the sensing curves towards temperature also enhanced the sensing functions, as shown in Fig. 15.9. A slight increase in resistance is observed only after each heating-cooling cycle, which may be related to the stability of the dopant ion and the rearrangement of the Ppy chain during annealing processes.

Humidity can also exert an influence on the conductivity of Ppy. The adsorbed moisture can promote the free motion of doping ions in the Ppy film, thereby increasing its conductivity.[20] Moisture can react with Ppy to decrease the conductivity.[26,61] It also shows a strong synergetic effect with oxygen on the degradation of Ppy.[61] Therefore, the stability of the sensor was further investigated by humidity ageing. Figure 15.10(a) shows the results of cyclic humidity sensing. Ppy-coated samples fabricated without DBS doping or thermal annealing shows a tremendous increase in hysteresis, at the 3rd cycle up to 30%. It can be seen in Figs 15.10(b) and (c), both DBS applications effectively eliminate the influence of humidity on the textile sensors. After heat treatment, the sensing curves to humidity reduce tremendously, particularly at the 2nd humidity cycle the hysteresis goes down from without heat treatment at around 85.5% to only 14.5% after heat treatment. Since it shows better linearity the sample is less sensitive to humidity. The results are further improved by application of DBS doping as reflected by the dropping of hysteresis between the wetting and drying cycles down to 3.8% for the sample with DBS and vacuum heat treatment. From the above results of cyclic humidity sensing, the findings show that the sample containing DBS doping has very little effect on the conductivity of the samples and is even insensitive to humidity change.

The thermal and humidity ageing tests show that the conductivity of the strain sensor fabricated from the Ppy-coated fabrics is only slightly dependent on the humidity and temperature, but it exhibits very high strain sensitivity. Therefore, it can be concluded that within the investigated ageing range of humidity and heating, the effect of the temperature and humidity on the

302   Intelligent textiles and clothing

conductivity of the strain sensor can be minimised, which is very important for its potential application. However, strain sensitivity will also be reduced after the storage of e-textile sensors in ambient condition for a long period of nearly one year as shown in Fig. 15.11.

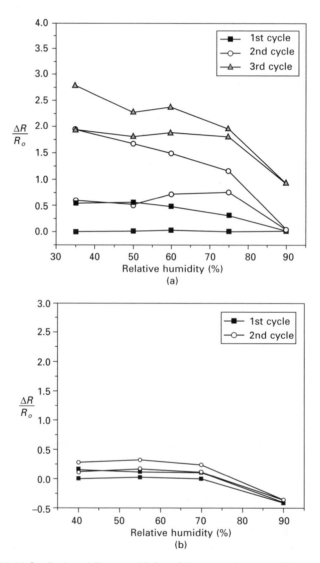

15.10 Cyclic humidity sensitivity of the samples (a) without any DBS doping and vacuum heat treatments, (b) without DBS doping but with vacuum heat treatment, and (c) with DBS doping and vacuum heat treatment.

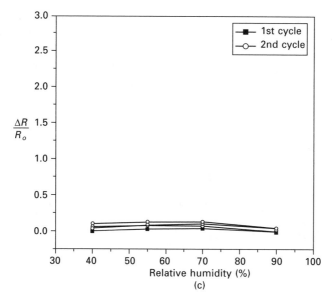

15.10 Continued

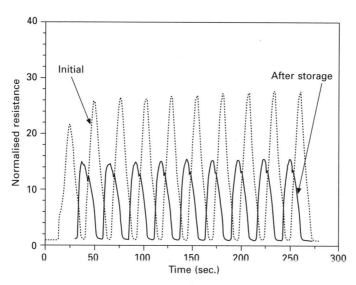

15.11 The conductivity change versus the strain of sensors prepared at room temperature before and after storage.

15.5 Conclusion

In summary, a flexible strain sensor produced from the polypyrrole-coated stretched fabrics and featured with both high strain sensitivity and good stability has been developed. It is proposed that the combination of printing and chemical vapour deposition of thin coatings of polypyrrole on the fabric surfaces will result in high strain sensitivity. The relatively lower temperature (60 °C) annealing treatment of the Ppy-coated fabrics and the introduction of bulky dodecyl benzene sulfonate anions into the polypyrrole backbone help improve the stability of the Ppy-coated sensors. The developed e-textile sensors based on the Ppy-coated fabrics are suitable for applying to sensing garments, wearable hardware and rehabilitation, etc.

15.6   Acknowledgement

This work is financially supported by the Innovative Technology Fund, The SAR Government of Hong Kong, and the Internal Research Fund, The Hong Kong Polytechnic University.

15.7   References

1. Wallace G. G., Spinks G. M., Kane-Maguire L. A. P. and Teasdale P. R., 2003, *Conductive Electroactive Polymers: Intelligent Materials Systems*, CRC Press, Boca Raton, 2nd edn.
2. Jin G., Norrish J., Too C. and Wallace G., 2004, Polypyrrole filament sensors for gases and vapours, *Curr. Appl. Phys.* 4, 366–369.
3. Otero T. F. and Cortès M. T., 2003, Artificial muscles with tactile sensitivity, *Adv. Mater.* 15, 279–282.
4. Spinks G. M., Xi B. B., Zhou D. Z., Truong V. T. and Wallace G. G., 2004, Enhanced control and stability of polypyrrole electromechanical actuators, *Synth. Met.*, 140, 273–280.
5. Wang L. X., Li X. G. and Yang Y. L., 2001, Preparation, properties and applications of polypyrroles, *React. Funct. Polym.* 47, 125–139.
6. Tessier D., Dao L. H., Zhang Z., King M. W. and Guidoin R., 2000, Polymerisation and surface analysis of electrically-conductive polypyrrole on surface-activated polyester fabrics for biomedical applications, *J. Biomater. Sci. Polymer Edn.* 11, 87–99.
7. Jiang X. P., Tessier D., Dao L. H. and Zhang Z., 2002, Biostability of electrically conductive polyester fabrics: An *in vitro* study, *J. Biomed. Mater. Res.*, 62, 507–513.
8. Lee C. Y., Lee D. E., Jeong C. K., Hong Y. K., Shim J. H., Joo J., Kim M. S., Lee J. Y., Jeong S. H., Byun S. W., Zang D. S. and Yang H. G., 2002, Electromagnetic interference shielding by using conductive polypyrrole and metal compound coated on fabrics, *Polym. Adv. Technol.* 13, 577–583.
9. Kim H. K., Byun S. W., Jeong S. H., Hong Y. K., Joo J. S., Song K., Park Y. H. and Lee J. Y., 2002, Environmental stability of EMI shielding PET fabric/polypyrrole composite, *Mol. Cryst. Liq. Cryst.* 377, 369–372.
10. Kim H. K., Kim M. S., Chun S. Y., Park Y. H., Jeon B. S., Lee J. Y., Hong Y. K., Joo J. S. and Kim S. H., 2003, Characteristics of electrically conducting polymer-coated textiles, *Mol. Cryst. Liq. Cryst.* 405, 161–169.
11. Kim S. H., Jang S. H., Byun S. W., Lee J. Y., Joo J. S., Jeong S. H. and Park M. J.,

2003, Electrically properties and EMI shielding characteristics of polypyrrole-Nylon 6 composite fabrics, *J. Appl. Polym. Sci.* 87, 1969–1974.
12. Kuhn H. H., Child A. D. and Kimbrell W. C., 1995, Toward real applications of conductive polymers, *Synth. Met.* 71, 2139–2142.
13. Oh K. W., Park H. J. and Kim S. H., 2003, Stretchable conductive fabrics for electrotherapy, *J. App. Polym. Sci.* 88, 1225–1229.
14. Scilingo E. P., Lorussi F., Mazzoldi A. and Rossi D. D., 2003, Strain-sensing fabrics for wearable kinaesthetic-like systems, *IEEE Sensors J.* 3, 460–467.
15. Rossi D. D., Santa A. D. and Mazzoldi A., 1999, Dressware: wearable hardware, *Mater. Sci. Eng.* C 7, 31–35.
16. Spinks G. M., Wallace G. G., Liu L. and Zhou D. Z., 2003, Conducting polymers, electromechanical actuators and strain sensors, *Macromol. Symp.* 192, 161–169.
17. Wallace G. G., 2002, *Synthesis and Structure of Polypyrroles, in Conductive Electroactive Polymers – Intelligent Materials Systems*, CRC Press, 51–88.
18. Rodriguez J., Grande H. J. and Otero T. F., 1997, Polypyrroles: from basic research to technological applications, in *Handbook of Organic Conductive Molecules and Polymers: Vol. 2. Conductive Polymers: Synthesis and Electrical Properties*, H. S. Nalwa, editor. John Wiley & Sons Ltd., 415–469.
19. Shirota Y., 1997, Electrically Conducting Polymers and Their Applications as Functional Materials, in *Functional Monomers and Polymers*, K. Takemoto, R. M. Ottenbrite, and M. Kamachi, editors., Marcel Dekker, Inc.: New York, Basel, Hong Kong.
20. Nowak C. K. J., 1995, Specific Interactions Between Electrically Conductive Polypyrrole and Nylon 6,6, in *Textiles, Fibers and Polymer Science.*, Clemson University.
21. Street G. B., 1986, *Handbook of Conducting Polymers*, T. A. ed., New York: Marcel Dekker, Inc.
22. Tsang H. Y. J., Li Y., Leung M. Y., Tao X. M., Cheng X. Y. and Yuen C. W. M. 2005, Enhancement of the Environmental Stability of Polypyrrole Coated Fabrics, working paper submitted to *Journal of Applied Polymer Science* for publication.
23. Wallace G. G., *Conductive Electroactive Polymers*. 2003, New York: CRC Press.
24. Cheah K., Forsyth M. and Truong V. T., Ordering and stability in conducting polypyrrole. *Synthetic Metals*, 1998, 94: p. 215–219.
25. Oh K. W., Park H. J. and Kim S. H., Electrical Property and Stability of Electrochemically Synthesized Polypyrrole Films. *Journal of Applied Polymer Science*, 2004, 91: p. 2659–2666.
26. Liu Y.-C. and Hwang B.-J., Mechanism of conductivity decay of polypyrrole exposed to water and enhancement of conductivity stability of copper(I) – modified polypyrrole. *Journal of Electroanalytical Chemistry*, 2001, 501: p. 100–106.
27. Gustafsson G., Inganas O., Nilsson J. O. and Liedberg B., 1988, Thermal undoping in poly(3-alkylthiophenes). *Synthetic Metals*, 26(3): p. 297–309.
28. Wang Y. and Rubner M. F., Stability studies of the electrical conductivity of various poly(3-alkylthiophenes). *Synthetic Metals*, 1990, 39(2): p. 153.
29. Thieblemont J. C., Brun A., Marty J., Planche M. F. and Calo P., Thermal analysis of polypyrrole oxidation in air. *Polymer*, 1995, 36(8): p. 1605–1610.
30. Samuelson L. A. and Druy M. A., *Macromolecules*, 1986, 19: p. 824.
31. Münstedt H., Ageing of electrically conducting organic materials. *Polymer*, 1988, 29(2): p. 296–302.
32. Inganas R. E. and Salaneck I. L. R., XPS and electrical characterization of BF4?- doped polypyrrole exposed to oxygen and water. *Synthetic Metals*, 1985, 10(5): p. 303–318.

33. Druy M. A., The role of the counterion in the reactivity of conducting polymers. *Synthetic Metals*, 1986, 15(2–3): p. 243–248.
34. Kuhn H. H., Child A. D. and Kimbrell W. C., Toward Real Applications of Conductive Polymers. *Synthetic Metals*, 1995, 71: p. 2139–2142.
35. Child A. D. and Kuhn H. H., Enhancement of the thermal stability of chemically synthesized polypyrole. *Synthetic Metals*, 1997, 84: p. 141–142.
36. Chen X. B., Devaux J., Issi J. P. and Billaud D., The stability of polypyrrole electrical conductivity. *European Polymer Journal*, 1994, 30(7): p. 809–811.
37. Cheah K., Forsyth M., Truong V. T. and Olsson-Jacques C., Mechanisms Governing the Enhanced Thermal Stability of Acid and Based Treated Polypyrrole. *Synthetic Metals*, 1997, 84: p. 829–830.
38. Wang Y. and Rubner M. F., Stability Studies of Electrically Conducting Polyheterocycles. *Synthetic Metals*, 1991, 41–43: p. 1103–1108.
39. Billingham N. C., Calvert P. D., Foot P. J. S. and Mohammad F., Stability and degradation of some electrically conducting polymers. *Polymer Degradation and Stability*, 1987, 19(4): p. 323–341.
40. Thieblemont J. C., Planche M. F., Petrescu C., Bouvier J. M. and Bidan G., Kinetics of degradation of the electrical conductivity of polypyrrole under thermal ageing. *Polymer Degradation and Stability*, 1994, 43: p. 293–298.
41. Thieblemont J. C., Gabelle J. L. and Planche M. F., Polypyrrole overoxidation during its chemical synthesis. *Synthetic Metals*, 1994, 66: p. 243–247.
42. Gustafsson G., Lundstrom I., Liedberg B., Wu C. R. and Wennerstrom O. I., The interaction between ammonia and poly(pyrrole). *Synthetic Metals*, 1989, 31(2): p. 163–179.
43. Cheung K. M. and Stevens D. B. C., Characterization of polypyrrole electropolymerized on different electrodes. *Polymer*, 1988, 29(9): p. 1709–1717.
44. Otero T. F., Tejada R. and Elola A. S., Formation and modification of polypyrrole films on platinum electrodes by cyclic voltammetry and anodic polarization. *Polymer*, 1987, 28(4): p. 651–658.
45. Ennis B. C. and Truong V. T., Thermal and electrical stability of polypyrrole at elevated temperatures. *Synthetic Metals*, 1993, 59: p. 387–399.
46. Ouyang J. and Li Y., Enhancement of the electrical conductivity of polypyrrole films on applying higher voltage. *Synthetic Metals*, 1995, 75: p. 1–3.
47. Salmon M., Diaz A. F., Logan A. J., Krounbi M. and Bargon J., *Molecular Crystals Liquid Crystal*, 1982, 83: p. 265.
48. Gao J. and Tian Z., Electrochemical and *in-situ* Raman Spectroscopic Studies of Dodecylbenzenesulfonate Doped Polypyrrole. *Journal of Xiamen University* (Natural Science), 1995, 34(2): p. 226–233.
49. Turcu R., Neamtu C. and Brie M., Effects of thermal annealing on the electrical conductivity of polypyrrole films. *Synthetic Metals*, 1993, 53: p. 325–332.
50. Singh R., Narula A. K., Tandon R. P., Rao S. U. M., Panwar V. S., Mansingh A. and Chandra S., Growth kinetics of polypyrrole, poly(N-methyl pyrrole) and their copolymer, poly(N-methyl pyrrole-pyrrole): effect of annealing on conductivity and surface structure. *Synthetic Metals*, 1996, 79: p. 1–6.
51. Kim D. Y., Lee J. Y., Moon D. K. and Kim C. Y., Stability of Reduced Polypyrrole. *Synthetic Metals*, 1995, 69: p. 471–474.
52. Rupprecht L., *Conductive Polymers and Plastics in Industrial Applications*. 1999, US: Plastics Design Library.
53. Gregory R. V., Kimbrell W. C. and Kuhn H. H., *Journal of Coat. Fabrics*, 1991, 20: p. 167.

54. Cassignol C., Olivier P. and Ricard A., Influence of the Dopant on the Polypyrrole Moisture Content: Effects on Conductivity and Thermal Stability. *Journal of Applied Polymer Science*, 1998, 70: p. 1567–1577.
55. Smela E., *Adv. Mater.*, 1999, 11, 1343.
56. Bay L., West K., Larsen P. S., Skaarup S. Benslimane M., *Adv. Mater.*, 2003, 15, 310.
57. Scilingo E. P., Lorussi F., Mazzoldi A. and Rossi D. D., *IEEE Sensors J.*, 2003, 3, 460.
58. Jager E. W. H., Smela E. and Inganas O., *Science*, 2000, 290, 1540.
59. Bay L., West K., Larsen P. S., Skaarup S. and Benslimane M., *Adv. Mater.* 2003, 15, 310.
60. Omastova M., Pavlinec J., Pionteck J. and Simon F., *Polym. Int.* 1997, 43, 109.
61. Truong V. T., Ennis B. C., Turner T. G. and Jenden C. M., *Polym Int.* 1992, 27, 187.
62. Omastová M., Košina S., Pionteck J., Janke A. and Pavlinec J., 1996, Electrical properties stability of polypyrrole containing conducting polymer composites, *Synth. Met.* 81, 49–57.
63. Omastová M., Pionteck J. and Trchová M., 2003, Properties and morphology of polypyrrole containing a surfactant, *Synth. Met.* 135–136, 437–438.
64. Xue P., Tao X. M., Tu T. X., Kwok K. and Leung M. Y., 2004, Electromechanical behaviour and mechanistic analysis of fibres cooked with electrically conductive polymer, *Textile Res. J.* 74(10), 929–936.

16
Electrical, morphological and electromechanical properties of conductive polymer fibres (yarns)

B KIM and V KONCAR, ENSAIT-GEMTEX Laboratory, France and C DUFOUR, Institute IEMN, France

16.1 Introduction

The textile industry has made considerable advances in the fields of intelligent and multifunctional textiles materials, mainly in the sectors of high-performance textiles, yarns and fibres. In addition, the demand of the electrically conductive fibres (or yarns) and textiles is rapidly growing up concerning industrial materials such as sensors, electrostatic discharge, electromagnetic interference shielding, dust and germ-free clothing, corrosion protecting, monitoring, data transfer in clothing and for military applications such as camouflage and stealth technology.[1–8] For all applications previously mentioned, the basic element is the textile fibre. The novel electrical properties can be obtained by modifying or integrating electronic devices into the traditional textile materials. An overview of methods related to the realisation of conductive textiles is given below.

Commercialised conductive spun or filament fibres are prepared by dispersion of the conductive particles such as silver, nickel, stainless steel, aluminium, graphite and carbon black into a common polymer matrix. In order to obtain the highly conducting blends, both high concentration of charges and homogeneous blending morphology are necessary. However, high concentration of charges in matrix polymer implies brittleness of blends and high producing cost. The modification of fibre (yarn) using inherently conductive polymers seems to be an interesting approach. Conductive polymers show the electrical properties due to their conjugated double bond chain structures, which derive both their conducting or neutral (non-conducting) forms. However, they are inherently insoluble or infusible due to their strong intermolecular interactions. The production of high quality conducting blends with conventional polymers by melt mixing or by solution casting is still at the development stage.[9–17]

Among the conjugated conducting polymers, polyaniline (PANI) and polypyrrole (PPy) have attracted much interest worldwide due to their tremendous environmental, thermal and chemical stabilities. Polyaniline exists in a variety of oxidation states from leucoemeraldine (completely reduced) to emeraldine (partially oxidised) to pernignaniline (completely oxidised). Most broadly studied forms of polyaniline are the emeraldine base form and the emeraldine salt form. With a simple doping process by charge transfer chemistry or by acid-base (protonation) chemistry, polyaniline converts the semiconducting emeraldine base form (10^{-5} S/cm) to the metallic emeraldine salt form (10^2 S/cm).[18-20]

16.2 Preparation of conductive fibres – overview

In this study, a doped polyaniline in salt form has been used in order to prepare the conductive fibres. The originality of our research approach is to prepare the conductive yarns which are able to transport information or to be used as textile sensors, preserving at the same time 'textile' mechanical properties such as light weight, elongation, bending, shearing and twisting. These textile properties are very important because the conductive fibres are transformed in textile structures by weaving, knitting or other manufacturing processes. Therefore, two different methods have been developed to synthesise conductive yarns with textile mechanical properties. The first method is based on conductive polymer composites (CPCs) using the 'melt mixing process' and the second method is performed using the 'coating process'.

16.2.1 Method 1: melt mixing process

One of the effective methods is dispersing conductive fillers in a thermoplastic polymer matrix and blending by mechanical mixing process. Shacklette *et al.* reported that a conductive form of polyaniline (Versicon,™ Allied-Signal, Inc.) is dispersible in polar thermoplastic matrix polymers such as polycaprolactone and poly (ethyleneterephthalate glycole).[21] Ikkala *et al.* reported on conducting polymer composites prepared by blending thermoplastic polymers such as polyolefin, polystyrene with Neste complex (polaniline salt complex developed by Neste Oy and UNIAX). In their report, 1–20 S/cm of conductivity has been obtained in the range of 1–30 wt% of Neste complex.[10] In addition, the morphological properties and percolation threshold of conductive filler on conductive polymer blends have been studied. At this point, the melt spinning process was our first approach for preparing conductive composites fibres. Polyaniline emeraldine salts form (PANI), PPy and graphite were melt mixed with polypropylene (PP) or low-density polyethylene (LDPE) using a co-rotating twin-screw extruder. However, the conductivity of these monofilaments was not satisfactory, even when 40% of charges are added in matrix polymer. This can be explained by the non-homogeneous morphological structures of conductive fillers in PP, as we can see in Fig. 16.1(a).[22] The particles of conductive polymers were not completely dispersed and formed

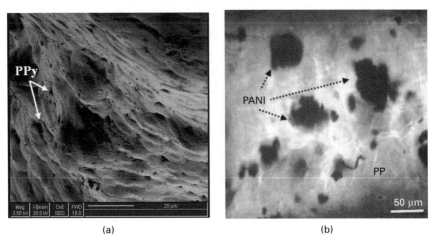

16.1 SEM photographs of (a) PP/PPy (20% in weight) blends and (b) aggregation of PANI charges in PP polymer observed by optical microscopy.

aggregates due to their strong intermolecular interactions (see Fig. 16.1(b)). The melt mixing process could be improved by changing the melting temperature, mixing time, or mixing speed. The other method, using solvent (solution blending) to prepare conducting polymer composites, is proposed in order to obtain highly conductive CPCs with better dispersion morphology.

16.2.2 Method 2: coating process

Even though conductive polymers can be electrochemically produced in fibre or film forms, they show weak mechanical properties precluding their application in traditional textile processes. Considering this difficulty, thin coating or polymerising using conducting solutions on the surface of textiles (fibres) should be a reasonable method to create conductive textile fibres. Since polypyrrole coated polyester textiles have been developed by Milliken Research Corporation, many research groups are actively involved in this field.[23] In addition, polyaniline coating or *in-situ* polymerisation on non-woven fabric, nylon 6, cotton, polyester fabric and Nomex fabric have been recently reported.[24–26]

The soluble conductive form of PANI may be obtained by protonation with functionalised protonic acid, denoted as $H^+(M^--R)$.[27] H^+M^- is a protonic acid group and may be sulfonic acid, carboxylic acid, phosphonic acid, etc. The proton of the protonic acid reacts with imine groups of polyaniline and the M^--R group serves as the counterion. R is an organic group that can be compatible with nonpolar or weakly polar organic solvents such as N-methyl-2-pyrrolidone (NMP), dimethylsulfoxide (DMSO), xylene and m-cresol.

Efficient doping methods of PANI with dodecylbenzene sulfonic acid (DBSA) or camphor sulfonic acid (CSA) were reported.[14–16,28] The PANI chains tend to form extended chain conformation in the presence of DBSA, and therefore electrons transported between polymer chains should be enhanced. In the polyaniline/solvent system, the film-like agglomerates of polyaniline take place on the solution-solid (substrate) interface when the solvent is removed. In our research work, this spontaneous molecular assembly has been used to coat conductive polymers on a fibre surface, and has been successfully applied to yarns and even fabrics. PANI coating was carried out during the impregnation of textile yarns in the PANI solutions. Electrical, morphological, electromechanical and electrical properties of the conductive yarns were investigated.

16.3 Experimental

16.3.1 Materials

Emeraldine salt form of polyaniline (PANI) was supplied by Sigma-Aldrich Chemical Company. Dodecylbenzene sulfonic acid (DBSA, Fluka) was used to obtain an homogeneous conducting PANI solution in xylene solvent. The chemical structures of used materials are shown in Fig. 16.2. The commercial polyethylene terephthalate (PET) spun yarn (111 dtex) and PET filament yarn (292 dtex) were supplied by Hyosung (Korea). Ultrahigh molecular weight polyethylene (UHMW-PE, Dyneema®, 220 dtex) yarn was supplied by the DSM Company (Netherlands).

16.3.2 Solution preparation and coating process

PANI and DBSA (1:0.5 w/w in ratio) mixtures were dissolved in xylene in order to prepare different PANI solutions ranging from 3–10 wt%. These

16.2 Chemical structures of used materials in this study.
(a) polyaniline salt form and (b) dodecyl benzene sulfonic acid.

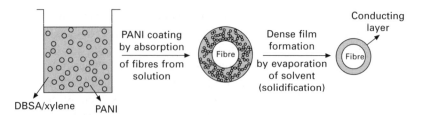

16.3 Overview of the coating process used in this study.

solutions were stirred vigorously for three hours at 90 °C and treated in an ultrasonic bath for two hours. The continuous coating process was performed using an experimental device which may be equipped with several baths filled with PANI solution. Textile yarns were dipped into the solution in the presence of a dry airflow around the coated yarns surfaces to enhance solvent evaporation and the solidification of a PANI layer on the fibre surface. The take-up speed of PANI-coated yarn in a bobbin was controlled taking into consideration the drying dynamics during the experimental procedure. The scheme of the coating process on the fibre surface using PANI solution is represented in Fig. 16.3. All the coated yarns were indicated using an abbreviation. For example, PANI/PET spun 3% means PET spun yarn coated with 3% of PANI solution.

16.3.3 Morphological and electrical characterisation

The electrical resistance of PANI-coated conductive yarns was measured with an Agilent 34401A multimeter at 25 °C, 55.56 HR%. Detailed coated fibre surface images and cross-sections were obtained using a scanning electron microscope (SEM, JEOL 100 CX model ASID4-D) and a focus ion beam system (FIB, Strata™ DB 235, FEI company). The SEM samples were gold-sputtered before observation.

16.4 Results and discussion

16.4.1 Morphological properties of PANI-coated yarn

Figure 16.3 shows the surface characteristics of PANI-coated PET spun and filament conductive fibres. The geometrical structures of spun and filament yarns are quite different. Spun yarns contain inherently many discontinuous fibres showing the bulky structures with many interstices (see Fig. 16.4(a) and (b)).[22] Most of these interstices are filled with PANI particles during the coating process. However, the structure of filament yarns is more uniform (fibres are often parallel). During the coating process, the bulky surface of spun yarns should influence the filling efficiency of PANI compared with

Electrical, morphological and electromechanical properties 313

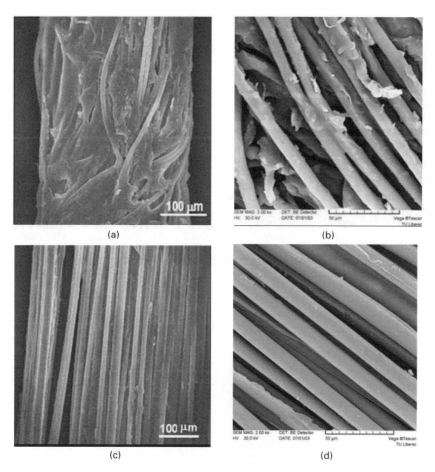

16.4 SEM photographs of PANI-coated polyester spun (a), (b) and filament (c), (d) conductive fibres

the filament one. This phenomenon is supposed to improve contacts among PANI chains therefore the conductivity increases. Figure 16.5 shows a cross-sectional SEM photograph of the PANI-coated PET filament fibres. A thin and compact PANI conducting layer (< 1 micron) was observed on the fibre surface.

16.4.2 Electrical resistance of conductive yarns

The electrical resistance (Ω) of PANI-coated conductive yarns decreases as the concentration of PANI solution increases and the yarn length decreases. It is obvious that the concentration of PANI solution is important to obtain good electrical properties. The electrical resistance (Ω) below 5% of PANI solution coated PE yarns is significantly higher than that of PE conductive

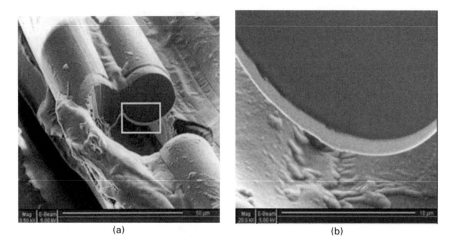

16.5 Cross-sectional SEM photographs of PANI-coated polyester filament conductive fibres.

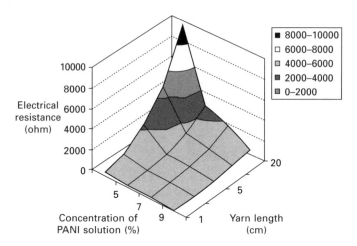

16.6 Electrical resistance of PANI-coated PE conductive yarns.

yarns coated with highly (7–10%) concentrated PANI solutions as can be seen in Fig. 16.6.[29] It was also observed in Fig. 16.7 that the electrical resistance decreases for PANI-coated PE conductive multiple yarns. However, the gelation of highly concentrated PANI solution occurs in short periods of time. The faster gelation occurs for over 8% of PANI solution compared to the 3~7% of PANI solution during the experimental procedure. The gelation of PANI solution is an irreversible reaction and is related to strong intra- and inter-chain bonding of hydrogen between secondary amine and tertiary imine groups of PANI. This gelation of PANI solution should be considered as an obstacle to the homogeneous coating of the fibre surface.[30] The temperature

Electrical, morphological and electromechanical properties 315

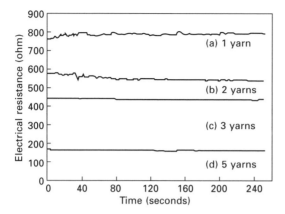

16.7 Comparison of electrical resistance for 5 cm of PANI-coated PE conductive multiple yarns.

of the PANI solution also influences the solidification of solution as well as the evaporation of solvent. In order to obtain a homogeneously coated conductive fibre using PANI solution, the temperature and the concentration of solution and the take-up speed of yarn from the solution have to be carefully controlled.

16.4.3 Environmental effects on the electrical properties

The environmental effects including times, thermal and ultraviolet (UV) radiation on the electrical properties of PANI-coated conductive yarns were studied. The ageing effect with time on the electrical properties of PANI-coated conductive yarns was studied under standard conditions (at 25 °C, 55.56 HR%). The electrical resistance for 5 cm of PANI/PE 7% conductive yarn increases for 12 weeks as we can see in Fig. 16.8.[31] The external protective layer on the conductive yarns could improve their electrical stability.

The influence of the temperature on the electrical resistance of PANI-coated PE conductive yarns is shown in Fig. 16.9. The electrical resistance increases slowly up to 55 °C. After passing this critical temperature, the electrical resistance increases more quickly. It is possible to notice that our conductive yarns may be used below 55 °C (critical temperature point) preserving their electrical properties.[31] This phenomenon may be explained by the combined actions of the fibre substrate and dopant used. Figure 16.10 shows the results from thermogravimetric analysis of the PE fibre, PANI powder, PANI-coated PE fibre and PANI/DBSA film. PANI powder is thermally stable up to almost 270 °C under nitrogen, and the decomposition of PE fibre is accelerated from 300 °C. PANI-coated PE conductive fibre is also thermally stable to up to 220 °C. However, thermal degradation of the PANI/DBSA

316   Intelligent textiles and clothing

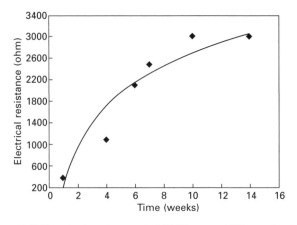

16.8 Electrical resistance of PANI-coated PE conductive fibres at time.

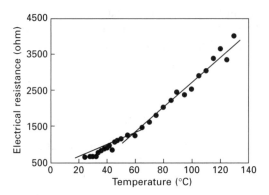

16.9 Influence of the temperature on the electrical resistance of PANI-coated PE yarns.

(dopant) film occurred faster. This result suggests that the dopant used strongly influences the thermal stability of conductive yarn coated with dopted PANI solution (see Fig. 16.10, line d).

Generally, incident UV light has sufficient energy to provoke a catalysed decomposition of the polymer. The active chain ends produced by photodegradation may react with the polymer structure, producing cross-links that generate the brittleness of the polymer.[32] The UV sensitivity study is important to determine the photo-degradation behaviour of PANI-PET conductive yarns for textile applications. PANI-PET 6% and 10% conductive yarns that have been exposed to UV radiation for 80 hours show an important decrease of conductivity as we can see in Fig. 16.11.

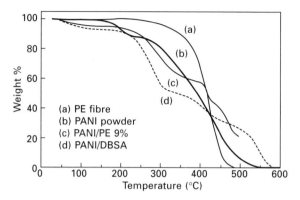

16.10 TGA scans of (a) PE fibre, (b) PANI powder, (c) PANI-coated PE fibre and (d) PANI/DBSA film.

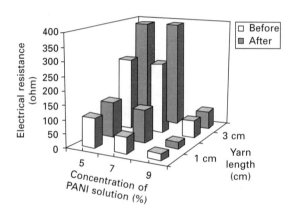

16.11 Electrical properties of PANI-coated PE conductive yarns before and after UV treatment.

16.4.4 Damage of power handling

It is important to study the specific electrical properties of conductive yarns such as damage of power handling and tension threshold of conduction. Power handling of PANI-coated PET spun, PET filaments and PE conductive yarns are studied in this section.[33] Figure 16.12 shows the intensity (I) according to the applied voltage (V) for 5 cm of PET spun conductive yarns coated with 10% of PANI solution. The threshold tension of conduction ranging between 1.5 and 2 V and the supported maximum voltage (V_{max}) is 24.5 V for an intensity of 0.0025 A. The maximum power handling is thus 0.6 W for 5 cm of PANI coated PET spun yarn.

On the other hand, PANI coated PET filament and PE conductive yarns show different behaviours of IV curves. The maximum voltage for 5 cm of PANI/PET 9% is approximately 20 V and the curve of intensity decreases

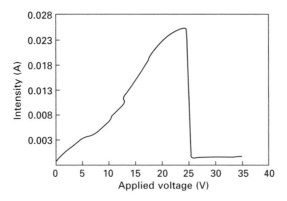

16.12 Intensity (I) as function of the applied voltage (V) for 5 cm of PANI/PET spun 10%.

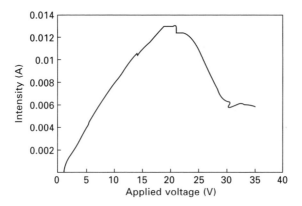

16.13 Intensity (I) as function of the applied voltage (V) for 5 cm of PANI/PET filament 9%.

irregularly from this point. The maximum power handling is approximately 0.25 W for 5 cm of PANI coated PET filament yarn, as shown in Fig. 16.13. In addition, the intensity of one yarn and two of PANI/PE 9% yarns increases linearly up to 15 V of applied voltage as we can observe in Fig. 16.14. At the rupture point of 15 V, the intensity for one fibre and two fibres is 0.027 A and 0.05 A, respectively. This result implies that the maximum power handing for one and two yarns should be 0.42 W and 0.75 W.

The electrical behaviour at maximum power handling of conductive yarns is not yet explained, but the local deformations of conducting layer at rupture are thought to play an important role. The morphological modification of the PANI conducting layers on the PET and PE fibre surface can be observed in Fig. 16.15. We suppose that the local temperature increases at maximum voltage and causes local deformation of the PANI layers which are responsible

Electrical, morphological and electromechanical properties 319

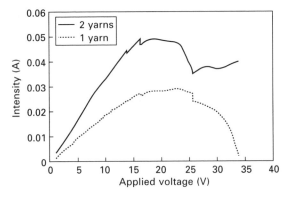

16.14 Intensity (I) as function of the applied voltage (V) for 5 cm of PANI/PE 9%.

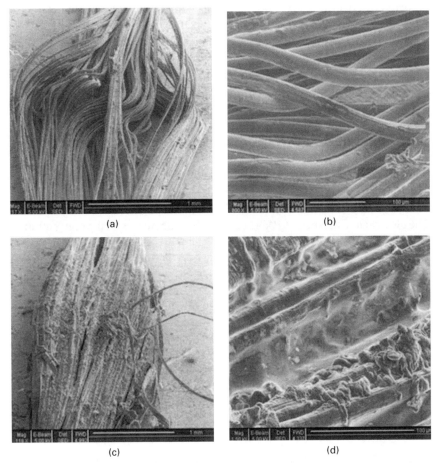

16.15 Morphological modification of PANI layers on the surface of PET (a), b) and PE (c), (d) conductive yarns at maximum power.

for conduction. The PANI chains on the conducting layers are locally disconnected and the conductivity decreases.

16.5 Applications: prototype

Since the prepared conductive fibres are based on the textile, the surfaces of conductive fibres have to be protected with an external layer in order to increase their stability and insulate them electrically. We have developed three different types of protecting layers for PANI-coated conductive fibres.[33] The first is covering the conductive fibres with a plastic film to protect fibres from abrasion. This prototype is flexible and lightweight as we can see in Fig. 16.16(a). These yarns can be used to transfer the electricity of approximately 3–6 V to light up the LED of 20 mA (see Fig. 16.16(b)). This prototype may also be used for transfer information. The second prototype is prepared by covering the conductive yarns with thermo retractable polyolefin tube (see Fig. 16.17(a)). The third prototype uses a weaving machine. PANI-coated conductive multiple yarns are protected by weaving a layer of Kevlar® fibres (see Fig. 16.17(b)). These prototypes show various possibilities for applications preserving the electrical and mechanical properties of textiles.

16.6 Conclusion

We have introduced the concept of conductive textile yarn preparation using two different methods, melt mixing and coating process. In this study, polyaniline (PANI) coated polyester and polyethylene conductive fibres have been prepared and analysed. From SEM photographs, we observe that approximately 1 micron of the PANI layer was deposited on the fibre surface. The electrical resistance of PANI-coated fibres decreases as the concentration

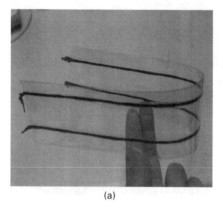 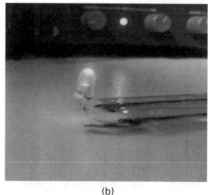

(a) (b)

16.16 (a) Prototype of conductive yarns covered with plastic films and (b) LED lighting test.

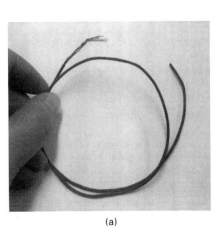 

16.17 (a) Prototype prepared using thermoretractable tube and (b) conductive yarns protected by Kevlar® fibers

of PANI solution increases. The concentration of PANI solution is important to obtain the homogeneous coating on the fibre surface. The threshold value of solution is between 6–9% of PANI concentration for our study. PANI-coated conductive fibres show sufficient electrical properties for various applications preserving their original strength and flexibility. Though PANI powder shows good thermal stability, the electrical properties of PANI-coated conductive fibres decrease at high temperature. Power handling and destruction of PANI-coated conductive yarns were also studied. It could be improved by using multiple yarns. Local deformation of PANI layers occurred at maximum current caused by increasing the temperature and the electrical properties have been modified. Three different prototypes of conductive multiple yarns have been developed in our laboratory using external protecting layers. These prototypes are expected to be used as conducting yarns, fibrous sensors, and connection devices in smart clothing or for electromagnetic shielding applications.

16.7 Acknowledgements

The authors wish to acknowledge the funding support for the Program of PREDIT 2002-2006 (No. 016-ISIC-VFIC) from the French Ministry of Transport and the Ministry of Research.

16.8 References

1. Park S, Jayaraman S, 'Adaptive and Responsive Textile Structures (ARTS)', *Smart fibres, fabrics and clothing*, Cambridge, Woodhead publishing Ltd, 2001.
2. Han M G and Im S S, 'Dielectric spectroscopy of conductive polyaniline films', *J App Polym Sci*, 2001 **82** 2760–2759.
3. Kraljic M, Mandic Z and Duic L J, 'Inhibition of steel corrosion by polyaniline coatings', *Corro Sci,* 2003 **45** 181–198.
4. Tan S T, Zhang M Q, Rong M Z, Zeng H M and Zhao F M, 'Properties of metal fibre filled thermoplastics as candidates for electromagnetic interference shielding', *Polym & Polym Comp*, 2001 **9**(4) 257–262.
5. Sastry U, 'An overview of worldwide developments in smart textiles', *Technical textile international*, International newsletters Ltd, June 2004, 31–35.
6. http://www.fitsense.com
7. Robert F, 'Electronic textiles charge ahead', *Science*, 2003 **301** 909–911.
8. Gragory Y C, Chap. 18. 'Solution processing of conductive polymers: Fibers and gels from emeraldine base polyaniline', *Handbook of conductive polymer*, second edition, New York, Marcel Dekker Inc., 1998.
9. Zilberman M, Siegmann A and Narkis M, 'Melt-processed electrically conductive polymer/polyaniline blends', *J Macromol Sci-Phy*, 1998 **B37**(3) 301–318.
10. Ikkala O T, Laakso J, Vakiparta K and Virtanen E, 'Counter-ion induced processibility of polyaniline: Conducting melt processable polymer blends', *Synth Met*, 1995 **69** 97–100.
11. Hosier I L, Vaughan A S, Patel D, Sutton S J and Swingler S G, 'Morphology and Electrical conductivity in polyaniline/polyolefin blends', *IEEE Trans dielectr electr insul*, 2001 **8** 698–704.
12. Yang J P, Rannou P, Planès J, Pron A and Nechtschein M, 'Preparation of low density polyethylene-based polyaniline conducting polymer composites with low percolation threshold via extrusion', *Synth Met*, 1998 **93** 169–173.
13. Zilberman M, Titelman G I, Siegmann A, Haba Y, Narkis M and Alperstein D, 'Conductive blends of thermally docecylbenzene sulfonic acid-doped polyaniline with thermoplastic polymers', *J Appl Polym Sci*, 1997 **66** 243–253.
14. Titelman G I, Zilberman M, Siegmann A, Haba Y and Narkis M, 'Thermal dynamic processing of polyaniline with dodecylbenzene sulfonic acid', *J Appl Polym Sci*, 1997 **66** 2199–2208.
15. Gettinger C L, Heeger A J, Pine D J and Cao Y, 'Solution characterization of surfactant solubilized polyaniline', *Synth Met*, 1995 **74** 81–88.
16. Heeger A J, 'Polyaniline with surfactant counterions: Conducting polymer materials which are processible in the conducting form', *Synth Met*, 1993 55–57 3471–3482.
17. Yang C Y, Cao Y, Smith P and Heeger A J, 'Morphology of conductive, solution-processed blends of polyaniline and poly(methyl methacrylate)', *Synth Met*, 1993 **53** 293–301.
18. Cao Y, Smith P and Heeger A J, 'Counter-ion induced processibility of conducting polyaniline', *Synth Met*, 1993 **55–57** 3514–3519.
19. Reghu M, Cao Y, Moses D and Heeger A J, 'Metal-insulator transition in polyaniline doped with surfactant counter-ion', *Synth Met*, 1993 **55–57** 5020–5025.
20. Aleshin A N, Lee K, Lee J Y, Kim D Y and Kim C Y, 'Comparison of electronic transport properties of soluble polypyrrole and soluble polyaniline doped with dodecylbenzene-sulfonic acid', *Synth Met*, 1999 **99** 27–33.

21. Shacklette L W, Han C C, and Luly M H, 'Polyaniline blends in thermoplastics', *Synth Met*, 1993 57 3532–3537.
22. Kim B, Koncar V, Devaux E, Dufour C and Vilallier P, 'Electrical and morphological properties of PP and PET conductive polymer fibers', *Synth Met*, 2004 **146/2** 167–174.
23. Kuhn H H and Child A D, Chap. 35. 'Electrically conducting textiles', *Handbook of conductive polymer*, second edition, Marcel Dekker. Inc., New York, 1998.
24. Gregory R V, 'Production and characterization of textile fibers made from intrinsically electrically conductive polymers', *NTC report*, March 1993.
25. Oh K W and Hong K W, 'Thermal stability of conductive polyaniline-nylon 6 composite fabrics doped by a mixture of protonic acids', *T R J*, 2001 **71**(8) 726–731.
26. Kim S H, Seong J H and Oh K W, 'Effect of dopant mixture on the conductivity and thermal stability of polyaniline/Nomex conductive fabric', *J Appl Polym Sci*, 2002 **83** 2245–2254.
27. Haba Y, Segal E, Narkis M, Titelman G I and Siegmann A, 'Polymerisation of aniline in the presence of DBSA in an aqueous dispersion', *Synth Met*, 1999 **106** 59–66.
28. Cao Y, Qiu J and Smith P, 'Effect of solvents and co-solvents on the processibility of polyaniline: I. Solution and conductivity studies, *Synth Met*, 1995 **69** 187–190.
29. Kim B, Koncar V and Dufour C, 'Electrical properties of PANI-Dyneema® conductive fibres – fuzzy logic modelling', *Third Industrial Simulation Conference*, Berlin, Germany, Eurosis, 2005.
30. Mattes B R, Wang H L and Yang D, 'Electrically conductive polyaniline fibers prepared by dry-wet spinning techniques', *Conductive fibers and plastics in industrial application*, New York, Plastic design Library, 1999.
31. Kim B, Koncar V, Devaux E and Dufour C, 'Characterization of polyaniline coated PET conductive fibres', *The international Polymer Fibre Conference*, Manchester, UK, 2004.
32. Warner S B, Chap. 6. 'Environmental effects: solvents, moisture and radiation', *Fiber Science*, New Jersey, Prentice-Hall, Inc., 1995.
33. Kim B, 'Creation and realization of conductive fibres – Application to the functional clothing', *Thesis*, University of Haute Alsace in France, April 2005.

17
Multipurpose textile-based sensors

C COCHRANE, B KIM and V KONCAR,
ENSAIT-GEMTEX Laboratory, France and C DUFOUR,
Institute IEMN, France

17.1 Introduction

In this chapter the classification of textile-based sensors has been carried out as a function of the measurement method which may be optically or electrically oriented. In the first case, frequently the Fiber Bragg Grating (FBG)[1] principle is used in order to determine the sensor dynamic stress[2-5] or temperature variations.[6-7] In the second case, the impedance following the same kind of physical modifications of the sensors' structure is measured. Usually, it is possible to measure only the electrical resistance but sometimes it is also necessary to measure the capacity and/or the inductance of the sensors. This depends mostly on the sensors' geometry.

This chapter is focused only on textile-based sensors using electrically oriented measurements, more precisely the electrical resistance. The most appropriate method to achieve this is to use electrically conductive polymers that are compatible with textile materials and structures. Therefore, two approaches are possible:

- coating the textile surfaces using conductive polymers also called 'organic synthetic metals'
- production of composites containing non-organic particles whose behaviour may be explained by metallic conduction theory.[8-11]

The common point of these two approaches is that polymers and metallic particles were used because of their good electrical properties (conductivity), facility of employment with temporal stability, oxidation resistance, and low cost.

Intrinsically conductive polymers (ICPs) have been considered as promising materials with their highly conducting properties in which polymer chains contain long conjugated double bonds.[12-13] These materials can be simply prepared using electrochemical or chemical oxidative methods but most ICPs are infusible or insoluble in common organic solvents. However, significant advanced solution processing of ICPs has been developed over the last decade to improve their low solubility. For example, chemical modification of monomers with dopants has been enhancing the solubility in the case of polythiophene (PT) and polyaniline (PANI).[14–18] Among the conductive polymers, polyaniline (PANI) was used in our first approach to obtain conductive polymer textile sensors. The mechanical and thermal influences on the electrical properties of PANI coated textile sensors are studied and discussed in Section 17.2.

On the other hand, among non-organic particles, carbon black[19] seems to offer the optimal compromise between electrical conductivity and price compared to other metal particles such as gold, silver and copper, etc. Electrically conductive polymer composites (CPCs) are obtained by blending (generally by melt mixing) an insulating matrix polymer (thermoplastic or thermosetting plastic) with conductive fillers like carbon black (CB), carbon fibres or metallic particles. However, adding the fillers to the matrix to impart conductivity generally modifies the mechanical properties of composites.[20–21] In addition, using alternative thermoplastic elastomeric (TPE) matrices seem to be more interesting because these composites do not always exhibit the best balance of properties for sensor applications. For pressure or deformation of CPCs sensors, the analogy with ferroelectrics suggests that the maximum sensitivity of electrical response to mechanical compression and elongation may be expected at conductive particles concentrations corresponding to a percolation threshold of electric conductivity.[22]

It is important to analyse the electrical conductivity of composite materials as a function of the filler concentration in order to identify the percolation threshold. It is well established that sensor sensitivity at this threshold point should be optimal.[23] In this case the global sensor conductivity is often not sufficient for measuring applications. In fact, this conductivity usually does not fit the measurement systems. For that reason it is necessary to find a good compromise between sensitivity and global conductivity which leads to optimal resistance of the sensor system. This resistance depends also on sensor geometry and dimensions. Finally, it means that the threshold point (equivalent to filler concentration) should not be selected if the global resistance of the sensor system does not fit the measurement system and the sensor will be made according to the compromise sensitivity ↔ global resistance. Moreover, other external parameters such as temperature, presence of solvent vapour and relative humidity (for example) may significantly affect a CPC electrical properties.[24–25] If there is a simple relationship between electrical conductivity and these external parameters and if the influence of each parameter can be identified, the composite has potential application as a sensor. For example, if the temperature influences strongly the electrical resistance and other parameters have non-significant influence, the composite material should be used as a temperature sensor.

17.2 Conductive polymer textile sensors

17.2.1 Sensors based on electrically conductive polymers – overview

Intrinsically conductive polymers (ICPs) such as polyaniline, polypyrrole, polythiophens and their derivatives show electrochemical reactions like the oxidation and the reduction of the polymer chains. They may also be sensitive to redox environment, temperature, humidity, and inorganic/organic vapours. One of the potential applications of ICPs has been extended from sensor devices to their use as transducer-active polymers in various sensing configurations (pressure, temperature, chemical and biological stimulus). Table 17.1 shows the various application areas and current commercial applications of ICPs as sensors.[26]

In this chapter, we have focused on the preparation of conductive polymer textile yarns for applications in the field of electro-mechanical fibrous sensors. PANI-based conductive yarns were prepared using a coating process that involves the transition system from a solution of PANI into a solid 'gel' phase on the fibre surface. We then investigate the mechanical and thermal effects on their electrical resistance.[27–28]

Table 17.1 Several potential applications of conductive polymers

General applications	Sensor configurations
• Electrostatic dissipation	• Potentionmetric chemical sensor
• Absorption of radar frequencies	• Amperometric sensor
• EMI (electromagnetic interference) shielding	• Conductimetric sensor
• Polymer electrolytes	• Voltammetric sensor
• Photovoltaics	• Electromembranc sensor
• Catalytic layer	• Combined conductive electroactive polymer membranes of polypyrrole and transistor (FET) device (CHMFETs)
• Redox mediation	
• Chemically and biologically sensitive on/off redox switch	• Artificial muscles as electrochemopositioning devices
• Chemically and biologically modulated resistor	
• Molecular recognition and perconcentration	
• Polyelectrolite gels as actuators or artificial	
• Electrochromic devices and smart windows	

17.2.2 Experimental

Materials

The following two important factors were carefully considered for the coating process. Firstly, the organic solvent has to be compatible with textile substrates without modifying their mechanical properties. Secondly, one of the challenges to improve the mechanical property of the PANI-coated textile yarns is to use high-modulus fibres like wholly aromatic polyamide fibre, carbon fibre, polyacrylate fibre or ultra high molecular weight polyethylene fibre.[29] Considering these two factors, the commercial polyethylene terephthalate (PET) spun (111 dtex, Hyosung) and ultra high molecular weight polyethylene (UHMW-PE, Dyneema®, 220 dtex, DSM Company) yarns have been used as fibre substrates because of their good solvent resistance and mechanical properties.

Sample preparation

The emeraldine salt form of polyaniline (PANI-ES, Sigma-Aldrich) and dodecylbenzene sulfonic acid (DBSA, Fluka) mixtures of 1:0.5 in weight ratio were dissolved in xylene to prepare PANI solutions ranging from 7 to 10 wt%. These solutions were stirred vigorously for three hours at 90 °C and treated in the ultrasonic bath for two hours. The continuous coating process was performed using an experimental device which may be equipped with several baths filled with PANI solutions. PANI-coated yarns were taken up in the presence of a dry airflow around coated yarn surfaces to enhance solvent evaporation. The thickness of the PANI layers on the fibre surface was approximately one micromet.

Electro-mechanical properties measurement

A tensile test was performed using a tensile tester (MTS®, Mechanical Testing and Simulation) with its software 'testworks' at 20–25 °C, 65 RH%. The sample length was fixed at 50 mm between two electrodes and the speed of traction was 12 mm/min. At the same time, the electrical resistance of the conductive yarns was measured with an Agilent 34401A multimeter. Figure 17.1 shows the electro-mechanical test set-up (middle) with two electrodes (20 mm × 20 mm) connecting the fibres to a multimeter and the tensile test machine.

17.2.3 Results and discussion

Mechanical sensor applications

In this section, we discuss several experimental results of conductive yarn sensors coated with 7%, 9% and 10% of PANI solutions. More detailed

17.1 (a) Electrodes in copper (b) head of electro-mechanical test.

study of the influence of the PANI concentration on the electrical resistance of coated yarns is given in Chapter 16. The dependence of the electrical resistance of conductive yarn on different elongations is shown in Fig. 17.2. The electrical resistance of the conductive PANI coated PET spun yarn increases slightly with an increase of the load. The increase of the electrical resistance is accelerated from the stretching yield point (approximately at 1.5 N) of used PET spun yarn. From the yield point of fibre substrate, the deformation of conductive PANI layers on the fibre surface is accelerated. The morphological structures and interactions among PANI chains have an important influence on conductivity. We suppose that with the increase of PANI molecular chain distances by micro-cracks in conductive layers, the number of connections between polymer chains decreases. In fact, the conductivity mechanism could not be explained only with a single chain theory.

Conductivity of polymers is rather a superamolecular phenomenon. The interactions of chains, defects at the molecular level, the direction and the

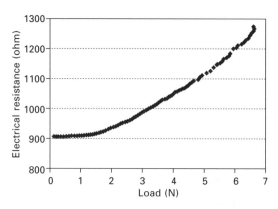

17.2 Electro-mechanical property for 5 cm of PET conductive yarn coated with 9% of PANI solution at different loads (N).

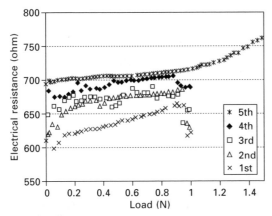

17.3 Repeated cycles of stretch recovery tests for PET conductive yarn coated with 9% of PANI solution.

arrangement of the chains and the chain distances are regarded as the important parameters for conductivity.[30-31] Therefore, models of solitons and polarons have been very popular to explain this mechanism in conductive polymers. The ability of conductive yarns to recover from deformation was also studied by means of the strain recovery test. The stretching and recovery cycles for PANI coated PET spun yarn were repeated four times up to 20% of the breaking load, and the fifth stretching was continued until breaking point. Electrical resistance decreases as a function of applied load for each cycle as we can see in Fig. 17.3. The loss of electrical resistance was approximately 12–15% of the initial value. The mechanical abrasion inducing this loss could be reduced using external protection of the sensor.

The electro-mechanical properties of PANI coated Dyneema® conductive yarns were also studied. The tensile test of conductive yarn was continued up to 90% of its breaking load and the electrical resistance was measured at the same time. The electrical resistance of PANI/Dyneema® conductive yarn increases about two times at 20 N of the load as we can see in Fig. 17.4. It can be explained that the PANI conducting layers on the fibre surface change their compact and continuous forms under strain. However, PANI coated Dyneema® yarns were not totally broken due to their important mechanical properties (resistance to break). The electro-mechanical properties are very important to confirm the possibility of applications of our conductive yarns as electromechanical fibrous sensors.

Temperature influence on electrical resistance

The thermal stability of conducting yarns is an important parameter. The study of thermal dependence of conductive yarns beyond a certain temperature should be interesting for their applications as temperature sensors. The influence

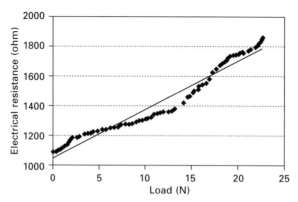

17.4 Electro-mechanical property at different loads for 5 cm of Dyneema® conductive yarns coated with 7% of PANI solution.

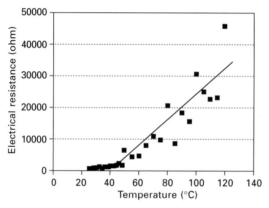

17.5 Temperature influence on the electrical resistance for 5 cm of PET conductive yarns coated with 10% of PANI solution.

of temperature on the electrical resistance of PET conductive yarns coated with 10% of PANI solution is shown in Fig. 17.5. The electrical resistance of conductive yarns increases slightly until 55 °C, then increases quickly at higher temperatures. It is possible to notice that our PANI coated PET conductive yarns may be used below 55 °C without important modification of the electrical properties and as a temperature sensor beyond 55 °C.

The temperature influence on the electro-mechanical properties of PANI coated Dyneema® yarns was also studied after treating the yarns in an oven at 70 °C for 24 hours in order to determine the thermal recovery properties of sensors. We can observe in Fig. 17.6, that the electrical resistance increases under strain, but sample sensitivity remains sufficiently for strain measuring applications after the thermal treatment. However, brief exposure at a higher temperature did not cause any serious deterioration of the electro-mechanical properties.

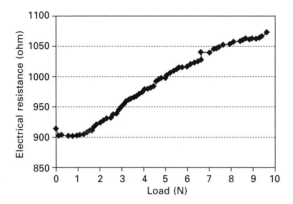

17.6 Temperature influence on the electro-mechanical properties for 5 cm of Dyneema® conductive yarns coated with 9% of PANI solution.

17.2.4 Discussion and conclusion

Polyaniline (PANI) based polyester (PET) and polyethylene (Dyneema®) conductive textile yarns were prepared by a coating process and their electro-mechanical properties and temperature influence on the electrical properties were studied. Our conductive yarns showed interesting properties for electrical applications preserving the original strength and flexibility of used textile substrates. The thickness and morphology of the coating conducting layer are considered as important parameters on the electrical response to stimuli. The coating layer properties can be modified by the concentration of conducting materials in solution, the applied tension and take-up speed of the yarns (fabric) and drying conditions of the coating process. The electrical stability to environmental damage (thermal, abrasion and humidity, etc.) of coated yarns should be controlled using an external protecting layer. Further improvement of processing is expected to enable the utilisation of conductive yarns in a wide range of potential applications such as fibrous sensors, flexible textile connections or for electromagnetic shielding (EMI).

17.3 Conductive polymer composites (CPCs) textile sensors

17.3.1 Materials and methods

Materials

For the realisation of the CPCs, ethyl-vinyl acetate (EVA) with density 0.94 g/cc (EVA 1080 VN5, ElfAtochem) and EVOPRENE with density 1.16 g/cc (EVOPRENE 007, AlphaGary) were used as matrix polymers. These

Table 17.2 Characteristics of Printex L6 (Degussa Corp.)

	Particle diameter (nm)	Structure (DBPA number)	Surface area (m^2/g)	Volatiles (%)
Printex L6	18	122	250	<1

thermoplastic elastomers (TPE) were selected because of their outstanding mechanical properties for various applications (weak Young's modulus and high elasticity). The filler (conductive particles) was a high structure carbon black (HSCB), Printex L6 supplied from Degussa Corp. The characteristics of this HSCB are given in Table 17.2.

A novel method via solvent at low temperature was developed in our laboratory as an alternative to the usual filling process (melt mixing) used to make the CPCs. The polymer (EVA or EVOPRENE) and the CB (Printex L6) were introduced, in adapted quantities, into a closed beaker with chloroform. The solution was heated between 35 °C and 60 °C then stirred for approximately 100 minutes. Then the mixtures were dried until all the chloroform evaporates (approximately 24 hours). The final result was a homogeneous bloc of CPCs. The filled polymer was cut into pellets. At least six samples of TPE with different values of filler contents of HSCB were prepared.

To undertake the characterisation (mechanical and electrical) of the CPCs, compression moulded plates were prepared. The CPC pellets were sandwiched between teflon frames heated at 170 °C (EVOPRENE) or at 200 °C (EVA) for two minutes without pressure followed by repeated pressure cycling at 50×10^5 Pa (50 bar) in order to remove air bubbles. Plates were then cooled to ambient temperature.

In addition to preparation of plate samples, the previous solution was used to prepare textile oriented samples by coating on a Nylon fabric. These samples were used to verify the compatibility of a new filling process with textile structures (fabrics, yarns). Moreover, these samples were used to measure the electrical resistivity-strain behaviour and the effect of external parameters on electric resistivity.

Methods

Stress-strain behaviour for uniaxial tension was measured at room temperature with sample gauges cut from the compression moulded plates. The distance between the grips was 35.0 mm and the thickness of a sample gauge was 1.0 mm. Gauges were deformed at the rate of 50 mm/min with a classical uniaxial tensile testing machine (MTS®). Electrical resistance was measured with a multimeter (Agilent 34401A). Each CPCs sample (width 20 mm, thickness 1.0 mm) cut from the compressed moulded plates was placed

Table 17.3 Characteristics of Nylon fabric used in this study

	Warp	Weft
Linear mass (dtex)	33	
Surface mass (g.cm^{-2})	42	
Young's modulus (Mpa)	930	200
Elongation at break (%)	24	46

between two copper electrodes. The measurement device was interfaced to a computer to record and process data.

For the textile-based sensor obtained by coating of 6.6 Nylon fabrics (Table 17.3), samples with dimensions 300 mm × 50 mm were used. The same multimeter and uniaxial tensile tester were used, but strain rate was fixed to 16 mm/min in order to have a sufficient number of points because of the low frequency of acquisition imposed by the measuring system. Both instruments were controlled by a computer. These samples were used to measure changes in conductivity during stretching.

The effect of external parameters (temperature and relative humidity) on electrical conductivity has been quantified by carrying out measurements on fabric samples (Nylon fabrics coated by CPCs) placed in a climatic chamber (FLONIC®). The electrical resistance for a given temperature and relative humidity was measured for at least two hours, which is the time necessary for the chamber's regulation. Finally, the mean value obtained during the last 30 minutes was recorded and used in the further processing.

17.3.2 Results

Mechanical properties

As a first step, the reinforcement effect of the CB nanoparticles for two thermoplastic elastomers (TPE) was investigated. The evolution of Young's modulus as a function of the CB content expressed in vol% was plotted in Fig. 17.7. It is obvious that Young's modulus of the composites increases with an increase of CB content. Since the Young's modulus of the inorganic fillers is usually much higher than those of polymers, this is typical behaviour for polymer/inorganic composites.

The influence of CB content on elongation corresponding to the break point of a compressed sample is shown in Fig. 17.8. The shape of these influences confirms the generally known fact that an elongation corresponding to the break point of polymeric composites decreases with an increase of the filler content. This behaviour is common for polymers filled with rigid inorganic particles. For EVA-CB composites, the curve suggests two different elongation phenomena. This result confirms the theory that the clusters formed by the

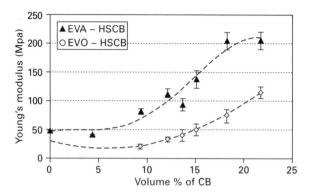

17.7 Young's modulus of the EVA-HSBC and EVO-HSBC composites.

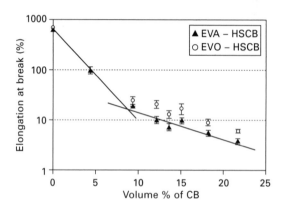

17.8 Elongation corresponding to the break point of compressed samples of the EVA-HSBC and EVO-HSBC composites.

conductive particles build a path not only for electrons but also for the formed microcracks that can grow easily into a crack of catastrophic size when the filler concentration is above a certain level.[32] For our composite EVA-CB, this 'catastrophic' filler concentration can be estimated to 8.4 vol% of CB.

Electrical properties without strain

The dependence of electrical conductivity of the EVA-CB and EVO-CB composites on the filler content is shown in Fig. 17.9. The resistivity as a function of filler content showed the typical S-shaped behaviour with three regions (dielectric, transition and conductive). Samples with low filler content were nonconductive. The step decrease in electrical resistivity was observed for different concentrations of fillers in spite of the fact that the CB used and

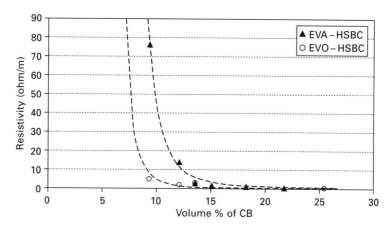

17.9 Electrical resistivity as function of filler contents.

the filling process are the same. Generally, electrical resistance depends on two parameters: the first is related to the concentration, the size of conductive CB particles and their 3D geometry configuration (dispersion, homogeneity...). The second is related to the physico-chemical compatibility between CB particles and matrix polymer. It is important to notice in Fig. 17.9 that for the same CB particles and the same concentration (for instance 9 vol%), the electrical resistance varies from 6 Ω.m for EVO-CB to 75 Ω.m for EVA-CB. Therefore, it is possible to conclude that EVO is more compatible with the CB particles that EVA. For higher concentrations, this difference is smaller. The concentration corresponding to a step decrease (threshold) is called the percolation concentration (Cp). It characterises the concentration of the filler where an internal network of particles is formed within the matrix and the material becomes electro-conductive. The percolation concentration was arbitrarily defined (identified) as an inflection point in the empirical fitting curve. For EVA-CB composite, the percolation threshold can be estimated to be lower than 9 vol% and for EVO-CB composite; it was roughly 7.5 vol%. The value obtained for the EVA-CB composite fits that obtained from Fig. 17.8. For this composite, mechanical and electrical percolations occur for around 9 vol%.

For applications as textile sensors, it is important to have a relatively high conductivity (< 1 Ω.m), high elasticity and high elongation at break at the same time. As recently shown, the mechanical properties decrease when the vol% of CB increases. It was thus necessary to fill the polymer with a minimum vol% of CB. However, for an identical resistivity (for instance 0.5 Ω.m) the necessary rate of CB in the EVA composite (21.7 vol%) was higher than the rate of CB in the EVO composite (15.1 vol%).

Electrical properties under strain of textile oriented sensor

These measurements were carried out on Nylon samples (300 mm × 50 mm) partially coated with EVA-CB composite, 21.7 vol%. The coating zone, positioned in the centre of Nylon samples, parallel to weft threads, has a dimension of 67 mm × 1 mm. The dependence of relative electrical resistivity of the EVA-CB composite on the elongation of the Nylon sample partially coated is shown in Fig. 17.10. The resistance of coating without elongation was arbitrary normalised to 1. The curve shows a very important increase of resistivity when the elongation increases slightly.

The variation of the resistivity was quasi-linear until 10% of elongation with the coefficient of correlation of 0.97. Unfortunately, the phenomenon was non-reversible because the yield stress of the Nylon fabric was approximately 3%. The sensitivity, K (see eqn 17.1)[28] of the pseudo-sensor was very good (~ 10).

$$K = \left(\frac{\frac{\Delta R}{R}}{\frac{\Delta l}{l}} \right) \qquad 17.1$$

For comparison the coefficient *K* of a classical metal gauge (copper-nickel) is 2.1. Moreover, a classical metal gauge has a range between 0.1 and 0.5% of elongation.[33]

Effect of external parameters on electrical properties

The dependence of relative electrical resistivity of EVO-CB (18.25%) composite on the temperature is shown in Fig. 17.11. Measurements were taken on Nylon fabrics partially coated by EVO-CB composite. The relative humidity

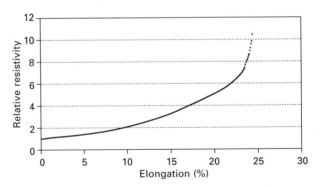

17.10 Relative electrical resistivity versus elongation of Nylon fabric coated with EVA-CB (21.25 vol%).

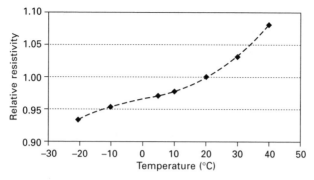

17.11 Temperature influence on the relative resistivity.

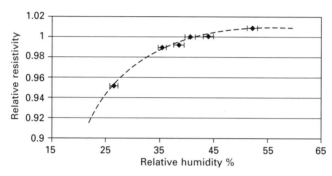

17.12 Influence of relative humidity on the relative resistivity.

was maintained at 40% ± 2.5% during measurements. The resistivity obtained at 20 °C was normalised to 1. The curve shows a non-linear behaviour. However, the relative resistivity variation was weak (0.2%/°C in the range between –20 °C and 40 °C) compared to the variation caused by lengthening (see p. 339). Therefore, it should be possible to make corrections according to the temperature; an approximate measurement of this one, would be sufficient.

The influence of the relative humidity on the relative electrical resistivity of EVO-CB (18.25%) composite is shown in Fig. 17.12. Measurements were taken on Nylon fabrics partially coated by EVO-CB composite. During measurements, the temperature was maintained at 20 °C ± 1 °C. The resistivity obtained at 40% of relative humidity was again normalised to 1. As for the effect of temperature on relative resistivity, the curve shows non-linear behaviour. However, the relative resistivity variation was weak (0.2%/1 RH%, in the range 25–55 RH%) compared to the variations caused by lengthening.

Taking into account the current results, two solutions may be envisaged: (i) bring corrections to data coming from the sensor using a calibration curve or (ii) realise a protection against moisture around the sensor. The first solution implies the realisation of cyclic measurements to detect the possible presence of a hysteresis. Therefore, the second solution seems to be simpler. However, we must be sure that this protection does not disturb the mechanical behaviour of the sensor.

17.3.3 Discussion

The electrical conduction of insulating polymers charged with conductive particles inducing a metallic behaviour is due only to physical phenomena. As shown earlier the resistivity as a function of filler content showed the typical S-shaped behaviour with three regions; dielectric, semi-conductive and conductive. In the dielectric region, the number of particles is not significant enough to form an electric driver network from one end to the other of the sample. On the other hand, if the number of conductive particles is significant and the network is continuous, electric conduction is assured. The transition between these two phenomena takes place at a critical rate of conductive particles called the 'percolation threshold'.

To create a sufficiently sensitive sensor, it is necessary that the rate of conductive particles of the composite should be slightly higher than the rate of percolation. Indeed, when the composite is lengthened, the number of electrical contacts between particles will decrease. If, in spite of this fall, the number of contacts remains high, electrical conductivity will not be affected. Lengthening a composite of concentration slightly higher than the percolation threshold will cause a significant loss in the number of contacts. The electrical sensitivity of the composite with respect to lengthening will thus be significant.

The polymer of the composite is useful to bind the carbon particles among them. The polymer should have the two following properties to avoid sensor deterioration during successive deformations:

1. The elastic range of the polymer should be larger than the sensor's potential extent of measurement.
2. It must be compatible (from the aspect of physicochemical adhesion) with the carbon particles. Thus, during deformations, the particles will not be exposed and electrical conduction will always be assured.

17.3.4 Conclusion

Realisation of a CPC solution at a low temperature was adequate to produce either composite pellets (useful for, for example, creating plates) or textile matter (fibres, yarns or fabrics) by coating. The results obtained with composite plates confirm that it is possible to detect the percolation threshold, either by electrical or mechanical measurements.

First results, concerning the behaviour as a sensor of the composite coated on a fabric, are encouraging. Below 10% of lengthening, the behaviour is quasi-linear and the sensitivity is four times more significant than for a traditional metal gauge. The composite is also rather sensitive to the moisture and the temperature. Calibration curves can be plotted to bring correction to the data according to these parameters. To improve the determination of the characteristics of the sensor, cyclic measurements (for lengthening, temperature and moisture) have to be carried out. The applications of a sensor as a textile-compatible lengthening gauge are numerous: intelligent boat's sails or aeronautic structures (parachutes), monitoring of deformable structures, etc. Moreover, if the composite is not deformed mechanically, a good sensitivity to moisture enables its use as a fast response moisture sensor.

17.4 Perspective

The use of intelligent materials reacting to external stimuli is growing in the field of technical textiles. Many products involving electrically conductive textiles exist on the market. In the majority of cases metal particles or carbon black were used to create conductive textile structures. In this chapter, the preparation of conductive materials based on textiles using two different methods, coating and solvent mixing process have been explained. The goal of our researches was to prove the feasibility of new textile materials (fibres and fabrics) for electronic sensing applications, electro-mechanical or thermal data acquisition or even for data transfer elements. Future research and development should be focused on the creation of new textile materials with semi-conductive properties enabling the 'weaving' or 'knitting' of electronic circuits and devices.

17.5 References

1. Hill K O and Meltz G, 'Fiber Bragg grating technology fundamentals and overview', *J Light Tech*, 1997 **15** 1263–1276.
2. Tao X M, Tang L, Du W C and Choy C L, 'Internal strain measurement by fiber Bragg grating sensors in textile composites', *Comp Sci Tech*, 2000 **60**(5) 657–669.
3. Du W, Tao X M, Tam H Y and Choy C L, 'Fundamentals and applications of optical fiber Bragg grating sensors to textile structural composites', *Comp Struct*, 1998 **42**(3) 217–229.
4. El-Sherif M, Chap. 6 'Integration of fibre optic sensors and sensing networks into textile structures', *Wearable electronics and photonics*, Cambridge, Woodhead Publishing Limited, 2005.
5. El-Sherif M, Fidanboylu K, El-Sherif D, Gafsi R, Yuan J, Richards K and Lee C, 'A novel fiber optic system for measuring the dynamic structural behaviour of parachutes', *J Intell Mater Syst Struc*, 2000 **11** 351–359.
6. Song M, Lee S B, Choi S S and Lee B, 'Simultaneous measurement of temperature and strain using two fiber Bragg gratings embedded in a glass tube', *Opt Fib Tech*, 1997 **3** 194–196.
7. Brady G P, Hope S, Lobo Ribeiro A B, Webb D J, Reekie L, Archambault J L and

Jackson D A, 'Demultiplexing of fibre Bragg grating temperature and strain sensors', *Opt Comm*, 1994 **111** (1/2) 51–54.

8. Flandin L, Brechet Y and Cavaillé J Y, 'Electrically conductive polymer nanocomposites as deformation sensors', *Comp Sci Tech*, 2001 **61**(6) 895–901.
9. Flandin L, Hiltner A and Baer E, 'Interrelationships between electrical and mechanical properties of a carbon black-filled ethylene–octene elastomer', *Polymer*, 2001 **42**(2) 827–838.
10. Knite M, Teteris V, Polyakov B and Erts D, 'Electric and elastic properties of conductive polymeric nanocomposites on macro- and nanoscales', *Mat Sci Eng C,* 2002 **19**(1/2) 15–19.
11. Knite M, Teteris V, Kiploka A and Kaupuzs J, 'Polyisoprene-carbon black nanocomposites as tensile strain and pressure sensor materials', *Sens Actuators A: Phys*, 2004 **110**(1/3) 142–149.
12. Heeger A J, 'Semiconducting and metallic polymers: the fourth generation of polymeric materials', *Synth Met*, 2002 **125** 23–42.
13. Kumar D and Sharma R C, 'Advances in conductive polymers', *Eur Polym J*, 1998 **34** NO 8, 1053–1060.
14. Jr A R, Soto A M G, Ello S V, Bone S, Taylor D M, Mattoso L H C, 'An electronic tongue using polypyrrole and polyaniline', *Synth Met*, 2003 **132** 109–116.
15. Ikkala O T, Laakso J, Vakiparta K and Virtanen E, 'Counter-ion induced processibility of polyaniline: Conducting melt processable polmer blends', *Synth Met*, 1995 **69** 97–100.
16. Heeger A J, 'Polyaniline with surfactant conterions: Conducting polymer materials which are processible in the conducting form'*, Synth Met*, 1993 **55–57** 3471–3482.
17. Haba Y, Segal E, Narkis M, Titelman G I and Siegmann A, 'Polymerization of aniline in the presence of DBSA in an aqueous dispersion', *Synth Met*, 1999 **106** 59–66.
18. Gettinger C L, Heeger A J, Pine D J and Cao Y, 'Solution characterization of surfactant solubilized polyaniline', *Synth Met*, 1995 **74** 81–88.
19. Donnet J B, Bansal R L, Wang M-J, *Carbon Black*, New York, Marcel Dekker, 1993.
20. Krupa I and Chodak I, 'Physical properties of thermoplastic/graphite composites', *Europ Polym J*, 2001 **37** 2159–2168.
21. Krupa I, Novak I, Chodak I, 'Electrically and thermally conductive polyethylene/graphite composites and their mechanical properties', *Synth Met,* 2004 **145** 245–252.
22. Knite M, Ozolinsh M and Sternberg A, 'Electrical and electrostrictive PLZT ceramics light modulators for infrared', *Ferroelectrics,* 1989 **94** 67–72.
23. Omastová M, Kosina S, Pionteck J, Janke A and Pavlinec J, 'Electrical properties and stability of polypyrrole containing conducting polymer composites', *Synth Met*, 1996 **81**(1) 49–57.
24. Feller J-F, Chauvelon P, Linossier I, Glouannec P, 'Characterization of electrical and thermal properties of extruded tapes of thermoplastic conductive polymer composites (CPC)', *Polym Test*, 2003 **22** 831–837.
25. Feller J-F, Linossier I and Levesque G, 'Conductive polymers composites (CPCs): comparison of electrical properties of poly (ethylene-co-ethyl acrylate)-carbon black with poly(butyleneTerephthalate)/Poly(ethylene-co-ethylAcrylate)-carbon Black', *Polym adv Tech*, 2002 **13** 714–724.
26. Anthony G E, Gordon G W and Tomakazu M, Chap. 34. 'Chemical and biological sensors based on electrically conducting polymers', *Handbook of conductive polymer*, second edition, New York, Marcel Dekker Inc., 1998.

27. Kim B, Koncar V, Devaux E, Dufour C and Vilallier P, 'Electrical and morphological properties of PP and PET conductive polymer fibers', *Synth Met*, 2004 146/2 167–174.
28. Kim B, 'Creation and realization of conductive fibres', 2005 PhD thesis in polymer and engineering, University of Haute Alsace in France.
29. Kim S H, Seong J H, Oh K W, 'Effect of Dopant Mixture on the Conductivity and Thermal Stability of Polyaniline/Nomex Conductive Fabric', *J Appl Polym Sci*, 2002 **83** 2245–2254.
30. Heeger A J, Kivelson S, Schrieffer J R, Su W P, 'Solitons in conducting polymers', *Reviews of Modern Physics*, 1988 **60** No. 3.
31. Wessling B, Chap. 19. 'Dispersion as the key to processing conductive polymers', *Handbook of conductive polymer*, second edition, New York, Marcel Dekker Inc., 1998.
32. Novak I, Krupa I and Chodak I, 'Investigation of the correlation between electrical conductivity and elongation at break in polyurethane-based adhesives', *Synth Met*, 2002 **131** 93–98.
33. Asch G, 'Les capteurs en instrumentation industrielle', fifth edition, Dunod, 2000.

18
Textile micro system technology

U MÖHRING, A NEUDECK and
W SCHEIBNER, TITV Greiz, Textile Research Institut
Thuringia-Vogtland, Germany

18.1 Textile micro system technology

During the last ten years micro system technology has not only focused on the size of the components but also on their flexibility. Importantly, the integrated electronic components are more and more based on flexible foils to avoid mechanical stress between the components and to be able to focus on new applications with new demands. In other words, the smaller the electronic devices the more mobile the final systems can be and the more flexible the electronic substrates the more wearable they become. That is why the focus is on:

- conductive structures on transparencies as new flexible electronic substrate
- printable conductive, semiconductive and insulating material to wire up the flexible electronic boards and to prepare printable electronic devices
- new technology to connect electronic devices on the flexible substrates.

But the electronic is only one component of a micro system. It needs to integrate power sources, sensors, actors, and input and output devices to communicate, to display and to transfer data.

Data transfer may occur first on a body area network, to collect the data from small sensor units via flexible data lines or wireless to a larger unit in the personal area network. This unit might be linked to a gateway to a wide area network or even with the internet. The Fraunhofer Institut für Photonische Mikrosysteme has already presented the first example of a flexible system based on electrodes placed on transparencies connected to a measuring unit which is located on the same transparency to measure the electrocardiogram (ECG) in a running project.[1] Importantly, for the development of such telemedical systems as well as for control units in hard- or even non-accessible units of machines and other technical devices, flexibility becomes more and more critical.

Nowadays we are able to integrate into our clothing such micro systems and electronic components due to their decreasing size and greater flexibility.

But clothing or, generally speaking, textile structures, are not simply flexible. Textiles are pliable. They have a non-rigid shape. They are pliable and depending on the structure more or less elastic. Furthermore, textile structures are permeable to air and moisture because of their inherent microstructure and that is much more than we can expect from a foil-based structure. So, why do we not use textiles to create micro systems?

Their advantage of being non-rigid is at the same time their disadvantage. Until now, no technology to create, handle and connect electronic or mecatronic micro components on such pliable microstructures existed, apart from textile technologies. The dilemma is that micro system technology as well as the electronic focus on lithographic structures and all technologies to generate, handle and to link such structures to prepare electronic devices or micro systems are focused on this basic technology and substrates.

Before we can start to use fabrics and textile structures as new substrates for electronic devices and as basic substrates to create micro systems it is necessary to have a closer look at the textile structures, their dimensions, the lateral resolution of such structures and the possibilities of textile technologies to design the structures needed to create textile-based electronics and micro systems.

18.2 Textiles are inherent microstructures

Let us examine more closely the classic textile technologies and their possibilities to create microstructures. A view of fabrics under the microscope show us the enormous variety of textile structures mainly designed to create new patterns, ornaments and decorations but also to have new functionalities or to improve textile properties. When we are going to use such structures for the applications mentioned above, we should know which kind of structures are available by weaving, knitting, warp knitting, braid and embroidery.

18.2.1 Weaving

The structuring possibilities of weaving can be illustrated with the example of the CMY Colour Weaving developed in the TITV Greiz.[2,3] The CMY Colour Weaving permits the creation of photographic images on inferlor floating Jacquard weavings on a uniform white or black warp beam together with defined coloured weft threads. The technology is based on a systematic analysis of weaves to obtain virtual colours on three basic colours (either RGB or their inverse colours CMY) together with a white and black thread. The procedure is based on a 4i × 4i matrix or 3i × 3i if the key thread (white in the case of RGB or black for CMY) is not used. The best result was obtained on a 8 × 8 matrix which permits over 28,000 colours with a woven virtual pixel size of about 1 mm^2. Furthermore, the systematic analysis of all binding possibilities yields several opportunities, to design identical colours on the same matrix size. Therefore the virtual colour binding sets can be subdivided into several classes. A systematic selection of binding of different classes enables us to give shape to the surface of the coloured weaving without loss of quality and colour impression. Finally, it permits the concealment of a code in each virtual pixel using bindings with the same virtual colour but from different classes. This is then readable with a commercial scanner.

Such a woven structure can be compared to a monitor and the hidden code shows that it is even possible to store permanent data. The minimum matrix size of 3 × 3 yields the best resolution of the resulting woven image of 10,323 pixels per square inch (ppsi) by use of a weaving machine with Satin quality (120 threads cm^{-1} = 305 threads per inch) or 2,567 ppsi by use of Taft quality (60 threads cm^{-1} = 152 threads per inch). These resolutions of more than 10,000 ppsi (Satin quality) and more than 2,500 ppsi (Taft quality) are much greater than the resolution of our TVs and comparable with a 17 inch monitor resolution in the different graphic modes. A 17-inch monitor driven with the lowest resolution 640 × 480 pixels corresponds to 2,581 ppsi with 800 × 600 pixels to 4,032 ppsi and with 1024 × 768 to 6,606 ppsi. Therefore the Taft quality would correspond to a 640 × 480 pixel monitor resolution and the Satin quality would be even superior to the 1024 × 768 pixel monitor resolution and additionally do not forget the possibility to store hidden information. An amount of 2*(1280/8)*1*2*..*8 bits = 1,032,192,000 bits = 129,024,000 Byte = 129 MByte » 89,600 pages » 896 books can be theoretically stored on one woven image using 1280 warp threads per image, the same number of weft threads as an 8 × 8 elementary binding set. This is a vast amount of information on a piece of fabric of 4 inch × 4 inch when woven in satin quality or 8 inch × 8 inch on a Taft machine. Nobody would recognise the woven image as a carrier of such a huge amount of information.

This means that as soon as we are able to weave conducting and semi-conducting threads together with non-conducting threads to create interactive weave crossings, then textile displays and electronic boards will be available. Furthermore, it can be shown that multi-layer weaving allows us to create electrical contacts or to avoid such a contact of conducting threads in the warp and weft directions. This is the basis to create the well-known structures used in micro system technology, e.g., electrode comb structures, double electrode comb structures also known as interdigital structures, meander and coil structures, as has been shown in refs 8–9. The interdigital structure is the most important structure to prepare gas sensors and in combination with piezoelectric materials to build actuators. Furthermore these structures with their double comb electrode structures can be used to design batteries, super capacitors and even light sources. Until now interdigital structures may only be prepared with comb electrode distances below 300 μm by lithographic techniques.

18.2.2 Knitting

Knitting structures are special mesh structures. If we look at Nylon stockings the size of a single mesh is in the micrometre range using monofilament yarns of 22 dtex (2.2 tex) or even less. (For those) who are not familiar with the units used for yarn size, it is given as the ratio of the weight and the length of a thread either in tex or numometric (Nm): 1 tex = 1 gkm^{-1} and 1 Nm = 1 mg^{-1}. A thread material density of 0.8 gcm^{-3} and a size of 2.2 gkm^{-1} yields a yarn diameter of the monofilament of 59.2 μm.

Recently, knitted structures were not used as electronic substrates apart from a few examples where textile heating was based on a knitted structure. The main interesting point of knitted structures is the possibility to get the thread back from such structures. This is called back-knitting and permits the modification of textile material in the shape of a fabric with a high surface. Hence, all textile finishing procedures known to handle knitting may be used to modify the surface of the thread and after back-knitting the modified threads are available. In this way each filament of the threads of the textile surface can be covered by conducting material. It can be done by use of carbon or conducting polymer dispersions or by chemical deposition of very thin metal layers. It will be shown later that it is possible to prepare highly conducting thread materials based on this well-known but partially forgotten technology. Furthermore, and this is the message of this chapter: remember the classic old textile technologies when trying to build flexible electronic and microstructures. One cannot be performed without the other.

18.2.3 Braiding

Braid is manufactured by interlacing three or more threads forming a flat or tubular narrow fabric. The difference from other textile technologies is that braiding allows the creation of microstructures without the need to make use of miniaturised tools. The yarn is directly fed from the bobbins to the braiding point. The dimension of a braid depends only on the count of the yarn, the number of threads, and the type of binding therefore it will be possible to produce braids that are much smaller in dimension. This makes braiding an ideal technology for textile micro-engineering.

18.2.4 Embroidery

In the traditional definition, embroidery is a technique of decorative needlework in which designs are created by stitching strands of some material onto a layer of another material. Embroidery has traditionally been used to decorate clothing and household textiles.

Up until now most embroidery used textile threads stitched onto a woven fabric, but recently embroidering has become ever more important in technical applications ranging from fibre reinforcement to electronic circuits. Stitches may be executed in wires and embroidery can be worked onto non-traditional materials such as plastic foils. One unique feature of embroidering is the possibility to place stitches in any desired direction forward, backward, and sideways. Very complex multi-layer patterns can be produced in this way.

Furthermore, we can construct a thread from two highly conducting Elitex® threads insulated against each other and coated with a paste containing copper doped zinc sulphide particles and insulated by a polyacrylnitril coating. Such a thread construction driven by an AC voltage of 110 V generated by an EL-converter from a low DC battery voltage can be illuminated and integrated in embroidery.

18.3 Goal of the application of compliant textile structures

All of these examples illustrate that textile structures provide a means to create compliant structures in the millimetre and micrometre range with many advantages over lithographic structures. These structures have the potential to combine micro- and nanostructuring. Especially noteworthy is the conducting textile microstructure base on conducting metallised threads (cf. Statex®, Shieldex® and Elitex®), which can be surface modified by use of electrochemical methods and allow the formation of thin structured layers in the 100 nm scale. The available methods to modify conducting thread surfaces and to create functional bonds to covalently link other molecules are already being used to modify electrodes to build chemical sensors (listed in Table 18.1).

Table 18.1 Overview of electrochemical surface modification technology applicable to metallised threads and textile structures

Electrochemical technique	Aim
Galvanic metal deposition	Tailor made conductivity of the thread material, protection against corrosion of the underlying metal layers and formation of alloys
Electro-deposition of paint	Deposition of insulation and protection layers in the micrometre scale
Electro-polymerisation	Deposition of conducting polymers in the micro- and nanometre scale on non-conducting nanolayers containing functional groups for further surface modification with enzymes, etc.
Anodisation	of the surface to protect against corrosion, create sensoric properties and textile electrode material for energy storing and electrochromic devices
Electro grafting (J. Pinson et al. 1995)	Covalent bonding of funtionalised molecules on electrode surfaces

Table 18.2 Interactions of the applications of smart textile structures in smart textiles and in new technical applications

	Smart textiles
Textiles with novel functions	Novel textile materials for technical applications
occupational wear	textile microsystems
leisure wear	textile-based electronic
welfare	textile microstructuring
home care/home monitoring	textile nanotechnology – micro-nano-coupling
medical care	
medical textiles	
business wear	
	Interactions on both directions

Such new conducting and sensory textile structures can be applied in two areas. On one hand they are technical textiles and may be used in fields where textiles are till now not in use like in the area of flexible electronic and to build new compliant microsystems but on the other hand the new compliant electronics and microsystems permit to integrate such devices into our clothing and other textile structures used in our cars, public transport in occupational clothing in the medicine or in the area of home care and home monitoring offering new possibilities for telemedical applications. These interactions are summarized in Table 18.2.

18.4 First attempt: textile electronic circuit technology based on copper wires in a lattice structure with interconnections and interruptions

An attempt to use a fabric as a substrate for electronic circuit technology has been described in ref. 10. The fabric consists of copper wires woven between conventional textile threads. The spatial arrangement of the copper wires is chosen to achieve appropriate features for chip mounting which was carried out using an electrically conductive glue. Some plain electronic circuits and a textile keypad[11] were built using this technology.

The most important advantages of these woven substrates in comparison with flexible foils and printed circuits or electronic boards are their mechanical non-rigidity, air permeability and low mass. The use of metal wires, however, is a limiting factor for the textile manufacturing process because of large differences in the tensile modulus of metal wires and conventional yarns.

18.5 Galvanic modification of yarns

In the preceding sections we have shown that textiles are inherent microstructures with fantastic properties. They are flexible and much more mechanically stable than foils and they are compliant structures. However, until now we did not have the right materials or we applied the wrong materials to use textile structures as electrodes, sensors and even as parts of a microelectronic structure. Metal wires, when they are integrated in textile structures or even when the threads are prepared from combinations of metal wires and conventional threads, have, on one hand, a negative influence on the properties of the resulting fabric and on the other hand, permit the highest conductivity. Applications of metal-containing pastes or other conducting materials do not permit the conductivity needed and cause many problems trying to reach the fastness expected from fabrics. The aim of this section is to give an overview of promising materials and composite structures as well as electrochemical finishing techniques to prepare them for textile substrates.

Chemical metallised threads are an interesting alternative and they are already available on the market.[12] They are used to produce textiles with antistatic properties or to shield electromagnetic waves. The metal layers are thin enough to be undamaged in the textile processing to form the final product, but they are too thin to achieve the conductivity necessary for the applications mentioned above. Furthermore, conducting inks are available to form a precursor structure on textile substrates. Based on both materials electrochemical techniques may be applied to increase the conductivity of such pre-structured slightly conducting textile structures or to improve the surface properties to avoid corrosion of the metal layers to reach the fastness expected. Electrochemical or galvanic deposition processes are the least expensive way to coat or modify a precursor structure.

The conductivity of such precursor structures is insufficient for electrodes, sensor structures or to build electronic devices, but sufficient for a further galvanic metal deposition by using special coating procedures and galvanic baths. The precursor structure can be electrochemically modified by metals and even noble metals,[7] as well as by electropolymerisation[15] and electrodeposition of paint. The metals can be electrochemically oxidised to form oxide structures with semiconducting properties. The textile precursor structure allows the coating of various zones with various metals, metal oxides and even new materials. Firstly, applications of textile substrates structured in this way will be presented.

A new and very interesting material which may be used to develop electrochemical finishing technologies and which allows textile microstructures to be created, are polyamide threads and yarns chemically coated with a silver layer of about 1 micrometre.[12] The thin metal layer remains stable on the polyamide filaments and does not significantly change the thread properties.

Such yarns can be processed by most textile technologies without serious problems and permit the preparation of textile microstructures by the use of the available monofilament Ag-PA-threads with a size of 22 dtex.

We tested the galvanic deposition of several metals including gold, platinum, copper, nickel, zinc as well as additional silver on different textile substrates. The microscopic images show reasonably smooth metal deposition on each filament of a pre-metallised thread by use of optimised galvanic baths. Furthermore, the threads galvanised by gold and platinum have been used as electrodes to record cyclic voltammograms in organic electrolytes using a well-known test redox system (ferrocene/ferrocenium). The cyclic voltammograms show a comparable potential window to that recorded on a disc electrode prepared from the same metal. The peak current dependence on the scan rate behaved as expected for an array of microcylindrical electrodes. This means that the galvanic noble metal layers on top of the thin silver metallisation on the polyamide filaments are tight and behave ideally even under such harsh conditions as in organic solvents. This offers the possibility to do further electrochemical modifications and to prepare modified thread electrodes for sensor and even actuator applications. On the basis of zinc depositions it is possible to get light-sensitive zinc oxide layers and by the use of the deposition procedures described by Schlettwein et al.[16] dye-modified solar electrodes may be prepared.

It is difficult, however, to control such electrochemical depositions on a textile substrate done on a laboratory scale. As soon as we think about a production technology plenty of problems have to be solved. That is why we should focus on easy applications to optimise the galvanic deposition on a textile precursor structure. Such easy applications are to prepare structures like antennas, textile electronic bus systems or as flexible electronic boards. We investigated the influence of the amount of galvanic-deposited metals on the basic structure with respect to conductivity and the loss of flexibility of the textile substrate. The conductivity available on thus prepared Elitex®-Materials may be even larger than 10^4 Scm^{-1}. That means the conductivity is increased by more than one order of magnitude compared to the pre-metallised thread material and is following the treatment at the same level as lead and mercury.

On the basis of these results, woven and embroidery structures have been prepared to create textile antenna coils for transponder applications. Antennas made from woven three-layer Jacquard weavings show the best properties. In the bottom layer the metallised warp threads are located. These threads are insulated by an intermediate non-conducting textile layer against the metallised weft threads in the top layer. By use of special weave bindings the metallised threads are connected to create a coil where two neighbouring weft threads are the start and end of the coil.[7] The transponder chip may be fixed on the coil structure ends. The coil resistance may be reduced to about

10–20 Ωm^{-1} and permits a reading distance with the Philips i-code System of about 50 cm. In a similar way a textile bus to pick up the signals from a textile keyboard (shown on the TechTex in 2000) was created.[6]

It is possible, furthermore, to weave interdigital structures (electrode structures with a shape of two merged combs) similar to the lithographic structures used in sensor applications. Interdigital structures may also be applied instead of optically transparent electrodes to build electroluminescence devices. Reference 7 shows the possibility to use woven interdigital structures to prepare textile electroluminescence devices. The distance of the thread electrodes in such woven structures is less than 300 μm. It has been shown that this is inferior to the critical distance to create electroluminescence.

All these structures look very promising but there is a considerable drawback when electrolytes containing cyanides are used. The removal of cyanide from the galvanic or electrochemical fabrics can only be done with an extremely high expenditure by a cascade of several washings and vacuum draining depending on the composition of the fabrics. As soon as the fabric contains natural fibres it became impossible. Therefore it was necessary to develop a new technology to produce highly conducting thread material. From the galvanic deposition on single threads it was known that threads with a silver layer of a thickness in the range of one up to two micrometres still behave like a textile thread and may be woven or used by other textile technologies. That is why we developed a galvanic bath connected with vacuum draining and washing to galvanise pre-metallised knitting. This knitting yields threads with a conductivity up to $5 \cdot 10^{-4}$ Scm^{-1} by back-knitting. That means it is possible to produce threads with a resistance of a few Ω per metre length depending on the thread size. When the size of the thread is larger than 23 tex the resistance is below 20 Ωm^{-1}.

18.6 Light effects based on textiles with electrically conductive microstructures

With the galvanically treated Ag/PA-Elitex® threads it is possible to weave double comb structures similar to the lithographic structures used in sensor applications. Interdigital structures may also be applied instead of optically transparent electrodes to build electroluminescence devices. Reference 6 shows the possibility to use woven double comb structures to prepare textile electroluminescence devices based on a woven structure (EL-weave) using 22 dtex Ag/PA-Elitex® threads as comb electrode material. The distance of the thread electrodes in such woven structures is less than 300 μm. It has been shown that this is below the critical distance to create electroluminescence. However, textile interdigital EL-weaves with an electrode distance larger than 200 μm show a different dependence on the stimulation frequency of the AC voltage. While EL foils show the highest light density of 60 Cdm^{-2}

in a frequency range of 400 Hz to 800 Hz, woven interdigital EL structures with an electrode distance larger than 200 μm have their highest light density at a characteristic frequency in the range of 1 kHz to 10 kHz depending on the size of the woven EL structure.

Optimal stimulation frequency decreases with the square root of the area of the woven interdigital structure depending on the inductivity of the secondary coil of the transformer of the EL converter. That means that in the case of woven EL structures with large thread electrode distances the resonance frequency of the oscillating circuit built from the capacity of the EL structure and the inductivity of the secondary coil of the transformer has to be found. The behaviour changes when the thread electrode distance becomes smaller than 200 μm. These new woven EL structures based on monofilament threads in the warp and weft direction with a weft density of 96 threads per centimetre (240 threads per inch) do not show such a sharp frequency dependence and may be stimulated as the EL foils in the frequency range of 400 Hz to 800 Hz yielding a light density of 40 Cdm^{-2} to 60 Cdm^{-2} and in the case of a stimulation with 150 V and 1 kHz, it reaches 80 Cdm^{-2}. Such woven EL structures can be easily screen printed with EL pastes illuminating in different colours to prepare luminescent/luminous images for advertisements, safety signs in buildings, cars, ships and aero plants.

18.7 Textile-based compliant mechanisms in microengineering and biomechatronics

Compliant mechanisms work by flexing all or some of their parts to transfer motion, force, or energy and, therefore, they are considered to be sophisticated flexible materials. Unlike their traditional rigid-body counterparts, compliant mechanisms do not contain moving parts that work together in order to carry out a function. Compliant mechanisms are made from one piece. This makes them ideal candidates for use as miniaturised components in microengineering. The application of textile materials and techniques in compliant mechanisms is the objective of ongoing research. Biomechatronics focuses on the development and improvement of mechatronic devices by transferring principles of medicine and biology into technical systems. Basically, it is the concept of taking ideas from nature and implementing them in another environment such as engineering and design. But first of all biomechatronics is an interdisciplinary system approach which involves biology, medicine, mechanical and electrical engineering, electronics, computer sciences, and system analysis. It leads to new designs which could not be realised in one of the disciplines alone. So biomechatronics becomes more than just a copy of nature.

18.7.1 Biological inspiration and technical implementation

The wings of insects and legs of spiders are two examples of compliant mechanisms where nature served as a pattern in textile microengineering of material-locked joints with fluidic drives. This development has been inspired by studies of the leg design and jumping in spiders. The spider leg represents a hollow structure with fluid drive which enables some species of spiders to jump over large distances. The whole spider leg, which is part of a rigid exoskeleton, is built from the same material, chitin polymer. The joints of the spider legs, however, consist of a flexible sealing membrane whose flexibility is due to a reduced degree of polymerisation within the chitin material. In this way a gradient of stiffness is produced along the spider leg. When applying pressure within the hollow tube the flexible membrane expands generating a torque and the spider leg is bent. A second example is the unfolding apparatus of the dragonfly's wing which can be understood as an arrangement of material-locked joints.

A bio-inspired technical system is a tube made from silicone rubber. Within a small region comprising the material-locked joint, the wall of this tube has an asymmetrical gradient of stiffness with respect to the tube's longitudinal axis. The artificial spider leg starts to bend in the same manner as its biological pattern does when pressure is exerted. The bending angle depends on the magnitude of the applied pressure. When testing some prototypes it became obvious that they were very imperfect replications of a real spider leg since the material-locked joint was blown up like a balloon. In contrast to the technical arrangement made from silicon rubber, the sealing up membrane of a spider expands only in one direction according to its internal structure. The study of this directionally dependent phenomenon was the starting point of efforts to use textile materials.[18]

18.7.2 Miniaturised material-locked joints based on textile materials

The construction principle of a material-locked joint based on textile materials may be illustrated by an elastic hollow cylinder with circular knit (or braid) and local stiffening. For its manufacturing, a flexible tube made from silicone rubber was enveloped with a circular knit. The local stiffening of the joint was effected by applying a small blot of glue resulting in an asymmetrical stiffening of the textile mesh structure. When applying pressure within the tube the joint undergoes a bending motion.

The most important advantage of this example is its simple construction and the possibility to produce it in an automated textile process as a running tape. The production process includes the following successive steps:

1. textile pattern design
2. manufacturing as an endless narrow fabric
3. local stiffening of the joints (adhesive, hot melt, welding)
4. making-up, i.e., tailoring to a given length and sealing of one end
5. attachment of hydraulic or pneumatic drive.

Taking into account that the material-locked joints are manufactured as endless textile cords it becomes obvious that they can be easily fitted to cascade systems of joints that can be tailored to any desired degree of flexibility and kinematics.

Most of the prospective applications of material-locked joints require a well-defined adjustment of their bending angle during operation. An appropriate means to equip flexible actuators with sensing capabilities is the integration of textile sensors. For this reason the knit which surrounds the flexible tube is made from electrical conductive threads. The knitted structure comprises a network of elementary electrical resistors whose magnitude changes systematically when the joint is deflected under application of pressure. So the resultant electrical resistance of the whole network, which can be easily measured, relates to the bending angle which depends on the pressure on its part. A flexible coating ensures a more stable operation and increased life expectancy of the sensor device. Some characteristic curves of electrical resistance plotted against pressure are also shown in Fig. 18.1. Using this effect, a simple pressure sensor with fast response was easily fabricated and could be used to study and to control motion sequences of material-locked joints. Combinations of flexible actuators with integrated textile sensors are straightforward to produce. The integration of sensors ensures mechatronic

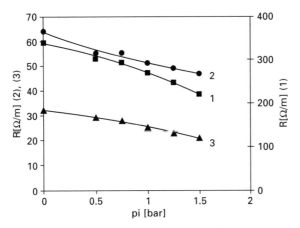

18.1 Characteristic curves (electrical resistance versus pressure) of a pressure sensor comprising a flexible tube enveloped with an electrical conductive knit.

feedback for automatically controlled motion systems. Future R&D efforts are focused on braiding technology because it allows the creation of microstructures without the need to make use of miniaturised tools.

18.7.3 Benefits of textile-based material-locked joints

- continuous manufacturing process as endless tapes
- made from inexpensive textile materials
- can readily integrate sensors
- chemically inert
- slip-stick-free operation
- completely non-metallic construction
- compliance helps correct for misalignment in robotics
- ability to be miniaturised
- operation without electricity, fluidic drive only.

18.7.4 Applications

- walking machines, robots, actuators, manipulators
- grasping apparatus for minimally invasive and microsurgery
- joint prosthesis with human-like response
- shape-change applications
- inflatable unfolding mechanisms.

18.8 References and sources of further information

Wearable microsystems

1. Holland, IPMS personal information, cf. http://www.ipms.fraunhofer.de/.

Textile structuring

2. A. Neudeck, D. Zschenderlein, D. Obenauf and U. Möhring, 'Übertragung digitaler Bildvorlagen und Designs in Jacquardgewebe', *Melliand Textilberichte* Heft 9/2002 and *Melliand Band- und Flechtindustrie* Heft 03/2002.
3. A. Neudeck, D. Zschenderlein and D. Obenauf, 'CMY Colour Weaving', DE 101 59 971 A1.

Smart textiles

4. S. Gimpel, U. Möhring, H. Müller, A. Neudeck and W. Scheibner, 'Textile Based Electronic Substrate Technology', *Proceedings of the Techtextil*, Frankfurt, 2003.
5. U. Möhring, W. Scheibner, A. Neudeck and S. Gimpel, 'Integration von Mikrosystemtechnik in textile Etiketten'/(Integration of microsystem technology in textile lables), *Melliand – Band- und Flechtindustrie*, 40/2 (2003) 63.

6. S. Gimpel, U. Möhring, H. Müller, A. Neudeck und W. Scheibner, 'Galvanische und elektrochemische Modifizierung von Textilien zur Integration von Mikrosystemtechnik, Sensorfunktionen und Elektrolumineszenz', *Melliand – Band-und Flechtindustrie*, 40/4 (2003) 115.
7. S. Gimpel, U. Möhring, H. Müller, A. Neudeck and W. Scheibner, 'Textile Based Electronic Substrate Technology', *Journal of Industrial Textiles*, 33/3 (2004) 179–189.

Sensory textiles

8. A. Neudeck and R. Reisch: 'Textile Elektronische Zunge', in H. Ahlers (Hrsg.), *Früherkennung der Diabetes*. Herausgegeben JENASENSORIC e.V. 2004, Kap. 4.6., S. 78–83.
9. R. Reisch, A. Neudeck and C. Wenk: 'Textile elektronische Nase', in H. Ahlers (Hrsg.), *Früherkennung der Diabetes*. Herausgegeben JENASENSORIC e.V. 2004, Kap. 4.7. S. 84–86.

Textile electronic substrates

10. H. Hieber, W. Scheibner and J. Schüler, 'Textiler Verdratungsträger für elektronische Baugruppen', GMM Fachbericht 37: *Elektronische Baugruppen – Aufbau- und Verbindungstechnik* DVS/GMM-Fachtagung 6.-7.2.2002, Fellbach.
11. W. Scheibner, H. Reichardt and H. Schaarschmidt, 'Textile switches made from electrically conductive narrow fabrics', *Narrow Fabric and Braiding Industry* 38 (2001) 2, 51.

Ag/PA threads

12. See for example: Statex GmbH (www.statex.de).

LEDs and chips on textiles

13. W. Scheibner, A. Neudeck, D. Zschenderlein und U. Möhring, 'Bandgewebe mit elektrischen Lichteffekten', *Melliand – Band- und Flechtindustrie*, 41/1 (2004) 16–17.

Electrochemical modifications

14. Andreas Petr, Andreas Neudeck, and Lothar Dunsch, 'On the magnetic susceptibility of polyaniline – an alternative approach', *Chemical Physics Letters*, (2004) in press.
15. P.-H. Aubert, A. Neudeck, L. Dunsch, P. Audebert and M. Maumy, 'Electrochemical Synthesis and Structural Studies of Copolymers Based on the Electrooxidation of Pyrrole and some Salen Compounds', *J. Electroanal. Chem.*, 470 (1999) 77–88.
16. See for example:
    T. Yoshida, K. Terada, D. Schlettwein, T. Oekermann, T. Sugiura and H. Minoura, 'Electrochemical Self-Assembly of ZnO/ Eosin Y Thin Films and Their Sensitized Photoelectrochemical Performance', *Adv. Mater.*, 12 (2000) 1214;

D. Schlettwein, T. Oekermann, T. Yoshida, M. Tochimoto and H. Minoura, 'Photoelectrochemical sensitisation of ZnO-tetrasulfophthalocyaninatozinc composites prepared by electro-chemical self-assembly', *J. Electroanal. Chem.*, 481 (2000) 42; T. Yoshida, M. Tochimoto, D. Schlettwein, G. Schneider, D. Wöhrle, T. Sugiura and H. Minoura, 'Self-Assembly of Zinc Oxide Thin Films Modified with Tetrasulfonated Metallophthalocyanines by One-Step Electrodeposition', *Chem. Mater.*, 11 (1999) 2657.

Biotechnological Application

17. P.J. Müller, A. Christner, M. Feustel, U. Möhring and A. Neudeck, 'Biotechnology and Textile Structures (part 1)', *Melliand Textilberichte*, 4 (2004) 283.

Biomechatronics

18. H. Witte, C. Schilling, 'Biomechatronics: How much biology does the engineer need'?, *Proceedings of the 2nd International Symposium on Adaptive Motion of Animals and Machines*, Kyoto, March 4–8, 2003.
19. L. Zentner, *Untersuchung und Entwicklung nachgiebiger Strukturen basierend auf innendruckbelasteten Röhren mit stoffschlüssigen Gelenken,* ISLE Verlag Ilmenau, 2003, ISBN 3-932633-77-6.
20. W. Scheibner, D. Zschenderlein, U. Möhring, T. Keil and H. Ahlers, 'Textile Riss- und Dehnungssensoren', *Band- und Flechtindustrie* 40 (2003) 4, S. 128–131.
21. W. Scheibner, M. Feustel, U. Möhring, L. Zentner, C. Schilling and H. Witte, 'Application of textiles in material-locked joints for biomechatronic engineering', *Narrow fabric and braiding Industry* 41 (2004) 4, 120.

Part IV

Applications

19
WearCare – Usability of intelligent materials in workwear

H MATTILA, P TALVENMAA and M MÄKINEN,
Tampere University of Technology, Finland

19.1 Introduction

Conditions in many industrial jobs are challenging regarding protective clothing. Traditionally, workwear fabrics are durable and may be water, oil or chemical repellent as well as flame retardant, but may not be especially comfortable. Temperature regulation, flexible ventilation or size and shape regulation has been possible only recently with so-called intelligent textile materials. In the healthcare sector the comfort of patients as well as medical personnel could be improved with intelligent textile applications, such as textile embedded sensors, which may be used for monitoring body functions like temperature and heartbeat.

Wearable technology was the first step towards intelligent textiles. By attaching electronic devices to pieces of textile or clothing electronics became wearable. This, however, was not very comfortable to the user, and several challenges remained unsolved, such as washability, cables that remained visible and heavy power sources. During the past four to five years interactive textiles or so-called intelligent textiles have emerged. Many of them are still at a very early stage of development, but new solutions are being developed continuously. This was well demonstrated by the third Avantex[1] symposium and exhibition held in Frankfurt in June 2005.

Intelligent textile materials are becoming more practical and their interactivity is being enhanced making them a real alternative for workwear and healthcare apparel. The extensive R&D efforts being carried out in this area will no doubt result in numerous breakthrough inventions and applications in the near future.

19.2 Objectives

'WearCare' was a joint project between the Department of Textiles & Fashion

---
1. www.avantex.messefrankfurt.com

Design of the Faculty of Art & Design at the University of Lapland[2] and the Institute of Fibre Materials Science at the Tampere University of Technology[3] between 2000 and 2003. As at the start-up of the project the intelligent textile concept was still rather new and not yet presented to the market, the objective of the project was to evaluate the usefulness of intelligent textile materials in general and particularly in the health sector and in the high-temperature working environment.

Secondly, the aim of the project was to analyse and define how intelligent garment properties could be illustrated and presented by means of 3D design techniques, as such properties are often invisible and difficult to demonstrate. Defining the usability of futuristic and conceptual designs is difficult due to lack of prototypes. The third objective of the project was to develop visual methods for presenting to potential users the ideas and properties of intelligent garments that may be made in the future.

19.3 Methodology

The project partners included the two earlier mentioned universities, textile producer Finlayson Forssa Oy,[4] a workwear manufacturer, Image Wear Oy[5] and Nokia Research Centre.[6] The University of Lapland concentrated on design and usability testing and the Tampere University of Technology on intelligent textile materials. Conceptual design leaned heavily on market and user research, which was carried out in two hospitals for the healthcare products and in a steel mill for the protective clothing products. Based on product designs, an extensive search for suitable interactive textile materials as well as communication devices was carried out.

The consumer-needs model FAE (functional, aesthetic, expressive) by Lamb and Kallal was chosen to be the testing method. Functionality in the model refers to the fit of the garment, the mobility it allows, comfort, protection and ease of dressing and undressing. Aesthetic properties refer to the design and art elements, such as line, pattern, colour and texture. Expressiveness means the communicative or symbolic characteristics of a garment. Computer animation was also used for visualisation of the product concepts. The animations built around hospital and high-temperature work environments demonstrated various intelligent properties, such as communications and transmitting of alarm and distress signals.

---

2. www.ulapland.fi
3. www.tut.fi
4. www.finlaysonforssa.fi
5. www.imagewear.fi
6. http://www.nokia.com/nokia/0,50249,00.html

19.4 Textile materials

There are specific standards for textile materials to be used in the healthcare sector and especially in patients' clothing, by which material composition, weaving structure, washing ability and colour fastness are defined. Polyester and cotton are the most commonly used materials. It would be possible to use phase change materials as well as colour change materials for improving user comfort and for signalling, but their use would be restricted by standards. Textile embedded sensors and electronics are therefore more feasible applications.

The requirements for protective clothing in high-temperature environments are defined by two standards, EN 340,[7] which outlines the general requirements and EN 531,[8] which presents the requirements for protective clothing against molten metal splatter. DuPont's aramid fibres (Nomex[9] and Kevlar[10]), glass fibre, silica fibres, and BASF's Basofil[11] fibre are among others fibres that are very heat resistant and flame retardant. In areas where radiating heat stress is high, it is necessary to use heat-reflective aluminium-based materials for protecting both the user and any embedded electronics. Such materials have traditionally been very rigid and uncomfortable. BASF has, however, developed two new approaches. One is a warp knit-based material with aluminium coating, which is flexible and more comfortable. The other one is a woven fabric made by using Basofil yarns and Nomex yarns as weft and Kevlar yarns as warp. Basofil-fibre has a high LOI value (Limited Oxygen Index[12]) and does not transfer heat well. The fibre is used in woven, knitted and non-woven structures-with various application area as shown in Table 19.1.

Phase change materials are suitable as underwear or as a jacket worn under the protective clothing, especially during winter in cold climates when the temperature may vary extensively. Heat radiation may produce very high temperatures while in other parts of the plant the temperature is below zero. Three-dimensional fabrics, for example, Spacetec by Heathcoat,[13] may be

---

7. http://www.idec.gr/ppe/en/EN340.htm
8. http://www.idec.gr/ppe/en/en531.htm
9. http://www.dupont.com/nomex/
10. http://www.dupont.com/kevlar/
11. http://www.basofil.com/
12. The Limiting Oxygen Index (LOI) indicates the relative flammability of polymeric materials. Index 'LOI' is defined as the minimum concentration of oxygen in an oxygen–nitrogen mixture, required for downward burning of a vertical test object. The higher the LOI value the more flame retardant the fabric is. The LOI test is widely used for determining the relative flammability of rubbers, textiles, paper, coatings and other materials. Standards: BS 2782 (Part 1, Method 141), ASTM D2863 and ISO 4589-2.
13. www.heathcoat.co.uk

Table 19.1 Basofil fabrics

Composition	Weight (g/m²)	Application
100% Basofil or 20–60% Basofil and 40–80% aramid	370–600	Firefighters' suit when coated with aluminium, protective clothing against splashing molten metal, protective industrial gloves, protective aprons, protection in welding
60% Basofil and 40% cotton	220–300	Industrial protective clothing, knitted protective gloves
60% Basofil and 40% flame retarded viscose	140–240	Linings, upholstery
40–70% Basofil and 30–60% para-aramid	220–290	Protective clothing, against splashing molten metal when coated with aluminium, firefighter's coverall
20–60% Basofil and 40–80% meta-aramid	150–220	Industrial protective clothing, automobile racing, knitted hoods for firefighters

used for knee areas and material composition can be chosen according to each purpose. Thermochromic materials for high-temperature colour change were not commercially available during the project.

19.5 Electronics

Embedding electronics in clothing is challenging. Beside washing and ironing problems electronics may not always function properly especially in demanding working conditions. Usability surveys have also shown that consumers do not appreciate solutions that are uncomfortable with bulky devices and visible cables. Instead of wearable technology future innovations and product applications focus on integrating electronics fully into textile structures. Nanotechnology may be one way of solving such problems, and with ultra-light and strong nano-tubes textile structures can be made conductive. During the project such solutions were, however, not commercially available.

Communications, music, localisation and monitoring of biosignals are normally the properties associated with textile-integrated electronic systems. Several prototypes have been developed and some have even been launched commercially, although real breakthroughs remain still to be seen. An example is the musical jacket developed by Philips and Levi's which has a communications system and an MP3 player.[14] In this jacket, as in all wearable

---

14. www.media.mit.edu/hyperins/levis

computer products, all electronics are removable and have to be removed if the jacket is washed.

Tele-medicine is an interesting area offering a wide range of possible applications that combine textile-based sensors, wireless communication and monitoring of patients' biosignals from a distance. Applications may range from pure monitoring of the patient to disease prevention and rehabilitation.

Batteries, the traditional power source, fit poorly in line with the objectives of making intelligent garments comfortable. Other solutions must be found. Current research and development focuses on tiny fuel cells, textile-integrated batteries and exploitation of excess body heat. A transducer, which converts body heat to electricity, would be a perfect solution.

Various types of natural-user interfaces and connections have been developed for intelligent textile products, ranging from press-buttons to Velcro tapes. Tapes, bindings and zippers can be used for integrating cables into the garment. Pressure-sensitive textiles are used for keyboards, and sensors may be laminated or embroidered on fabrics. Metal and carbon yarns and conductive polymers can be integrated into the knit or woven fabric and used for signal transfer. An example is DuPont's Aracon fibre, which has been applied to heart-rate-monitoring sensors. One of the problems with conductive yarns is how to produce a yarn with an even level of resistance. Thüringen Institute[15] is currently developing a conductive cellulose-based filament by using the lyocell[16] process. Their aim is to produce a fibre with a wide scale of resistance controlled by the relative quantity of carbon (35–100%). With a high proportion of carbon in the fibre resistance increases up to a point, which makes it possible to use a fabrics made from this fibre as heating elements. Gorix[17] is an example of commercially available carbon-based fabrics that can be used for heating elements, but it cannot be applied for all purposes as it still requires a powerful energy source.

Encapsulation of electronic devices is necessary in order to make them washable and to isolate them from high thermal radiation. Thermoplastic encapsulation techniques have been developed against heat, shocks, water and chemicals and it can be applied to microchips, connectors, microsensors, switches and keyboards. This kind of protection is necessary in workwear for high-temperature environments. Encapsulation must, however, not block sound waves, for example, when microphones or loudspeakers are attached to the garment.

Optical fibres are useful for illumination and signal transfer when attached to textile materials. Fibre optic sensors are able to measure temperature, pressure, gases and smells and they can function like nerves when attached

---

15. www.titk.de
16. www.lyocell.net
17. www.gorix.com

to textile structures. Such sensors are of very light weight and flexible, they need no electric insulation and do not cause electromagnetic interference. Light from an LED source can be transferred through optical fibres to areas where light is not sufficient. This light could also be used for alarm signals. France Telecom presented a fibre optic display prototype at the Avantex exhibition in 2002. It was a rather robust display attached to a backpack woven of optical fibres and controlled by mobile phone. This was an invention produced by France Telecom's Studio Créatif within their special research programme called 'Communicating Clothes'.[18] Further development of this idea has resulted in a bluetooth-operated LED screen, which displays text, drawings and animations transferred by a multimedia messaging service.

Project WearCare focused on communications, monitoring of body functions and temperature regulation. The aim was to find rather than develop suitable textile-based solutions for making such functions possible. There are already suitable applications that could be used and are commercially available but ideal solutions were not found. Furthermore, extensive prototype testing in actual working environments (steel mill and hospital) should be undertaken in order to ensure that the solutions function properly and maintain their durability for a longer period of time. This, however, was to be done in the second phase of the project.

19.6 Usability testing

Standards EN 531 and EN 340 define the requirements for textile materials to be used for workwear in high-temperature conditions. Several other standards define the requirements for testing electronics parts. Testing temperature varies usually between 125 °C and 150 °C. The most efficient method for testing the behaviour of intelligent materials are physiology tests carried out with real persons or simulated with a thermal sweating manikin.

Current test methods are not very suitable for testing the thermoregulating properties of phase change materials, as the traditional testing method is based on a static quantity of air in the clothing. Intelligent materials are dynamic thermoregulators and they interact with skin and environmental temperature changes. Differential scanning calorimetry (DSC)[19] is used for measuring the thermocapacity of a garment or the enthalpy of a single microcapsule of the fibre containing PCM capsules. This method is normally used for measuring the melting and crystallising temperatures and heat absorbing and releasing potential of different materials. Thermogravimetric analysis (TGA) is a method for estimating thermal resistance of a microcapsule,

---

18. http://www.studio-creatif.com/Vet/Vet02Prototypes05Fr.htm
19. www.msm.cam.ac.uk/phase-trans/2002/Thermal2.pdf

as the capsule must endure the process temperatures in fibre production, spinning, weaving and fabric finishing.

At temperatures above the fusion point of carbon hydride, steam pressure rises inside the capsule causing weight loss. Normally there are two weight loss points: at 100 °C water starts to vaporise and at 280–310 °C carbon hydride escapes as capsule walls collapse. In order to prevent considerable weight loss, the temperature at which the capsule walls collapse must be clearly higher than the textile process temperatures. Doctor Doug Hittle of Colorado State University has developed a method called TRF for testing the performance of phase change materials. The equipment and method simulates physiology tests with human beings and measures the temperature-regulating ability of fabrics.

Usability testing methods applied within the project consisted of background study, virtual reality produced by 2D and 3D images and animation, analysis of users' impressions and finally evaluation of the test method. The empirical testing method is described in Fig. 19.1. Two collections of patients' clothing and three collections of high-temperature workwear were designed and visual images as well as animations were produced. Prototypes were, however, not produced, as the objective was to be able to evaluate which properties are preferred by users before launching expensive prototype production. Furthermore, this testing method allows for futuristic ideas to be tested before actual technology for producing them is available. Both patients' clothing and workwear contained non-visible intelligent solutions, such as tele-monitoring and communications.

Background research was carried out by interviewing patients and personnel in two hospitals and workers in a steel mill. Information collected through these interviews was used as a guideline for designing the collections. Both 2D and 3D images of each garment in the collections were produced. The animation demonstrated how the intelligent functions were performed and what help and assistance they would provide to the user. Altogether 13 patients, 5 nurses and 10 steel mill workers carried out the test. The test

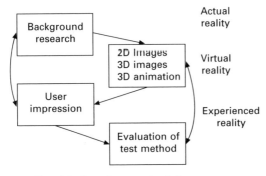

19.1 Empirical testing methodology.

person was first given background information and told about the project objectives. Then he or she watched the multimedia animation, was shown the product images, and was interviewed according to a pre-set questionnaire. According to the Lamb and Kallal's conceptual model the usability evaluation was based on functional, aesthetic and expressive properties. Functional properties are further defined by comfort, ease of dressing and undressing, protection, fit and mobility.

The patients regarded modifiability of clothing and colours as most important issues regarding comfort, followed by pocket applications and protective properties. Materials, however, were not regarded very important in terms of comfort. Two kinds of intelligent applications were attached to the patient's clothing, an interface that could be used, for example, for calling a nurse and for adjusting the bed angle and maternity pants for monitoring body functions of a fetus with pregnant women. The patients rated calling for nurse and adjusting bed angle as most important functions in patients' clothing. The hospital personnel regarded calling for nurse and monitoring of biosignals as most important features. Monitoring of fetus biosignals was also rated important. For details see Table 19.2. In addition to the intelligent properties patients and hospital personnel both regarded expressive and aesthetic aspects of clothing important in supporting patients' self-esteem and mental well-being.

High-temperature protective clothing collections contained such intelligent solutions as hands-free communications with textile-embedded microphones and loudspeakers operated with a pull-aid user interface, heating panels for back, textile-embedded wrist lights and heat radiation reflecting body shield. Heating panels for back, hands-free communication system, pull-aid user interface and hearing protectors made of textile fabrics were rated as most important intelligent properties by steel mill workers as shown in Table 19.3.

Table 19.2 Importance of intelligent properties in hospital clothing

n = 13 patients, 5 maternity patients, 5 personnel	Patients	Personnel
Calling for nurse	69%	80%
Dinner menu selection	23%	20%
Clothing selection	31%	20%
Control of massaging socks	15%	20%
Adjusting bed angle	54%	20%
Monitoring of biosignals	38%	60%
Music selection	15%	20%
Maternity pants for monitoring fetus biosignals	40%	40%

Table 19.3 Importance of intelligent properties in high temperature work wear

n = 10 steel mill workers	Steel mill workers
Heating panels for back	100%
Hands-free communication system	83%
Pull-aid user interface	80%
Hearing protection made of textile fabric	80%
Speciality knee protection	77%
Heat radiation reflecting body shield	50%
Wrist lights	45%

19.7 Conclusions

Setting standards and developing testing methods is a long process, and existing standards or test methods may not be suitable for testing totally new functional textiles, as was the case in this project with phase change materials. Researcher creativity may be the only solution for deciding how such materials should be tested until proper standards and test methods are developed. As well as textile functions, the functionality of electronics, communications, signal monitoring and transfer must also be tested. When electronics are embedded to textile structures flexible encapsulation against moisture and heat, as with high-temperature workwear, becomes necessary. Such properties may not be required from conventional electronic devices and therefore traditional testing methods may not be suitable.

The number of intelligent textile innovations and their applications in clothing will no doubt increase in the future. At the moment numerous research institutes as well as industrial companies are busy developing such innovations and applications. Intelligent textile innovation projects are very cross-scientific requiring skills and knowhow not only regarding textiles, but often electronics, signal transfer, communications and medicine. The aim is to develop something new, which we have no prior experience of and whose usability and functionality may be questionable. Therefore, concept planning with usability testing is a rational first step before producing prototypes or bringing the product to the market.

One of the main objectives of project WearCare was to investigate how useful 3D modelling and animations are in usability testing, specifically when we need to find out the impressions and opinion of the potential user. Such systems, when successful, would enable the research team to analyse their product concepts and ideas at a very early stage and would make it possible to eliminate errors and handicaps that might cause the product to be unusable. The results of WearCare clearly demonstrated that 3D modelling and animations can be successfully used for early-stage usability testing. Patients, hospital personnel and steel-mill workers were able to identify which features were desirable and which in their opinion were not so important. The results of the usability test proved to be very valuable and they produced hands-on guidelines for further research and development.

19.8 Bibliography

Anderson K and Seyam A (2002), 'The road to true wearable electronics', *Proceedings of the Textile Institute 82nd World Conference, Cairo,* March 2002.

Hittle D C and Tifani L (2002), *A new test instrument and procedure for evaluation of fabrics containing phase change materials,* ASHRAE, Atlanta, GA, USA.

Lamb J and Kallal M (1992), 'A Conceptual Framework for Apparel Design' *Clothing and Textile Journal* 10 (2), 42–47.

Uotila M, Hilden M, Matala R, Pursiainen M, Ruokanen M, Mäkinen M and Talvenmaa M (2002), 'WearCare, intelligent materials in workwear', *Textile Asia,* October 2002, 33–35.

Internet websites

http://www.ravistailor.com/customtailor/Tomorrowacutes_Ewardrobe_._.custom_Clothing_Online.htm
www.tut.fi/units/ms/teva/projects/intelligenttextiles/
http://www.tie-lock.com/general-article-6.htm
http://computer.howstuffworks.com/computer-clothing2.htm
http://www.herenorthere.org/msquare/aesth.htm

20
Intelligent textiles for medical and monitoring applications

J-S SOLAZ, J-M BELDA-LOIS,
A-C GARCIA, R BARBERÀ, J-V DURA,
J-A GOMEZ, C SOLER and J-M PRAT
Instituto de Biomecanica de Valencia, Spain

20.1 Introduction

Healthcare is a key market for the textile industry. In the year 2000, over 1.5 million tons of medical and hygienic textile materials, with a value of 5.4 billion dollars, were consumed worldwide. It is estimated that this figure will increase in volume by 4.5% per annum, so that by 2010 it will have reached 2.4 million tons, with a market value of 8.2 billion dollars (David Rigby Associates, 2002). Thus, the healthcare and medical sector offers the greatest opportunities for developing the most sophisticated high value textiles for niche applications (Czajka, 2005) and is one of the most important areas to foster growth. Within the textile industry there is ever-increasing competition from manufacturers worldwide that brings the need to find developed markets for specialised products, in a sector that is changing from traditional production methods to advanced technology. This consideration is especially important for countries with high labour costs such as those of western countries.

Intelligent textiles combine various technologies from different fields, mainly from the traditional textile industry and from research in technologically advanced smart materials. The incorporation of these technologies into textiles and vice versa will open new opportunities for intelligent textiles in medicine and medical monitoring. Among the possible applications of intelligent textiles, wearable systems and ambient intelligence are two of the most important. Wearable systems integrate information and communication technologies in garments. Integration into clothes and textiles is one of its major objectives. Ambient intelligence is a way of making interfaces between humans and computers disappear (Gaggioli et al. 2003). Ambient intelligence implies an intelligent environment, in which computers disappear or become transparent. Integration of an interface in clothing is one of the latest trends in this line of research. One example of this is the development by TU Berlin of the technology for embedding active devices (chips) in the inner layers of flex laminate (Vanfleteren and Lutz-Günter, 2005). BAN (body area network), PAN (personal area network), smart clothes and other devices placed in the body that will be able to detect patients' vital signs and retransmit them to sorting nodes, in real time (Gaggioli *et al.* 2003) are some examples of the potential for this technology in medical applications. Within ambient intelligence, intelligent textiles can be considered as a transparent interface incorporated in clothes, or as sensors for recording the user's activities. Both approaches are highly suitable for medical and monitoring applications.

The incorporation of interfaces and sensors into clothes is a highly attractive option for medical and monitoring applications as it facilitates the automatic sensorisation of the activity. If sensors are included in the garments, wearing the garments equals to sensorisation. This is superior to other approaches to e-health. Also, wearable medical devices are often rejected by the user as they highlight the existence of pathologies or medical problems. If the devices are integrated in clothing this problem disappears. The advantage of intelligent textiles for medical applications is due to two reasons: they are in contact with the skin, and they follow the user's movements and therefore make possible the sensorisation of physiological and biomechanical parameters. However, there are two drawbacks that have to be dealt with: the high impedance of skin contact for recording physiological signals and the draping movements of garments (Martin *et al.*, 2004), which mean that the follow-up of movements is not precise. This consideration should lead to new techniques for analysis and treatment of the signals provided by intelligent textiles. New approaches and techniques are necessary instead of the established algorithms and procedures.

These approaches will lead to new concepts in health, with a better quality of life for patients and lower social costs, changing the model from a system in which citizens have to go to a centre where they receive medical attention (hospital-centred health systems), to a system in which the attention is individualised and delivered to the user wherever he may be (citizen-centred health) (Lymberis and Olson, 2003). This chapter focuses on the applicability of intelligent textiles to medical and monitoring uses from a practical perspective. The importance of intelligent textiles in medical and monitoring applications is described in section 20.2. Section 20.3 analyses the potential applications of medical textiles grouped into diagnosis/complementary explorations, secondary prevention and healing/medical treatment. In section 20.4 the existing intelligent textiles technologies applied to monitoring and medical treatment are described. Finally, in section 20.5 future trends and applications are considered.

20.2 Importance of intelligent textiles for healthcare

Major social changes, progress in science and technology and increased medical knowledge are driving the evolution of healthcare and health delivery

Intelligent textiles for medical and monitoring applications 371

(Lymberis and Olsson, 2003). Present-day society is undergoing changes such as ageing of the population, further integration of people with disabilities and an increase in chronic diseases. The control of these diseases is critical, as seven major diseases (heart disease, cancer, diabetes, arthritis, chronic bronchitis, influenza and asthma) accounted for 80% of deaths in the US in 1990. For many of these health problems, early systematic intervention would be highly beneficial (Park and Jayaraman, 2003). However, this intervention has to be carried out keeping in mind the direct annual medical cost, which leads to the search for alternative solutions. The healthcare industry must meet the challenge of balancing cost containment with maintenance of desired patients' outcomes (Park and Jayaraman, 2003). In this context, wearable electronics and intelligent textiles may become key elements in ambient intelligence (Catrysse *et al.*, 2004). In fact, the possibilities of unobtrusive long-term monitoring from intelligent textiles could benefit, among others, patients with spinal cord injuries, the cognitively impaired, elderly people living at home and infants at risk of Sudden Infant Death Syndrome (Van der Loos *et al.*, 2001; Gibbs and Asada, 2005; Korhonen *et al.*, 2003). Although the costs are low and there is a high degree of integration with the user, which makes them very attractive, there are some aspects that should be pointed out regarding these potential treatments.

20.2.1 Levels of integration of the devices in textiles

There is no one single 'intelligent textile' or 'wearable device' able to gather and process all the signals generated by the human body. Each application needs to acquire different physical parameters or to act on different parts of the body. Also, the devices can be integrated to different degrees in textiles, depending on the cost, desired complexity and application. At the first level, solutions can adapt to clothes (i.e. a pocket to hold a mobile phone). At the next level, electronics and/or microsystems can be integrated into clothes (Fig. 20.1) or into textiles with connectable modules (e.g. with textile conductors). The third level of integration consists of embedding active structures in the textile fibres (e.g. woven displays). There is still another level; active fibres are the micro-electronic building blocks, such as transistors or diodes (Lee and Subramanian, 2005). However, substantial developments of flex technologies (Strese *et al.*, 2005) are still needed to convert these applications into commercial products. At this level, textronics appears as a new interdisciplinary branch of science with synergic connections between textile science, electronics and computer science. The objective of textronics is to obtain multifunctional, smart textile products of complex inert structure but with uniform functional features (Gniotek and Krucinska, 2004).

20.1 Microelectronics triaxial gyroscope. (Instituto de Automática Industrial.)

20.2.2 User requirements for intelligent textiles

The use of intelligent textiles as wearable devices (sensors or actuators) must satisfy a series of user requirements in order to guarantee results and the appropriate response in use (Korhonen *et al.*, 2003):

- Reliability, robustness, and durability. The environments in which the sensors need to operate vary, and the users are nonspecialists not necessarily aware of the technical limits of usage.
- Look/unobtrusiveness. Intelligent textiles must be integrated into the everyday life of the users. Hence, their appearance should either fit in with the individual's preferences or they should 'disappear', i.e., be as unobtrusive as possible. As textiles are used by everybody in their daily lives, the integration of intelligent textiles will not present serious problems.
- Communication. The sensors and measurement devices should be capable of transferring their information to some central data storage, preferably by means of a fully automatic system, or at least so easily that it does not pose a burden to the user.
- Zero maintenance and fault recovery. An important issue in the case of health monitoring is self-calibration and drifting avoidance, i.e., finding ways to guarantee that the sensor performance does not deteriorate over time. Calibration and accuracy are two issues that are obviously fundamental (Teller, 2004).
- Customization. Sizing of the wearable sensorised systems and

confidentiality of the obtained data are two aspects in regard to the user that must be considered in the development of intelligent textiles applications.

20.3 Potential applications of intelligent textiles

The range of intelligent textiles applications in medicine is fairly wide, ranging from clinical monitoring, in which a continuous flow of data is produced during long periods of time, to a drug-administering treatment. Doctors need tools that permit them to improve the performance of treatments and subsequently increase the quality of life of their patients. Several models can be employed to classify intelligent textiles in health and monitoring applications but, if the patient's welfare is taken as the criterion, there are two approaches that can be considered as adequate. The first classification involves the participation of the patient in interaction with the system embedded in the textile. According to this criterion, medical applications of intelligent textiles could be classified into disease management and remote monitoring. The second classification criterion considers the role that intelligent textiles may play in each clinical phase; diagnosis, prevention or treatment. This last classification will be used in the present section to describe in detail the potential application of intelligent textiles.

According to the first classification criterion, medical applications of intelligent textiles can be carried out in two different modes:

Mode 1: In disease management the individual is actively participating in the management process. An example of this type of application is a diabetic subject who may be monitoring blood glucose values, storing these results in a database, and receiving feedback. This will be used in evaluating the success of a pharmacological treatment or a diet to improve the blood glucose balance or the regime of activity, sport, etc. In this model it is the individual who is actively willing to receive feedback regarding his wellness or disease status and to participate in his own care (Korhonen *et al.*, 2003). The role of the physician is to support and advise the individual in his daily activities affecting the management of the disease. The capability of the system integrated in the textile to communicate in a comprehensible manner with the patient is a fundamental factor in maintaining motivation for continuing use of technical devices.

Mode 2: Remote monitoring is often related to independent living. In this model, the main (and often only) consumer of the measurement data is the caregiver, and the individual is not interested in or capable of interpreting the measurements (Korhonen *et al.*, 2003). The health status of an elderly subject living alone might be remotely monitored to detect possible deterioration in the subject's health status, or to detect sudden problems such as falls or health hazards (Perry *et al.*, 2004). Telemonitoring ECG (Bonato, 2003) could be useful in patients with unstable cardiac diseases in certain phases. If the patient complains of heart-related symptoms, by using the device, he/she may send the ECG to the care centre for review and judgement of whether immediate help is needed.

The ease and automation of use of the technical devices is strongly emphasised as the users may have limited technical skills, compromised capabilities to attach wearable sensors on a daily basis, and/or reluctance to accept any new technical device to be worn on their body. Besides, since their independence may be entirely due to the monitoring system, the subject will be highly motivated to accept the possible inconvenience caused by using the system. Integration is also a key factor for user acceptance. Most of the reasons for rejecting current approaches are social exclusion for wearing visible devices that may be a sign of disease or a health problem. If the applications can be part of the clothes or are an in-depth integration with user garments, acceptance (and therefore commercial success) will increase.

According to the second criterion, potential applications can be classified as follows (Fig. 20.2):

Diagnosis/complementary explorations: Monitoring and recording of signals (i.e. Holter monitors, body temperature, sweating, etc.).
Secondary prevention: Detecting associated risks and communication of physical changes (i.e. biofeedback) to avoid health problems.
Healing/medical treatment: Acting on an existing problem and modulating the response (i.e. intelligent orthosis).

Inside these general groups, there exist particular cases of especial interest due to the prevalence of the disease or the seriousness of the illness. In the following subsections the main potential applications in complementary exploration, prevention and treatment are described.

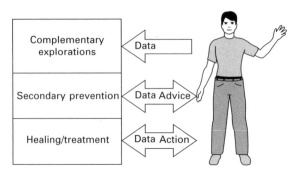

20.2 Classification of medical application of intelligent textiles according to the clinical phase.

20.3.1 Diagnosis/complementary explorations

Clinical monitoring of physiologic signals, such as electrocardiogram or blood pressure, provides only a brief window on the physiology of the patient. The interaction of variability in heart rate and blood pressure supplies information about higher regulatory systems, such as the systemic arterial baroreceptors. Insight into such higher-order control systems permits even greater understanding of disease pathogenesis and, potentially, disease therapy. However, the brief periods of clinical monitoring limit the potential of these data, mainly because of three major shortcomings (Binkey *et al.*, 2003):

1. They are likely to fail sampling rare events that may be of profound diagnostic, prognostic or therapeutic importance.
2. They fail to measure physiological responses during normal periods of activity, rest and sleep, which are more realistic indicators of the health of the patient.
3. Brief periods of monitoring cannot capture the circadian variation in physiological signals that appear to reflect the progression of disease.

To overcome these problems, devices that readily supply this information in an integrated fashion over a complete 24-hour interval in an outpatient setting would be the solution. However, even though there are several current research projects that are trying to solve this problem, a complete application is still not available (Binkey *et al.*, 2003).

In this section, the main monitoring applications in which intelligent textiles may play a fundamental role are described. Heart rate and blood pressure are basic in medical diagnosis and their control will reduce significantly the number of serious health problems. Patterns of movement and EEG are used in diagnosis of neurological diseases and as an input to decide treatment. Obtaining these data in a non-invasive manner will increase the health status of the patients and subsequently their quality of life, and will permit them to adapt to the treatment in real time.

Heart rate

A classical application in which ambulatory measurements are required is in the exploration of cardiovascular diseases. It is known that variability in heart rate over a 24-hour period is an important indicator of disease evolution and progression. A lack of diurnal heart rate variability is characteristic of patients with congestive heart failure (CHF) and cardiomyopathy and is likely to be a result of the profound abnormalities in autonomic function that characterise these patients (Binkey *et al.*, 2003). Patients with CHF have in common an imbalance of the autonomic nervous system, which may contribute to the progression of circulatory failure and influence survival. The sustained imbalance of the autonomic tone over a 24-hour period may promote the progression of circulatory failure and predispose these patients to malignant ventricular arrhythmias and sudden cardiac death (Panina *et al.*, 1995). Similar observations have been made in patients with ischemic cardiomyopathy (Binkey *et al.*, 2003). Diurnal variations in heart rate variability may indicate specific times of day that a patient may be more responsive to a given therapeutic intervention and may be a helpful tool for indicating the optimal timing of effective drug administration.

For all the above reasons, ambulatory systems for ECG monitoring (Bartels and Harder, 1992) (Panina *et al.*, 1995) have been part of the routine evaluation of cardiovascular patients for almost three decades (i.e. Holter monitors). However, these systems are not suitable when monitoring has to be accomplished over periods of several weeks or months, as in a number of clinical applications (Bonato, 2003) and their poor signal quality limits the accuracy of arrhythmia classification algorithms to fully monitor and manage rhythm disorders. Although impressive advances have been made in recent years in the field of ECG, there are still some problems that remain unsolved (Lobodzinski and Kuzminska, 1998) (Catrysse *et al.*, 2004).

The capacity of intelligent textiles to non-invasively detect physiological signals makes them the ideal candidate to solve these problems. Also, as they have the capacity to implement the equivalent of servo-controlled administration of therapy, they may also greatly improve the efficacy of treatment for cardiovascular disease (Binkey *et al.*, 2003).

Blood pressure

Patients with hypertension who fail to experience a drop in blood pressure during nocturnal intervals have been found to have a greater risk for events such as stroke and greater end-organ damage, such as renal dysfunction (Binkey *et al.*, 2003) (Engin *et al.*, 2005). Persistent rises in arterial pressure imply disturbance in the complex and multifactorial cardiovascular control mechanisms. In this context, neurohumoral disturbances could play a special role, in view of the fact that they have demonstrated that elevated sympathetic drive seems essential in hypertensive patients. Parameters obtained by spectral analysis of heart rate variability (HRV) might furnish useful information on autonomic normal and abnormal nervous system regulation (Pagani and Lucini, 2001).

Patterns of movement

Neurological disorders such as Parkinson's disease (PD) or hemiparetic stroke severely affect motor functions. PD is the most common disorder of movement affecting at least three per cent of the population over the age of 65. The characteristic motor features are development of a rest tremor, bradykinesia, rigidity and impairment of postural balance.

Several lines of research provide preliminary evidence that motor function in humans can be enhanced after a stroke through exercise and perhaps with medication. Dopamine therapies are often successful for some time in alleviating abnormal movements, but most patients eventually develop motor complications as a result of the treatment.

A reliable quantitative tool for evaluating motor complications in PD patients would be valuable both for routine clinical care of patients as well as for trials of new therapies. In routine care, it would be very useful to obtain information on a patient's motor pattern during the course of several days and then relate this to the timing and dose of medications (Binkey *et al.*, 2003).

EEG

Ambulatory electroencephalography (AEEG) monitoring allows prolonged EEG recording in the home setting. Its ability to record continuously for up to 72 hours increases the chance of recording an ictal event or interictal epileptiform discharges (Waterhouse, 2001) and is a useful tool in the study of sleep disorders, etc.

20.3.2 Secondary prevention

Secondary prevention aims at the avoidance of health problems in people at risk. The scope of secondary prevention is the analysis of clinical risks and its communication to the user or the carer to avoid adverse consequences. The possibility of inclusion of sensors and advice devices into the garments of the users should lead to a broad new range of applications in this field.

Detecting clinical risks and communicating physical changes to the patient or the carer may save children's lives in the prevention of sudden infant death syndrome (SIDS) (Park *et al.*, 1999), reduce deaths of elderly people caused by falls, increase the quality of life of people with dementia or Alzheimer's and detect critical situations when monitoring patients in post-operative recovery.

Sudden infant death syndrome

SIDS is the sudden death of an infant under one year of age that remains unexplained after a thorough case investigation. In the last 20 years the incidence of SIDS has been steadily diminishing (since 1983, the rate of SIDS has fallen by over 50%), but there are still about 2500 deaths per year in the US, and thousands more throughout the world. One of the methods suggested for preventing SIDS is a home monitoring system (American SIDS Institute, 2005). Online monitoring of heart and breath rates of infants and their integration into children's clothing are an area of research interest (Hertleer et al., 2002).

Prevention of falls

Other critical cases in which the life of the patient may depend on the transmission of information to the carer are falls in elderly people and potentially dangerous situations due to agitation in people suffering from dementia. From 1992 through 1995, 147 million injury-related visits were made to emergency departments in the US. Falls were the leading cause of external injury, accounting for 24% of these visits. Emergency department visits related to falls are more common in children less than five years of age and adults 65 years of age and older. Compared with children, elderly persons who fall are ten times more likely to be hospitalised and eight times more likely to die as the result of a fall. Trauma is the fifth leading cause of death in persons more than 65 years of age, and falls are responsible for 70% of accidental deaths in persons 75 years of age and older (Fuller, 2000).

There are several biomechanical parameters that can be used as options for the assessment of the risk of falling. Inertial sensors (accelerometers or gyroscopes) are good candidates because of their light weight and size. A change in the pattern of movements of the trunk of a user at risk of falling can activate an alarm system to advise the user to take additional safety measures to avoid falling (i.e. when sitting or leaning on a wall).

Detection of states of agitation

According to the United Nations, the population of the developed countries was 1143 million in 1990, with 143 million of these being aged 65 or over. Applying the prevalence rates for various age groups given above, we arrive at an estimate of 7.4 million persons with dementia (Prince and Jorn, 1999). Given that Alzheimer's disease generally makes up the majority of cases in the developed countries, more than half of these people would have Alzheimer's disease (i.e at least 3.7 million people). The agitation that can occur in association with dementia greatly complicates patient care, poses a risk to the patient's health and safety, and significantly increases the burden for families and carers. Alarm systems and medical bracelets are solutions nowadays used to control patients. But the information provided by these systems does not include the health or agitation level of the patient. As harmful or violent behaviour towards self or others is possible in some patients, they can easily break the bracelet or any other wearable device. Integration of healthcare systems into textiles is the only feasible solution

Intelligent textiles for medical and monitoring applications 379

for controlling the situation and health status of this group of patients. There are many possible strategies for the analysis of a state of agitation; heart rate, breath rate, EEG or galvanic skin conductivity. Inertial sensors can also be useful as changes in the length of stride can be indicative of a state of agitation.

Biofeedback

Biofeedback is a therapeutic technique in which an individual learns to consciously control involuntary responses such as heart rate, brain waves, or muscle contractions. Information about a normally unconscious physiological process is relayed back to the patient as a visual, auditory, or tactile signal. These responses are electronically monitored and noted through beeps, graphs, or on a computer screen, which are seen and heard by the participant (Newman, 2003).

Biofeedback has been playing a significant role in the prevention of health problems such as urinary incontinence (UI) in women through the monitoring of pelvic floor muscles (PFM) contraction during therapeutic exercises with surface EMG (Tries and Eisman, 1995). Other applications of biofeedback are the control of attention deficit hyperactivity disorders (ADHD) in children through monitoring with digital polygraphs, or the prevention of repetitive strain injuries in workplaces using surface EMG (Pepper and Gibney, 2000). Biofeedback can be greatly improved through the integration of electronic monitoring systems in garments, to allow feedback monitoring when and where the measurements are required; in the workplace for injury prevention, at home for daily activities, etc.

Patients undergoing a physical therapy regime in post-operative recovery need to find out if they are doing their exercises correctly and if their range of movement is increasing. Intelligent textiles integrated into clothing could be a tool to inform patients and carers about the recovery process (Gould, 2003). In daily life, injured people may need information about the potential risk of an action (i.e. lifting a weight or reaching a limit position). Devices like the Intelligent Knee Sleeve, provide audible feedback with respect to changes in knee flexion angle during movements (Munro *et al.*, 2003) in order to prevent an excessive stress that may worsen the problem. The generalisation of this application to other parts of the body could be a useful tool for traumatologists and physiatrists.

20.3.3 Healing/medical treatment

Ambulatory medical treatments are, in most cases, based on drugs for different purposes or, in orthopaedic surgery patients, may require an orthosis.

Delivery of drugs

Medical treatments based on the administration of drugs may be a problem in some patients (i.e. in children) due to the concentration of the dosage and, in the case of dependent people (i.e. people with dementia), since a carer may be necessary to supply the proper dose of the drug at the right time. In other cases, such as in external treatments (i.e. ointments), some patients may have difficulties in reaching certain parts of the body or the dose may be inadequate. For these cases, materials that release substances open up a huge number of applications as drug supply systems in intelligent textiles (Van Langenhove and Hertleer, 2004).

Wearable orthoses

The development of wearable ambulatory devices (orthoses) that mechanically suppress upper-limb tremor while preserving, as far as possible, natural movement, could vastly increase the quality of life of patients. Tremor can be a significant problem to sufferers of essential tremor, multiple sclerosis, Parkinson's disease and other neurological problems resulting in movement disorders, when engaged in the activities of daily living, requiring particular dexterity such as eating, dressing and writing (Manto *et al.*, 2003). The categories of tremor to be addressed are those resulting from progressive neurological disorders, such as Parkinson's disease and multiple sclerosis as well as tremor resulting from cerebellar trauma, and atypical essential tremor. The DRIFTS project has allowed the development of a wearable device that filters out undesirable movements, allowing the patient to carry out his daily activities (Manto *et al.*, 2003; Belda-Lois *et al.*, 2005; Loureiro *et al.*, 2005; Rocon *et al.*, 2005). However, the DRIFTS results are only a first approach to improving quality of life, as its weight and invasivity are still a problem. The use of intelligent textiles as force actuators may significantly improve the device, increasing comfort and wearability (Fig. 20.3).

Although wearable solutions are being developed, there are still many integration problems to be solved in order to transform the prototypes into really usable products. Most of the wearability problems may be solved by means of intelligent textiles used as sensors and actuators, reducing the weight and volume of the device. Thus, intelligent textiles and garments that provide an appropriate degree of comfort have a high market potential (Bartels, 2005).

20.4 From medical needs to technological solutions

In section 20.3 a series of medical needs were described, ranging from sensors for EMG to drug delivery systems and alarm systems for dependent

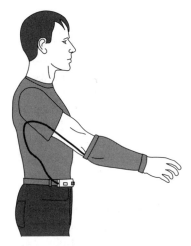

20.3 Prototype of DRIFTS orthosis. (DRIFTS QKL6-CT-2002-00536).

people. All these needs have something in common; the patient's independence and quality of life will be dramatically increased if it is possible to integrate the systems into a textile. A comfortable shirt or suit or a sheet for patients that have to remain in bed would be a less invasive solution and would alleviate the user's sensation of being ill and controlled, besides the clinical advantages.

All components of interactive electromechanical systems (sensors, actuators, electronics for data processing and communication and power sources) can be made from polymeric materials, to be woven directly into textile or printed or applied onto fabrics (Engin *et al.*, 2005). In this section, the existing intelligent textiles used as sensors, actuators and in communication are described. Some of them are existing commercial products and others are still under development.

Integrating the aforementioned components into textiles is not an easy task, and several problems have to be solved. Some questions, such as getting textile-shaped energy supply systems (Seyam, 2003; Powerpaper, 2005) or washability (Kallmayer *et al.*, 2005; Martin *et al.*, 2004) are common to all the applications, but others are specific to a certain case. Before designing a system based on intelligent textiles with a definite purpose, design criteria based on studies of existing wearable electronic and medical devices similar to the new product should be carried out (Teller, 2004). However, one thing is fundamental to guarantee the success of the intelligent textile application; user needs (Buhler, 2000) must be considered from the initial phases of the development, otherwise the final product will be useless.

20.4.1 Intelligent textiles as sensors

To fully qualify as a smart textile, a sensor and actuator have to be present in the textile structure (Engin *et al.*, 2005), but nowadays most of the experimental and commercial intelligent textiles act as 'sensors + communicators' or 'sensors + data storage'. Various body signals (see Fig. 20.4) can be acquired by means of the specific properties of smart materials capable of being embedded into textile structures. In fact, there already exist, both as prototypes and also as commercial products, several applications of intelligent textiles that are being used to obtain data from the subject's physical response.

In this subsection, sensoring applications have been classified, according to the type of acquired data, into physiological signals, biomechanical signals and other parameters. However, the most interesting applications are achieved by combining several measurement techniques (Van der Loos *et al.*, 2001) to obtain a more complete subject diagnosis that overcomes the disadvantages of the classical methods (guardians and clinical polysomography).

Physiological signals

Physiological signals may supply critical information to doctors and carers about the health of the patient. In fact, the proper identification of patterns or occasional phenomena in these data could avoid a high percentage of deaths. Heart rhythm, breath rhythm, temperature, galvanic skin response and pH sensors are described in the following paragraphs.

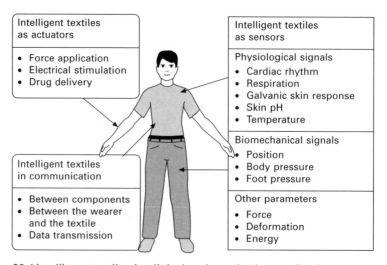

20.4 Intelligent textiles in clinical and monitoring applications.

Heart rhythm

Several initiatives have been developed in order to integrate cardiac rhythm sensors in textiles. The 'smart shirt' wearable motherboard, which can be used in persons who have known disorders, permits a constant monitoring of their physical condition by medical personnel in a non-invasive manner (Park and Jarayaman, 2003). This shirt includes special sensors and interconnections to monitor an individual's vital body signals, and provides a systematic way of controlling the vital signs of humans in an unobtrusive manner (Park et al., 2002).

Wealthy is also a health monitoring system based on a wearable interface with integrated fabric sensors. Its aim is to set up a fully integrated garment system that is capable of simultaneously acquiring a set of physiological parameters in a 'natural' environment. The system targets the monitoring of patients suffering from heart diseases during and after their rehabilitation (Paradiso and Wolter, 2005). Similarly, Lifeshirt is a commercial product that collects, analyses and reports on the subject's pulmonary, cardiac and posture data. It also correlates data gathered by optional peripheral devices that measure blood pressure, blood oxygen saturation, EEG, periodic leg movement, core body temperature, skin temperature, end tidal CO_2, and coughing (Vivometrics, 2004).

The development of integrated heart rate measurement textiles is still a topic of interest and research. There have been several national projects like VTAMN (French National Funded-RNTS, 2000; Lymberis and Olsson, 2002) and currently there are still some others under the Sixth European Framework Programme to investigate the potential of body-worn electronics. The project 'MyHeart' focuses on prevention of cardiovascular diseases with intelligent biomedical clothing (Locher et al., 2005). Its goal is to integrate system solutions into functional clothes with integrated textile sensors. The combination of functional clothes with integrated electronics and on-body processing, creates a sort of intelligent biomedical clothing. The processing consists of making diagnoses, detecting trends and reacting to them (MyHeart Project, 2005).

All these shirts use several techniques for measuring ECG by means of textile embedded sensors in direct contact with the skin. Steel fibres are an appropriate solution for making these sensors as they feel good, have the right conductivity, low toxicity to living tissue, little or no danger of contact allergies because of the very low content of nickel, can be easily washed without losing their properties, and can be manipulated as a textile material (Van Langenhove and Hertleer, 2004). Steel fibres have been used in different ways, wound round acrylic yarns, with a layer of acrylic/cotton fabric coupled with a layer containing stainless steel threads (De Rossi et al., 2003; Engin et al., 2005) or twisted around a viscose textile yarn and knitted (Paradiso and Wolter, 2005). Fibres are woven and knitted, generating a family of sensors called 'textrodes'.

Nonmetallic solid state conductive polymers can also be used as ECG electrodes. A possible electrode structure can be made of cross-linked poly N-vinylpyrrolidone. The electrode (Lobodzinski and Kuzminska, 1998) comprises a thin strip of polyester film coated on one side with silver/silver chloride. Mühlsteff and Such (2004) describe dry electrodes based on silicone rubber produced using a thermal moulding process and integrated by a sandwich design (rubber–textile–rubber) to measure ECG signals. These electrodes can use sweat produced by the glands for a conductive bridge from the skin to the electrode and have a very good long-term stability due to the chemically inert material.

These systems based on electrodes show signal artefacts induced by the subject's motion, electrode motion, patient's skin and cable motion. The influence of motion depends on how the electrodes are fixed to the body and can be greatly reduced by increasing contact pressure. Finding appropriate electrode positions on the body and ensuring appropriate contact pressure will be major tasks in the system design process. Also, the implementation requires that electrodes have sufficient electrical properties, show long-term robustness and should not require any interaction with the end user. Easy integration techniques and low production costs can establish a mass market in functional textiles (Mühlsteff and Such, 2004).

Breathing rate

Breathing rate is commonly measured by means of magnetometers, strain gauges or inductance plethysmography (Martinot-Lagarde et al., 1998). Respibelt (Hertleer et al., 2002) is a fabric sensor made of a stainless steel yarn, knitted in a Lycra® belt, providing an adjustable stretch. By wrapping the Respibelt around the abdomen or thorax, changes in the circumference and length of the Respibelt caused by breathing cause variation in both inductance and resistance. Most studies report that, for long-term monitoring, when only respiration rate is required, there is no significant difference between measurement by strain gauges and inductance plethysmography (Catrysse et al., 2004).

The Wealthy system also contains a prototype of a respiration sensing device that uses impedance to derive the respiration of the wearer (Wealthy, 2002). Lifeshirt® monitors respiration by means of thoracic and abdominal inductive plethysmography bands sewn into a Lycra® vest. In this case, the shirt is functionalised with carbon-loaded rubber (CLR) piezoresistive fabric sensors, used to monitor respiration trace (Grossman, 2003).

Temperature

Body temperature can be measured by textile-embedded thermocouples (Van Langenhove and Hertleer, 2004) or thermistor-based sensors (Teller, 2004; Locher *et al.*, 2005). Temperature-sensitive polymers (amphiphilic block copolymers) offer a change of phase at a tuneable temperature. In these copolymers, the different parts of a molecular chain have different water-attracting properties; in other words, parts of the molecule are hydrophilic and other parts hydrophobic. This polymer can reversibly alternate between a gel structure and a micelle structure (Williams, 2005).

Galvanic skin response

The wearable armband sense wear system measures the galvanic skin response (GSR). GSR represents the electrical conductivity between two points on the wearer's arm. The Armband's GSR sensor includes two hypo-allergenic stainless steel electrodes connected to a circuit that measures the skin's conductivity between these two electrodes. Skin conductivity is affected by the sweat from physical activity and by emotional stimuli. GSR can be used as an indicator of evaporative heat loss by identifying the onset, peak, and recovery of maximal sweat rates (Teller, 2004). This system could be implemented in the textile by means of the previously described textile-embedded ECG sensor techniques.

Skin pH

Skin pH is related with cutaneous diseases and its value can be determined by means of embedded pH responsive polymer fibres. A dramatic alteration in the degree of ionisation of the polymer can result in a profound change in the molecular arrangement. This structure is effectively 'coiled' under one condition and 'uncoiled' under another; this is manifested by significant changes to volume, shape and properties (Williams, 2005).

Biomechanical signals

Biomechanical signals are very varied in type and include joint motion, body and foot pressure signals among others.

Joint motion

Monitoring body kinematics has a fundamental importance in several biological and technical disciplines. In particular, the possibility of exactly determining posture may furnish a useful aid in rehabilitation topics (Togneti *et al.* 2005).

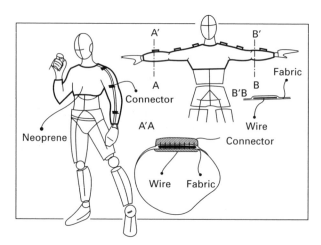

20.5 Mobility measurement by integrating accelerometers in the textile. Source: IBV.

Four different approaches have been employed to date for the kinematic analysis of joint motion: attachment of discrete inertial sensors to the textile, inclusion of thin piezoelectric films, manufacturing of piezoresistive fibres and coating of yarns and fibres with CLRs. Accelerometers and gyroscopes are inertial sensors very suitable for attachment to a textile substrate (Fig. 20.5), as they are small, lightweight and reliable (Belda-Lois *et al.*, 2005).

Piezoelectric films can act as shape or angle sensors (Martin *et al.*, 2004). These sensors exhibit excellent motion-sensing capabilities (Edmison *et al.*, 2002), low power consumption, and can be found in a broad range of voltages based on the type and magnitude of the applied stimulus. Piezo-resistive textiles (a combination of conducting polymer and fabric) are useful due to their elasticity, ideal conformability to the human body, and high piezoresistive and thermoresistive coefficients (De Rossi *et al.*, 1999). However, conducting polymer-based sensors are not easily amenable to textile technology (Carpi and De Rossi, 2005). A new generation of high-performance strain sensors has been obtained by coating yarns and fabrics with CLRs, typically consisting of a silicone matrix filled with carbon black powder (Carpi and De Rossi, 2005; De Rossi *et al.*, 2003; Engin *et al.*, 2005). CLRs show, as do PPy sensors, thermoresistive properties.

Interface pressure

Pressure values between body and an interface (i.e. seats or beds) have been widely studied in order to improve users' comfort (Buckle and Fernandes, 1998) and reduce pain or sores in elderly patients. For this purpose, measuring devices integrated into a sandwich-like structure covered by textile layers have been used in practice (Babbs *et al.*, 1990). Following this line of thought, many devices have been patented to measure body position, breathing rate and heart rate using force sensitive resistors, capacitive sensors, piezoelectric sensors and microphones, but in most cases, they were expensive and wired to other devices (Van der Loos, 2001).

The approach to measuring pressure by means of intelligent textiles has taken the form of pressure sensitive mats consisting of a spacer fabric with embroidered electrically conductive patch arrays on both sides (Locher *et al.*, 2005). Each opposing patch pair in the arrays forms a plate capacitor whose capacity changes with the compression force on the spacer fabric. Technological production is becoming ever cheaper and nowadays it is possible to find pressure-sensing textiles such as the Softswitch (Softswitch, 2001), which uses a so-called 'quantum tunnelling composite' (QTC). This composite is an insulator in its normal state and changes into a metal-like conductor when pressure is applied to it. The pressure sensitivity can be adapted for different applications. Using existing production methods, the active polymer layer can be applied to every type of textile structure, knitted fabrics, and woven or nonwoven fabrics (Van Langenhove and Hertleer, 2004).

Foot pressure

Foot pressure is related to comfort (Chen *et al.*, 1994) and is extremely important in patients with Diabetes Mellitus (DM) and other foot pathologies. Diabetics are prone to foot ulcerations due to neurological and vascular complications. Ulcers appear as a combination of physiological, structural and biomechanical changes and several wearable piezoelectric insoles have been developed in order to study how to reduce plantar pressure (Sarnow *et al.*, 1994) such as Biofoot/IBV (Fig. 20.6).

Other parameters

Closely related to biomechanical signals, but in a more general manner, some other sensors should be mentioned because of their potential interest, namely, force sensors (general applications) and optical deformation sensors.

Force

The properties of piezoelectric materials mean that they can also be used as force sensors (Lane and Craig, 2003; Gould, 2003) in the form of extensometric gauges (De Rossi, 2003). These materials respond to almost any type of magnitude of physical stimulus including, but not limited to, pressure, tensile force, and torsion (Edmison *et al.*, 2002).

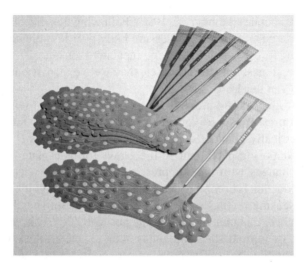

20.6 Biofoot/IBV wearable pressure-sensing insoles. Source: IBV.

Deformation

Fibre Bragg grating (FBG) sensors convert energy from mechanical into optical and electrical. FBG sensors look like normal optical fibres, but inside they contain a diffraction grid that reflects the incident light of a certain wavelength (principle of Bragg diffraction) in the direction of origin of the light. The value of this wavelength linearly relates to a possible elongation or contraction of the fibre. In this way, the Bragg sensor can function as a sensor of deformation (Van Langenhove and Hertleer, 2004). A possible use of these fibres as intelligent textiles is integration into flexible orthoses to determine the requirements of the patient or even generating feedback.

20.4.2 Intelligent textiles as actuators

Actuators can perform movements, release substances, make noise, and many other actions. Shape memory materials are the best-known examples; they transform thermal energy into motion (Van Langenhove and Hertleer, 2004) and have a notable importance in the thermo- and hygro-controllability of textile products (Gniotek and Krucinska, 2004). In the following subsections, the three main types of actuating applications in health and monitoring (force application, electrical stimulation and drug supply) will be described.

Force application

Three types of electroactive materials are being used nowadays as actuating fabrics: electroactive polymers (EAPs), dielectric elastomers and carbon nanotubes. EAPs mainly comprise inherently conductive polymers and conductive plastics. The latter are traditional plastics, almost exclusively thermoplastics that require the addition of conductive fillers such as powdered metals or carbon (usually carbon black or fibre). Conducting polymers show a drastic change in electrical conductivity and in the dimensions associated with changes in ionic doping inside the polymer. Conjugated electroactive polymers can exert high forces, much greater than those of natural muscle, which makes them appropriate as orthotic systems. They also undergo volume changes with noticeable variations of elastic moduli when ionic species are forced to penetrate inside their network by electrodiffusion (De Rossi et al., 2003; Engin et al., 2005).

Dielectric elastomers generate strains proportional to the square of the electric field applied between two compliant electrodes, showing high strains, large force densities, low response times and long lifetimes. They suffer the disadvantage of requiring high driving electric fields (De Rossi et al., 2003) that limit their use in clothing. However, it is possible to integrate them into applications such as sheets or in cushion covers to avoid bedsores.

The other feasible option for applying forces, carbon nanotube actuative fibres, is a sheet of carbon atoms rolled up into a tube with a diameter of around tens of nanometres. Their projected superior mechanical and electrical properties (high actuating stresses, low driving voltages and high energy densities) suggest that superior actuating performances can be expected. However, the manufacture of carbon nanotube fibres needs to be much improved in order to produce fibres that can demonstrate all their actuating potentialities (De Rossi et al., 2003).

Electrical stimulation

Electrical stimulation is defined as the use of an electrical current to transfer energy to a wound. The type of electricity that is transferred is controlled by the electrical source. Capacitatively coupled electrical stimulation involves the transfer of electric current through an applied surface electrode pad that is in wet (electrolytic) contact with the external skin surface and/or wound bed. When capacitatively coupled electrical stimulation is used, two electrodes are required to complete the electric circuit. Electrodes are usually placed over a wet conductive medium, in the wound bed and on the skin at a distance from the wound (Sussman, 2005).

Belgium's Centexbel is studying the textile transposition of a known technology for the rehabilitation of paralysed limbs. This process is known as functional electrical therapy. It is based on the stimulation of certain motor functions through electrodes that transmit electrical microwaves. The aim is to integrate functional electric stimulation (FES) motor aid into a textile structure that would offer the patient a degree of user comfort on a completely different level to present devices (European Commission, 2005).

Drugs delivery

From a pharmaceutical perspective, intravenous drug infusion may be the ideal. But for the patient, the downside is pain and inconvenience. Through one single patch on the skin, transdermal drug delivery can be customised to deliver the appropriate medication. Because this method ensures sustained release over time, it provides a steady flow of medication, reduces the risk of side effects and ensures higher patient compliance. For this reason, a material that releases substances and that can be embedded in textiles opens up a huge number of applications. Controlled drug supply systems in intelligent suits can also make an adequate diagnosis. However, active control of the release is not easy (Van Langenhove and Hertleer, 2004).

Microencapsulation technology (microcapsules of phase-change materials – PCMs – e.g. nonadecane) is being used nowadays in thermal clothing to reduce the impact of extreme variations in temperature. As well as being designed to combat cold, textiles containing PCMs also help to combat overheating, so overall the effect can be described as thermoregulation. They can also be used for medical purposes. Textiles that 'interact' with the consumer, reducing stress, promoting comfort and relaxation, are possible through active delivery from microcapsules. For example, encapsulated glycerol stearate and silk protein moisturisers can be embedded into bandages and support hosiery. The materials maintain comfort and skin quality through extensive medical treatment where textiles are in direct contact with the skin. These systems can also be developed to deliver measured dosages of chemical to combat muscle pain or other more serious injuries (Nelson, 2002). Currently the research group Medical Textiles/Biomaterials at the ITA (Institut für Textiltechnik der RWTH Aachen) focuses on the optimisation of a drug release system using degradable non-woven structures and microspheres. Natural hollow fibres and certain man-made fibres are possible microenvironments for bacterial bioreactors. Polydimethylsiloxane (PDMS) cellular microenvironments can be incorporated into yarns and non-woven fabrics to create bioactive fabrics (Wang *et al.*, 2001). PDMS is the most widely used silicon-based organic polymer, and is known for its particularly unusual rheological (or flow) properties.

20.4.3 Intelligent textiles in communication

Several levels of communication can be defined in an intelligent textile, depending on the emitter and the receiver: communication between components (within the textile), communication between the textile and the user (interfaces) and communication between the textile and a data storage system at a certain distance from the user (data transmission). Any kind of communication system must be compatible with comfort, durability and resistance to regular maintenance processes.

Communication between components

Electronic components in intelligent textiles (Gniotek and Krucinska, 2004) can be integrated at three levels. The first level of integration is the use of freely available electronics, connected by inserting the devices into the fabrics. At a deeper level, are the special miniaturised constructions and their composite-like installation in the textile structure, for example between the layers of the product. Finally, the direct installation of p–n junctions on fibres (FE-Fibre electronics) overcomes the limitations (low flexibility, fault-tolerance and cost-effectiveness) of the embedded silicon cores connected by fixed wiring. Weave-patterned organic transistors offers a novel approach to provide interconnection in intelligent textiles (Lee and Subramanian, 2005). In most cases, microelectronic devices are not really integrated into textile structures; they are 'stitched' on the fabric or hidden in the textile structure. Textiles can have 'floating' wires at regular locations across the fabric to allow for attachment of e-Tags (Lehn *et al.*, 2004). e-Tags are small printed circuit boards with connectors specifically designed to be attached to textiles (Martin *et al.*, 2004).

When electronics are added to the textile, the most obvious thing to do is to integrate the connection wires of the different components into the textile. To this end, conductive textile materials are used. In the second generation, the components themselves are transformed into full textile materials (Van Langenhove and Hertleer, 2004). Communication within the suit can be currently possible by either optical fibres (Park and Jayaraman, 2002) that are insensitive to electro-magnetic radiation and do not heat, but signals need to be transformed into an electrical signal, or conductive yarns (Hertleer *et al.*, 2002; Gimpel *et al.*, 2005; Locher *et al.*, 2005). Both have a textile nature (Van Langenhove and Hertleer, 2004) and are used in several prototypes and products.

Several frameworks for the incorporation of sensing, monitoring and information processing devices, such as Georgia Tech Wearable Motherboard (GTWM) have been developed and are still under research (Park *et al.*, 1999, 2002). Some of them are based on integrating wires (Gould, 2003) such as conductive fibres wrapped in a polymer coating in the textile and others are developed using printed circuit board-like conductive layers (Lobodzinski and Kuzminska, 1998).

Communication between the user and the textile

Communication between the wearer and the suit is necessary for passing instructions from the suit to the wearer and vice versa. The level of sophistication of the communication system varies from a simple buzzer to a flexible colour screen. The conducting polymers integrated into a knee sleeve developed by the University of Wollongong (Munro *et al.*, 2003) act as a strain gauge whereby the sensor is stretched when the wearer bends his knee and resistance within the sensor decreases. At a predetermined threshold resistance an audible tone is emitted to alert the wearer that the desired knee flexion angle has been achieved.

More complete information is offered by the WEALTHY system. The WEALTHY Central Monitoring System is a software module interpreting physical sensor data received from the portable patient unit (PPU) and representing them in simple, graphical forms (Paradiso and Wolter, 2005). In this case, although the information supplied is very detailed, it is transmitted via an antenna and it is not visualised on the textile.

The highest degree of integration of communication from the textile to the patient is achieved by the optical fibre flexible screen developed by France Telecom that is completely integrated in the textile (Hatcher, 2002) and the PolyLED (plastic LED by Phillips) (Engin *et al.*, 2005) which can offer the patient complete information. Regarding the transmission of information from user to textile, conductive textile materials and conductive polymers offer a highly integrated solution like that used in Softswitch (Softswitch, 2001) flexible keyboards.

Data transmission

Integrating antennas into clothing is easy because a large surface can be used without the user being aware of it. In the summer of 2002, a prototype of this application was presented by Philips Research Laboratories, UK and Foster Miller, USA on the International Interactive Textiles for the Warrior Conference (Boston, USA) (Van Langenhove and Hertleer, 2004). Regarding communication protocols, different technical solutions have been introduced, ranging from non-standard custom solutions (Polar Electro Oy, 2005; IST International Security Oy, 2005) to standards such as IEEE 802.11 (also known as WLAN) and Bluetooth. These solutions may be used to implement a personal area network (PAN) or home area network (HAN) (Korhonen *et al.*, 2003).

Not only information can be transmitted to the textile but also energy in order to reduce to the minimum the batteries to be worn by the user. An inductive link that provides wireless power transmission and bi-directional data transmission has been developed by ESAT-MICAS and Ghent University.

It consists of two coils and is able to transmit bidirectional data at a bit rate of 60 kbits/s. Simultaneously, 500 mW can be wirelessly transmitted from the base station to the suit at maximum coil separation of six centimetres (Catrysse et al., 2004).

20.5 Summary and future trends

A revolution in the form of how healthcare is delivered to citizens is on the way. Welfare organisations require new approaches that keep the current standards of services while reducing the associated costs. eHealth applications and wearable systems are at the forefront of new technologies to achieve this revolution. The future trend in wearable computing is to integrate electronics directly into textiles for better performance.

There are a large number of medical applications that will clearly benefit from intelligent textile technologies. These applications cover a broad range in complementary explorations (i.e. diagnosis of heart disease), secondary prevention (i.e. SIDS, biofeedback) and treatment/healing (i.e. wearable orthoses, drug delivery). Future medical applications may include sensors that will non-invasively measure blood gases (CO, SO_2, CO_2) and vital signs. The time required for these applications to be on the market depends to a large extent on the available technologies. The level of integration of technology in textiles is a good indicator of the time required to arrive at commercial solutions:

- Level 1 (adapted solutions) is ready to be incorporated in commercial solutions.
- Level 2 (module interconnection within textiles) should appear in the next few years.
- Level 3 (embedding solutions in fibres) will be ready in the mid-term.
- Level 4 (textronics) is still in the primary phase of development, but is clearly the trend in the technological development of smart fabrics.

In order to become technological realities there are still a number of issues that applications based on intelligent textiles must face including:

- Problems arising from long-term use: washability, deformation and interconnections.
- Energy supply: there are currently two trends that may combine to produce integrated solutions to overcome the problems related to energy supply: the increase in performance, and size reduction of current batteries and energy harvesting of the body (Kymissis et al., 1998). A third approach may be the use of distributed power sources in the textile that use electrochemical reactions.

In conclusion, intelligent textiles offer a new range of possibilities for healthcare. These possibilities depend on the development of the technologies, but they are fast becoming a reality. The first applications are already on the market as specialised products, but the development of technologies will widen the range of possible users. Citizens and welfare systems will be the main beneficiaries of this trend. Patients can improve their degree of independence and quality of life through the use of these technologies. Welfare systems can lower their costs associated with hospitalisation and nursing and provide sustainable care in the future.

20.6 Acknowledgements

We would like to thank the R+D+i Linguistic Assistance Office at the Universidad Politécnica de Valencia for their help in revising this chapter.

20.7 References

American Sids Institute (2005) [Internet], Available from: <http://www.sids.org/> [Accessed 2nd September 2005].

Babbs C F, Bourland J, Graber G P, Jones J T, Schoenlein W E (1990), 'A pressure-sensitive mat for measuring contact pressure distributions of patients lying on hospital bed', *Biomed Instr Tech*, 24 (5), 363–370.

Bartels V T (2005), 'Physiological function and wear comfort of smart textiles', *Int newsletter on micro-nano integration.* 2 (05), 16–38.

Bartels A and Harder D (1992), 'Non-invasive determination of systolic blood pressure by heart sound pattern analysis', *Clin Phys Physiol Meas*, 13 (3), 249–256.

Belda-Lois J M, Rocon E, Sánchez-Lacuesta J J, Ruiz A F, Pons J L (2005), 'Estimation of biomechanical characteristics of tremorous movements based on gyroscopes', in *AAATE2005. Assistive technology From virtuality to reality. 8th European conference for the advancement of assistive technology in Europe*, Lille, France.

Binkey P, Frontera W, Standaert D G, Stein J (2003), 'Predicting the Potential Impact of Wearable Technology', *IEEE Eng Med Biol Mag* 22 (3), 23–27.

Bonato P (2003), 'Wearable sensors/Systems and their impact on biomedical Engineering', *IEEE Eng Med Biol Mag* 22 (3), 18–20.

Buckle P and Fernandes A (1998), 'Mattress evaluation-assessment of contact pressure, comfort and discomfort', *Appl Ergonomics*, 29 (1), 35–40.

Buhler C (ed.) (2000), *FORTUNE Guide: Empowered participation of users with disabilities in projects*, Wetter, Evangelische Stiftung Colmarstein – Forschungsinstitut Technologie – Behindertenhilfe.

Carpi F and DeRossi D (2005), 'Electroactive Polymer-Based Devices for e-Textiles in Biomedicine', *IEEE Trans Inf Tech in Biomed*, 9 (3), 295–318.

Catrysse M, Puers R, Hertleer C, Van Langgenhove L, van Egmond H, Matthys D (2004), 'Towards the Integration of Textile Sensors in a Wireless Monitoring Unit', *Sensors and Actuators A: Physical*, 114, (2–3), 302–311.

Chen H, Nigg, B M, Koning J D, (1994), 'Relationship between plantar pressure distribution under the foot and insole comfort', *Clin Biomech*, 9 (6), 335–341.

Czajka R, (2005), 'Development of Medical Textile Market', *FIBRES & TEXTILES in Eastern Europe*, 13, 1 (49), 13–15.

David Rigby Associates (2002), *Technical Textiles and Industrial Nonwovens: World Market Forecast to 2010*, Manchester, David Rigby Associates.

De Rossi D, Della Santa A, Mazzoldi A (1999), 'Dressware: wearable hardware', *Materials Science and Engineering C*, 7 (1), 31–25.

De Rossi D, Carpi F, Lorussi F, Mazzoldi A, Paradiso R, Pascuale Scilingo E, Tognetti A (2003), 'Electroactive Fabrics and Wearable Biomonitoring Devices', *AUTEX Research Journal*, 3 (4), 180–185.

Edmison J, Jones M, Nakad Z, Martin T (2002), 'Using Piezoelectric Materials for Wearable Electronic Textiles', in *Proc. 6th Int. Symp Wearable Computers 2002*, 41–48.

Engin M, Demirel A, Engin E Z, Fedakar, M (2005), 'Recent developments and trends in biomedical sensors', *Measurement*, 37 (2), 173–188.

European Commission (2005), 'Inwoven Intelligence', *RTD Info*, 45, (14–15).

Fuller G F (2000), 'Falls in the Elderly', *American Family Physician*, 61 (7), 2159–2173.

Gaggioli A, Vettorello M, Riva G (2003), 'From Cyborgs to Cyberbodies: The Evolution of the Concept of Techno-body in Modern Medicine', *PsychNology Journal*, 1 (2), 75–86.

Gibbs P T and Asada H H (2005), 'Wearable Conductive Fiber Sensors for Multi-Axis Human Joint Angle Measurements', *J NeuroEng Rehab*, 2 (1), 7.

Gimpel S, Möhring U, Neudeck A, Scheibner W (2005), 'Integration of microelectronic devices in textiles', *Int newsletter on micro-nano integration*, 2 (05), 14–15.

Gniotek K and Krucinska I (2004), 'The Basic Problems of Textronics', *Fibres & Textiles in Eastern Europe*, 12 (1), 13–16.

Gould P (2003), 'Textiles gain Intelligence', *Materials Today*, 6 (10), 38–43.

Grossman P (2003), 'The lifeshirt: A multi-function ambulatory system that monitors health, disease, and medical intervention in the real world', in *Proc Int Workshop. New generation of wearable systems for eHealth: Towards a revolution of citizen's health and life style?*, Lucca, University of Pisa, 73–80.

Hatcher M (2002) 'France Telecom debuts fiber screen' [Internet], Available from: <http://optics.org/articles/news/8/7/1/1> [Accessed 7th September 2005].

Hertleer C, Van Longenhove L, Catrysse M, Puers R, Van Egmond H, Matthys D (2002), 'Intelligent Textiles for Children in a Hospital Environment', in *Proc. 2nd Autex Conference: Textile Engineering at the Dawn of a New Millennium*, Bruges, 44–48.

IST International Security Oy (2005). [Internet], Available from: <http://www.istsec.fi> [Accessed 15th September 2005].

Kallmayer C, Linz T, Aschenbrenner R, Reichl H (2005), 'System integration technologies for smart textiles', *Int newsletter on micro-nano integration*, 2 (05), 42–43.

Korhonen I, Pärkkä J, Van Gils M (2003), 'Health monitoring in the home of the future', *IEEE Eng Med Biol Mag*, 22 (3), 66–73.

Kymissis J, Kendall C, Paradiso J, Gershenfeld N (1998), 'Parasitic Power Harvesting in Shoes', *in Proc 2nd Int Symp Wearable Comp*, Pittsburgh, 132–139.

Lane R and Craig B (2003), 'Materials that sense and respond: an introduction to smart materials', *The AMPTIAC Quarterly*, 7, (2), 9–14.

Lee J B and Subramanian V (2005), 'Weave patterned organic transistors on fiber for e-textiles', *IEEE Trans on electron devices*, 52, (2), 269–275.

Lehn D I, Neely C W, Schoonover K M, Martin T L, Jones M T (2004) 'e-TAGs: e-Textile Attached Gadgets', in *Proc of the Communication Networks and Distributed Systems Modeling and Simulation Conference*, San Diego.

Lobodzinski S, Kuzminska M (1998), 'Silicon whole body sensors for medical applications

– A glimpse of the future', in *Bioeng and Biotech – Applications Conference WESCON/ 98* Anaheim, 286–291.

Locher I, Kirstein T, Tröster G (2005), 'From smart textiles to Wearable Systems', *Int newsletter on micro-nano integration*, 2 (05), 12–13.

Loureiro R C V, Belda-Lois J M, Rocon E, Pons J L, Sanchez-Lacuesta J J, Harwin W S (2005), 'Upper Limb Tremor Suppression in ADL via an Orthosis Incorporating a Controllable Double Viscous Beam Actuator', in *ICORR2005 IEEE 9th Int Conf Rehab Robotics. Frontiers of the Human-Machine Interface, Chicago*.

Lymberis A and Olsson S (2002), 'Smart biomedical clothes promising way to keep the European citizen healthy', Virtual Medical Worlds [Internet], Available from: <http://www.hoise.com/vmw/02/articles/vmw/LV-VM-08-02-35.html> [Accessed 14th September 2005].

Lymberis A and Olsson S (2003), 'Intelligent biomedical clothing for personal health and disease management: State of the art and future vision', *Telemed J and e-Health*, 9 (4), 379–386.

Manto M, Topping M, Soede M, Sanchez-Lacuesta J, Harwin W, Pons J, Williams J, Skaarup, S, Normie L (2003), 'Dynamically responsive intervention for tremor suppression', *IEEE Eng Med Biol Mag*, 22 (3), 120–132.

Martin T, Lockhart T, Jones M, Edmison J (2004), 'Electronic Textiles for *in situ* Biomechanical Measurements', *Proc 24th Army Science Conference, Orlando*.

Martinot-Largarde P, Sartene R, Mathieu M, Durand G (1998), 'What does inductance plethysmography really measure?', *J Appl Physiol*, 64 (4), 1749–1756.

Mühlsteff J and Such O (2004), 'Dry electrodes for monitoring of vital signs in functional textiles', in *Proc IEEE Eng Med Biol Soc*, San Francisco, 1 (3), 2212–2215.

Munro B J, Steele J R, Campbell T E, Wallace G G (2003), 'Wearable textile biofeedback systems: are they too intelligent for the wearer?', in *Proc Int Workshop. New generation of wearable systems for eHealth: Towards a revolution of citizen's health and life style?*, Lucca, University of Pisa, 187–193.

MyHeart Project (2005) 'Fighting cardio-vascular diseases by prevention and early diagnosis' [Internet], Available from: < http://www.hitech-projects.com/euprojects/myheart/> [Accessed 2nd September 2005].

Nelson G (2002), 'Application of microencapsulation in textiles', *Int J Pharm*, 242 (1–2), 55–62.

Newman D K (2003) 'Diagnostic Ultrasound. Glossary of terms' [Internet], Available from: < http://excellence.dxu.com/Glossary.htm> [Accessed 21th September 2005].

Pagani M, Lucini D (2001), 'Autonomic dysregulation in essential hypertension: insight from heart rate and arterial pressure variability', *Autonomic Neuroscience*, 90 (1), 76–82.

Panina G, Khot U N, Nunziata E, Cody R J, Binkley P F (1995), 'Assessment of autonomic tone over a 24-hour period in patients with congestive heart failure: relation between mean heart rate and measures of heart rate variability', *Am Heart J*, 129 (4), 748–53.

Paradiso R, Wolter K (2005), 'Wealthy – A wearable health care system: New frontier on E-Textile', *Int newsletter on micro-nano integration*, 2 (05), 10–11.

Park S, Jayaraman S (2003), 'Enhancing the quality of life through wearable technology', *IEEE Eng Med Biol Mag*, 22(3), 41–48.

Park S, Gopalsamy C, Rajamanickam R, Jayaraman S (1999), 'The wearable motherboard: a flexible information infrastructure or sensate liner for medical applications', *Stud Health Technol Inform*, 62, 252–258.

Park S, Mackenzie K, Jayaraman S (2002), 'The wearable motherboard: a framework for personalized mobile information processing (PMIP)', in *Proc 39th Design Automation Conference*, New Orleans, 170–174.

Pepper E, Gibney K H (2000), *Healthy computing with muscle Biofeedback*, Woerden, Biofeedback Foundation of Europe.

Perry M, Dowdall A, Lines L, Hone K (2004), 'Multimodal and ubiquitous computing systems: supporting independent-living older users', *IEEE Trans Inf Tech in Biomed*, 8 (3), 258–270.

Polar Electro Oy (2005) [Internet, Available from: <http://www.polar.fi> [Accessed 2nd September 2005].

Powerpaper (2005). [Internet] Available from: <http://www.powerpaper.com> [Accessed 29th July 2005].

Prince M and Jorn A F (1999), 'The prevalence of dementia', *Alzheimer's Disease International. Factsheet 3*.

Rocon R, Ruiz A F, Pons J L, Belda-Lois J M, Sanchez-Lacuesta J J (2005), 'Rehabilitation Robotics: a Wearable Exo-Skeleton for Tremor Assessment and Suppression', *Proc IEEE Int Conf Robotics and Automation*, Barcelona, 2283–2288.

Sarnow M R, Veves A, Giurini J M, Rosenblum B I, Chrzan J S, Habershaw G M (1994), 'In-shoe foot pressure measurements in diabetic patients with at-risk feet and healthy subjects', *Diabetes Care*, 17 (9), 1002–1006.

Seyam A M (2003), 'Electrifying Opportunities' [Internet], *Textile World, Billian Publishing*. Available from: < http://www.textileworld.com/News.htm?CD=1294&ID=3416> [Accessed 2nd September 2005].

Softswitch (2001), [Internet] Available from: <http://www.softswitch.co.uk/> [Accessed 7th September 2005].

Strese H, John L G, Kaminorz Y (2005), 'Technologies for Smart Textiles', *Int newsletter on micro-nano integration*, 2 (05), 6–9.

Sussman C (2005) Electrical Stimulation, [Internet] *Wound Care Information Network*. Available from: <http://wwww.medicaledu.com/estim.htm> [Accessed 28th September 2005].

Teller A (2004), 'A platform for wearable physiological computing', *Interacting with computers*, 16, 917–937.

Tognetti A, Lorussi F, Bartalesi R, Quaglini S, Tesconi M, Zupone G, De Rossi D (2005), 'Wearable kinesthetic system for capturing and classifying upper limb gesture in post-stroke rehabilitation', *J NeuroEng Rehab*, 2, 8.

Tries J, Eisman E (1995) 'Urinary incontinence: Evaluation and biofeedback treatment', *Biofeedback: A practitionner's guide*, New York, Guilford Press, 597–632.

Van der Loos H F M, Kobayashi H, Liu G, Tai Y Y, Ford J S, Norman J, Tabata T, Osada T (2001), 'Unobtrusive vital signs monitoring from a multisensor bed sheet', *Proc RESNA Ann Conf*, Reno, 218–220.

Van Langenhove L, Hertleer C (2004), 'Smart clothing: a new life. International Journal of Clothing', *Sci Tech*, 16, (1/2), 63–72.

Vanfleteren J, Lutz-Günter J (2005), 'Smart High-Integration flex technologies', *Int newsletter on micro-nano integration*. 2 (05), 40–41.

Vivometrics (2004). [Internet], Available from: <http://www.vivometrics.com/> [Accessed 2nd September 2005].

Waterhouse E J (2001) 'Ambulatory Electroencephalography (EEG)' [Internet], Available from: <http://www.emedicine.com/neuro/topic445.htm> [Accessed 25th July 2005].

Wealthy (2002) 'Wealthy – Wearable Health Care System' [Internet], Available from: <http://wealthy-ist.com/> [Accessed 14th September 2005].

Williams D (2005) 'Environmentally Smart Polymers', *Med Device Technol*, 16 (4), 9–13.

Wang W, Zhao Z, Fowler A, Warner S, Ellis D, Toner M, Morgan J (2001) 'Development of bioactive fabrics' [Internet], Available from: < http://www.mne.umassd.edu/faculty/alexbio.html> [Accessed 14th September 2005].

21
Context aware textiles for wearable health assistants

T KIRSTEIN, G TRÖSTER, I LOCHER and C KÜNG, Wearable Computing Lab, ETH Zürich, Switzerland

21.1 Introduction

The integration of electronic functionality into clothing offers new possibilities for medical monitoring. Our approach is to develop context aware textiles. Context aware means that textiles sense the state of the user and the environment and recognize situations and events. We combine electronic textile technologies with context recognition and wearable computing technologies in order to achieve an intelligent and at the same time wearable system. In the first part of this chapter we explain the vision of a wearable health assistant and our system concept. Then we describe the recent developments and achievements in the area of electronic textiles, context recognition and wearable technologies. We show how these three research fields can be combined. In the last part we present concrete applications of our wearable health assistants.

21.2 Vision of wearable health assistant

An increasingly important issue for many of today's devices and services is mobility. In particular there is a growing interest in mobile healthcare services such as portable health monitoring systems. The vision sketches personalized health services for everyone in a trusted and natural way, anywhere and at any time. Healthcare is not restricted to clinics or a stationary environment (like at home) but extended to our whole life. With this approach the current issues can be addressed:

- For many patients it is difficult to manage health problems in daily life. Generally there is a lack of motivation and advice for a continuously healthy lifestyle. Widespread problems that could be avoided are, for example, back pain, obesity and stress-related diseases.
- Physicians have only limited tools to assess patients' health status during their daily activities. Diagnosis is restricted to brief contacts with the patients.

- The costs of healthcare are increasing. The focus is on extensive professional treatment instead of illness prevention.

The wearable health assistant could help people to fight diseases by a preventive lifestyle and early diagnosis. Users could take control of their own health status and adapt a permanent healthier lifestyle. This self-management of health makes people more independent, improves their quality of life and at the same time, reduces healthcare costs.

Three main features characterize our vision of the wearable health assistant: monitoring of the physiological parameters, detection of the user's context and giving feedback to the user. Concentrating on non-invasive measuring methods, physiological parameters comprise heart rate and ECG, respiration, EMG, blood pressure, blood oximetry, skin conductance and temperature. But the meaningful assessment of these vital parameters requires the consideration of the current context of the user. For example, rapidly increasing heart rate could naturally be provoked by jumping up a staircase, but if the user has not been moving, it could indicate a dangerous health status. Context awareness includes the user's motion, activity, gestures and also the affective and emotional state like stress and depression. The user's location, both indoor and outdoor, time, weather, the illumination and noise define the environmental context. Apart from the user's activity and environment it is also important to determine the user's social context that means his contact and communication with other people.

The combination of the vital parameters with the wearer's context, the activity and the sleep patterns together with social interactions paint a picture of the user's health status. To facilitate the feedback and interface between the individual user and the wearable health assistant we propose a 'life balance factor' (LBF) as a plain health measure and generally understandable indicator, especially for medical laypersons. The LBF summarizes the current health status; it indicates changes and calls on a consultation if health parameters are moving to a critical range.

Weiser's visionary view (Weiser, 1991) of an invisible and pervasive computing world is now coming to fruition, where tiny autonomous systems, consisting of sensors, signal processing and transmitting units, possibly as small as a grain of rice, are scattered in the environment. Radio Frequency Identification (RFID) tags are the forerunner of this vision; attached to a variety of daily artefacts, these electronic markers enable the detection of their location and also provide information about the objects to which they are attached. The impact of a computerized environment on personal healthcare varies from monitoring of people with cardiac risks (Gouaux *et al.*, 2002) to home care for elderly living alone (Korhonen *et al.*, 2003).

21.3 Approach

For the realization of continuous health monitoring we need on-body electronics (Kirstein, 2004; Lukowicz *et al.*, 2004). Figure 21.1 shows different approaches to on-body electronics, handheld electronics, electronics in accessories, electronics in clothing and finally electronic textiles. In recent years advances in miniaturization, wireless technology and worldwide networking have enabled the development of many portable (hand-held) devices such as cell phones, organizers and laptops. But until now electronic devices on the market are still bulky and inconvenient to use and especially in the medical field rather home-based than truly mobile. The first step to wearability has been made by embedding electronics into accessories like watches and belts. The HealthWear armband by BodyMedia (www.bodymedia.com) monitors, for example, the calorie balance of the user. The next step is to use clothing as a platform for electronics. This idea offers many advantages especially in the medical field because of the direct contact and continuous interaction between the garment and the user. Another important aspect is the comfort of wearing, as humans prefer to wear textiles rather than heavy and hard boxes. Clothing allows integrating the system unobtrusively and conveniently into the daily life of the user. Simply hiding electronic components in pockets or seams is a possible solution. The Lifeshirt by Vivometrics (www.vivometrics.com) is an example of medical clothing where the fabric acts as a carrier of conventional cables and electronic devices. Using the textiles themselves as electronic components goes one step further and is a new approach to the next generation of on-body electronics.

We believe that a wearable assistant as described in our vision can be achieved only by context sensitivity as described above and by a modular system concept. The modular system concept means that we use different integration methods to embed the system into the user's outfit depending on the functionality and the cost of the components. People wear many different

Mobile					
Portable	Wearable				
		Electronic clothing			
				Electronic textiles	
Handheld electronics	Electronic accessories	Electronics embedded in clothing	Electronics attached to textiles	Textiles with electronic functionality

21.1 Approaches to on-body electronics.

402   Intelligent textiles and clothing

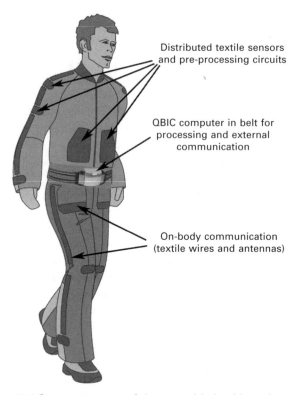

21.2 System concept of the wearable health assitant.

clothes and select their outfit according to their activities. Only cheap components with task-specific functionality (e.g. sensors) should be permanently mounted into the garment whereas more expensive general-purpose devices such as processors should be detachable and usable with different outfits. For distributing sensor functionality all over the body and for having direct contact with the skin, electronic textiles offer the best solution. Therefore we embed textile sensors as well as pre-processing circuits and communication facilities into the clothing. Processing devices and components for external communication can be centralized and embedded into accessories (such as belts and watches) or hand-held devices (such as mobile phones). This combination of smart textiles with miniaturized electronics is depicted in Fig. 21.2.

21.4   Electronic textile technology

Textile technological developments have created a whole range of so-called 'smart fabrics' for many applications. The concept of smart materials describes the ability of materials to sense and react to external stimuli. However, most of the advanced textile materials like breathable, fire-resistant or substance-releasing textiles cannot be considered as smart because they do not adapt their functionality to the environment. Hence they are not context-aware as required for the wearable health assistant. Textile materials that can store heat when it is warm and release the heat again when it gets cold (phase change materials) react to a change in environment and therefore possess a low level of context sensitivity, but they have no active control. It is, for example, not possible to regulate the temperature of the clothing according to the user needs. Such an active control is necessary for healthcare applications but it requires an electronic system that processes the sensor data and 'decides' about the reaction. This need for electronics in textiles induced a new research field 'e-textiles'. Considering the opposed properties of electronics and textiles a merging of both seems to be impossible. Nevertheless, the first results show the potential of this idea (Kirstein et al., 2005).

There are two possible ways to create textiles with electronic functionality. Miniaturized electronic components can be attached to fabrics if their size does not reduce comfort. Using electrically conductive fibres and fabrics is the second approach. In this case the textiles do not just act as a substrate but as electronic components themselves. In the following, we describe the latest developments in e-textile research.

21.4.1 Textiles for communication

Some early approaches to use textiles for communication are described in Marculescu et al. (2003). One of the biggest problems was that the fabrics lost much of their typical textile properties due to the embedded thick wires. Our aim was to achieve high-performance signal lines made from conductive textiles that have the same look and feel as conventional fabrics. Several types of conductive fabrics already exist and are applied mainly for shielding and antistatic applications. By developing measurement and simulation methods those textiles can now be optimized for data transmission.

The first systematic studies of the electrical properties of textile transmission lines were carried out by Cottet et al. (2003). The proposed textiles are fabrics with copper fibres in one or two directions and with different polyester yarn fineness. The variety of fabrics opens a wide range of possible transmission line topologies and allows finding a configuration that fits potential target applications. Using wire pair configuration the achievable characteristic impedances lie between 120 Ω and 320 Ω. To study the influence of fabrication tolerances, the textiles were modelled with an EM-field simulation tool. The simulation results showed that with the given geometry variations an accuracy of ± 5% to ± 10% for the characteristic impedances is achievable. High-frequency network analyser measurements were performed up to 6 GHz. The extracted frequency characteristics revealed that the dielectric and ohmic losses do not determine the line insertion loss. The loss is mainly influenced by a non-uniform impedance profile along the lines up to the half-wavelength and by coupling to parasitic modes above this frequency point. This results in cut-off frequencies of 1 GHz for 10 cm long lines. Good signal transmission for a 100 MHz clock signal was proved through 20 cm textile lines. Experiments showed also that a grounded copper fibre between two neighbouring lines reduced crosstalk from 7.2% to 2.8%. To conclude, conductive textiles provide potentials in signal transmission in addition to EMI shielding and power supply. Textile transmission lines can be used to create a network infrastructure in clothing and to connect different distributed components of a wearable assistant.

Another important ingredient of a wearable assistant is the connection to a wireless network. For this purpose, textile antennas were developed that guarantee flexible and comfortable embedding into clothing. Wearable antennas presented by Salonen et al. (2000) and by Massey (2001) are partially based on textiles possessing an inverted-F shape that results in a stiff structure. Other textile antennas described by Tanaka et al. (2003) and by Salonen et al. (2003) are designed as rectangular patches with a protruding probe feed and only linear polarization. Antennas such as presented in Salonen et al. (2004) utilize fabrics only as substrate whereas the patches and ground planes are copper foils.

We developed antennas that are purely textile and flat (Klemm et al., 2004). Those textile patch antennas are designed for Bluetooth in the frequency range from 2400 MHz to 2483.5 MHz. The design of this antenna is inspired by the build-up of printed microstrip antennas and consists of a three-layer structure; electrically conductive fabrics act as ground plane and antenna patch and are separated with a fabric substrate. The conductive fabric should have a homogeneous resistance below 1 Ω^2. Therefore we used a metallized fabric that was plated before weaving or knitting. Using a knitted fabric leads to a highly bendable and deformable structure. From a manufacturing point of view, the knitted structure is a drawback because precise shaping as well as assembly of the antenna without warpage are difficult. The manufactured antenna shapes finally achieved a geometrical accuracy of about ±0.5 mm. Another undesired effect of the knitted fabric is a change of the sheet resistance when the structure is stretched. Conductive fabrics that are woven possess better electrical performance, but bending of such an antenna is limited. The textile substrate provides the dielectric between the antenna patch and the ground plane and needs to have a constant thickness and stable permittivity. We chose a spacer fabric with a thickness of 6 mm and performed humidity measurements covering a range from 20% to 80% relative humidity within a temperature range of 25 °C to 80 °C. The measurements showed that permittivity variations are negligible compared to measurement uncertainty. Alternative substrate materials are felts and foams.

Context aware textiles for wearable health assistants

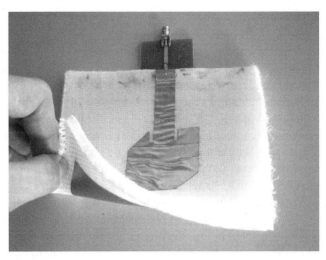

21.3 Textile Bluetooth antenna.

During the design process of the antenna it was important to achieve a flat and wearable structure; that also means a planar antenna feed. We designed a microstrip feedline and applied insets in order to adjust the antenna's input impedance and to avoid losses due to mismatch between feedline and antenna. Additionally, a microstrip feedline does not increase the height of the patch antenna and maintains wearing comfort when integrated into clothing. We designed linearly and circularly polarized textile antennas and proved a good directivity that minimizes unnecessary radiation exposure to the human body and radiation losses. The textile antennas feature a 10 dB bandwidth of 200 MHz on average. Even when bent around a radius of 37.5 mm resembling a mounting on a human upper arm, Bluetooth specifications can be assured. One of the textile patch antennas is shown in Fig. 21.3.

21.4.2 Textiles for signal pre-processing (System-on-Textile)

We believe that conductive textiles offer an even greater potential than just being used as cables or antennas. Such fabrics manufactured with high precision allow complex wiring structures. Along with the proper assembly technology for electronic components and sensors, entire electrical circuits can be embedded into the fabric. We call this technology 'System-on-Textile' (SoT). Using fabrics as substrates for electronic circuits instead of rigid circuit boards enables the placement of small circuits for signal pre-processing close to the sensors. In this way we can distribute sensing functionality all over the body without affecting wearing comfort.

Suitable textile substrates must support electrical routeing structures. The newly developed woven fabric with thin insulated copper fibres provides our platform for electrical circuits. This fabric is manufactured by Sefar Inc., a producer of precision filters. From an electrical point of view, precise yarn distances within the fabric are required in order to achieve satisfactory electrical performance. Secondly, yarn distances need to be small to meet the pitches of the electrical components. On the other hand, the fabric should be fine, light and maintain typical textile properties. Nevertheless, manufacturability of the fabric with the desired materials has to be considered.

After several iterations, we achieved a hybrid fabric consisting of woven polyester yarn (PET) with an exact diameter of 42 microns and copper alloy wires with a diameter of 50 microns. The hybrid fabric with a mesh opening of 95 microns (+/– 10 microns) and an opening area of 44% is shown in Fig. 21.4. Each copper wire itself is coated with a polyurethane varnish as electrical insulation. The copper wire grid in the fabric features a spacing of 0.57 mm (mesh count in warp and in weft is 17.5 cm^{-1}). The combination of PET yarn and copper wires requires a special weaving technology, which includes two yarn systems in warp and weft direction (3 PET wires and 1 copper wire) with separate tensioning systems. We positioned our hybrid fabric with its weight of 74 g/m^2 as interlining. Its application field is therefore very versatile. The fabric represents a compromise between preserving textile properties and copper wire density, i.e., electrical connectivity. To our knowledge, such a precise hybrid fabric consisting of PET yarn and copper wire is unique.

In order to build circuits on the fabric, we need the technology to mount electrical components and interconnect them through the fabric utilizing embedded copper wires. The desired wiring structure can be established by connecting crossing copper wires at their intersections and by cutting the wires at certain locations. Since the wires are insulated against each other, the insulation needs to be removed at these intersections to enable electrical connection. Altogether, three manufacturing steps, as shown in Fig. 21.5, are required to create such an electrical connection.

1. coating removal and cutting of the copper wire using laser light at defined locations
2. assembly of the electrical components and interconnecting of the skinned wire sections with conductive adhesive
3. adding epoxy resin to the electrical components and intersections as mechanical protection.

These three steps form the building block for defined wiring structures in fabrics. They can be manufactured in automated processes using equipment for printed circuit board (PCB) fabrication. This assembly technology is described in more detail in Locher *et al.*, (2004, 2005).

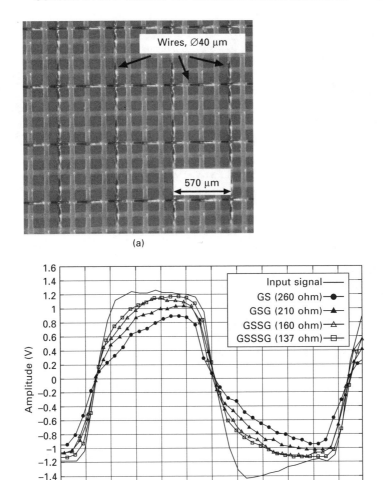

21.4 Conductive textile (by Sefar Inc.).

Electrical components usually feature a solid and rigid body whereas fabrics are soft and drapable. Thus, the placement of components onto the fabric requires special attention resulting in an additional trade-off between textile and electrical properties. From a textile point of view, the placement should be done such that drapability and softness are preserved. In other words, the components must not be placed too close together. On the other hand, the electrical wiring should be short in order to avoid electrical losses and noisy signals. In contrast to earlier developments by Virginia Tech and Infineon (Marculescu *et al.*, 2003), we were able to mount electrical components directly onto the fabric and interconnect them through the fabric over an

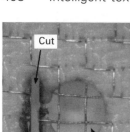  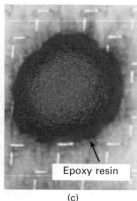

21.5 Manufacturing of wiring structures in fabrics.

arbitrary wiring structure. Utilization of this technique is only enabled by the high precision of our hybrid fabric.

The Sefar Petex hybrid fabric combined with our assembly technology opens a promising new perspective for flexible electrical circuits in the field of e-textiles. The technology enables textiles to be used for processing and sensing tasks and for displays such as those needed in applications for medical monitoring, smart interior fabrics and drapable advertising media. Additionally, the textile properties can be adapted to the application requirements by deploying of different finishes.

21.4.3 Textiles for sensing

The next step in e-textile research is to use fibres and fabrics not just to transmit but also to transform signals. Conductive textiles that change their electrical properties due to environmental impact can be used as sensors. Typical examples are textiles that react to deformation like pressure sensors and stretch sensors and measure body movements, posture or breathing. Further parameters that can be measured with textiles are, e.g., humidity and temperature. Textile electrodes can replace conventional electrodes for heart monitoring or electrical stimulation (Kirstein *et al.*, 2003). Apart from measuring biometric or environmental data, textile sensors can also act as an input interface as, for example, textile touchpads.

Textile pressure sensor

Pressure sensors that are made from textiles have many attractive features for wearable applications. They can cover a large three-dimensionally shaped surface area and detect pressure without reducing wearing comfort. Apart from acting as input interfaces they can measure pressure distribution during sitting or lying and even detect body movements due to pressure changes in the garment.

ElekTex (www.eleksen.com) is a laminate of three textile layers. Conductive fibres in the central layer are locally compressed allowing conductive contact between the top and bottom layer. Softswitch (www.softswitch.co.uk) is made of conductive textiles with a thin layer of elastoresistive composite material (Quantum Tunnelling Composite QTC) that reduces resistance when compressed. The Sensory Fabric (patent US2003119391) contains conductive and insulating yarns in a woven structure that can create electrical contact at yarn crossing points when compressed. All these structures act as switches and do not deliver pressure values. We developed a pressure sensor mat consisting of a spacer fabric with embroidered electrically conductive patch arrays on both sides. Sitting posture and risk of bedsores (decubitus) can be detected using such a mat on seats and beds. The textile pressure sensor is shown in Fig. 21.6.

Each opposing patch pair in the arrays forms a plate capacitor whose capacity changes with compression force on the spacer fabric. Although the capacity is reciprocal proportional to the distance between the patches, the capacity versus pressure is highly nonlinear. Firstly, the compression force nonlinearly depends on the compression distance. Secondly, the permittivity of the spacer fabric (dielectric) increases when compressed since air becomes displaced. Additionally, relaxation of the spacer fabric shows a hysteresis effect as depicted in Fig. 21.7. Using the Preisach model this behaviour can be described, so that the measurement of pressure distribution over a fabric area becomes feasible.

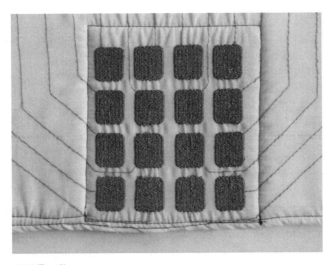

21.6 Textile pressure sensor.

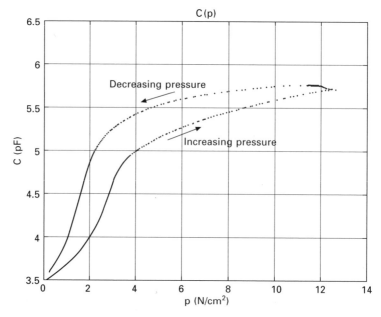

21.7 Capacity measurement hysteresis of textile pressure sensor.

Textile elongation sensor

Possible applications of wearable elongation sensors are, among others:

- posture and motion analysis in sports, medicine, rehabilitation or daily life
- artificial sensor skins for humanoid or animal-like mobile robots
- replacement of acceleration sensors used for context or gesture recognition

The high range (including pre-stretch) of up to 30% strain which is needed for wearable applications as, for example, body posture monitoring, sets wearable elongation sensors clearly apart from other industrial elongation sensors. Additional requirements are comfort of wearing, the need for overstress tolerance and the fact that high elongations have to be measured without stiffening the textile too much.

The most wearable option of making elongation sensors is to use a knitted electroconductive fabric which is stretchable (Fig. 21.8). When such a fabric is stretched, the interconnect topology in the garment changes, and hence resistance changes. Some work on such fabrics can be found in the literature (Pacelli *et al.*, 2001; Scilingo *et al.*, 2003; Oh *et al.*, 2003; Wijesiriwardana *et al.*, 2003; Farringdon *et al.*, 1999; Bickerton, 2003). Mainly fabrics polymerized with conductive polymers, or fabrics made of conductive threads are mentioned. From what is currently published, the following is apparent: these sensors do not perform well because either the response is rather weak

21.8 Conductive knitted fabric.

(polypyrrole coated threads/fabrics) or the range is too low (carbon filled rubber coated threads/fabrics). High temperature dependence and dependence of resistance on elongation rate are further problems mentioned for almost all published conductive fabrics.

We evaluated three highly stretchable electroconductive fabrics; two metallized knitted fabrics and a knitted fabric with activated carbon fibres. The general observations are the following: the fabrics show high transient times, peaks when movements start and high hysteresis. Electroconductive fabrics used as elongation sensors are wearable, but the transducer qualities are poor (see Fig. 21.9).

Another kind of wearable elongation sensors can be made by utilizing the piezoresistive effect of electroconductive elastomers. The following options were evaluated: carbon filled silicone rubber coated onto a fabric (Fig. 21.10) and carbon filled thermoplastic elastomer fibers. Carbon filled silicone rubber coated onto stretchable fabrics are extensively used at the University of Pisa (Tognetti *et al.*, 2005). Thereby, a commercial electroconductive silicone rubber product is used. This commercial product was extensively evaluated. It was found that the resistance of this material behaves in a very complex manner when the material is strained. The problems are that there is insufficient repeatability (Fig. 21.11) because the resistance depends on the strain history of the material, there is high electrical hysteresis, the resistance depends on the speed of deformation, the transient time is very high (due to the resistance recovery behaviour), and the general resistance level of the material depends on the applied strain. Resistance recovery is an effect observed in all particle filled elastomeric materials. After a movement, resistance is not constant, but decreases very slowly.

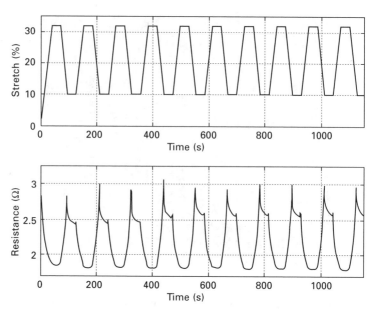

21.9 Periodic elongation and resistance change of conductive knitted fabric.

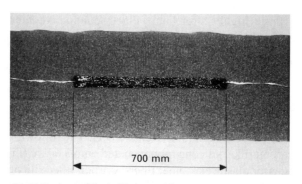

21.10 Carbon filled silicon rubber sensor.

As an alternative to the commercial conductive rubber described above, six different compounds of carbon filled thermoplastic elastomer fibres were tested. Most of the problems mentioned above can also be found in these materials. However, for some of the compounds, the problems are much less pronounced. The hysteresis problem seems to have been removed completely, but resistance recovery is present as well. It is very important to understand that these kinds of materials are sensitive to all kinds of deformations. Resistance changes not only in response to elongation, but also in response to pressure, shear and bending. Therefore, it is unlikely that these materials can be used as pure elongation sensors. If these materials are integrated into a garment,

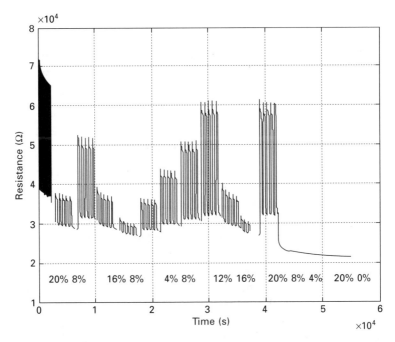

21.11 Repeated cycling of carbon filled silicon rubber sensor (with different strain levels, always starting from the pre-stretched zero position).

they will most likely act as general deformation sensors. This is in fact how such materials are used in (Tognetti *et al.*, 2005). In these papers, some work on modelling of these materials and on how signal processing techniques can be applied to compensate the excessive transient times is presented. Summarizing, electroconductive elastomers show complex behaviour and can therefore not be applied as elongation sensors without advanced signal processing. Furthermore, it is questionable if all the unfavourable effects can be compensated through modelling and signal processing techniques.

An alternative approach to using purely textile sensors is to mount non-textile structures onto a fabric. A sensor was produced which utilizes an electroconductive fluid as the transducer medium (Küng, 2005). The fluid is filled into a rubber tube which is glued onto the textile. For the experiments an electrolyte was used. The problem with this fluid is that it diffuses from the tube very quickly. Another type of sensor makes use of the changing light transmission through a fabric when it is stretched (Schultze, 2003). A prototype was built and characterized (Küng, 2005). The results of the evaluations done with the prototype are quite promising. There are no transient times, no hysteresis, and also no dependence on elongation rate. The drawbacks of this sensor are the relatively high obtrusiveness compared to the other sensors and the possibly high power consumption due to the requirement of a light source.

Summarizing, the sensors that are made by attaching a structure to the fabric are not as unobtrusive as the sensors that are made of pure textile material. Yet, their transducer performances are better compared to the other sensors.

21.5 Context recognition technology

The recognition of context, that is the activity of the user and the status of his environment, relies on continuously measured sensor data. In Lukowicz *et al.* (2002) recommendations are made about which sensor or which combinations of sensors are appropriate to detect specific context components. The signals of the sensor data have to be pre-conditioned, e.g., converted, amplified and filtered, before the characteristic features like signal energy or moments are extracted. Several methods and tools have been proved for the fusion of the features. The Bayesian decision theory, for example, offers a fundamental approach for fusion and assignment of predefined classes like motion, sleep, etc. Frequently used methods are the kNN-approach, Kalman and particle filter as well as the Hidden Markov Models and Neural Networks. Combining basic context classes affords the classification also of more complex user contexts like stress and depression. As described in Piccard *et al.* (2001), four wearable sensors (muscle activity EMG, blood oxygen SpO2, skin conductance and respiration) have been applied to detect and to classify eight different motions like anger, grief, joy or hate.

21.6 Wearable components

21.6.1 Embedded microsystems

Recent developments in microtechnology have paved the way to embedding microsystems, either directly in fabrics, or in clothing components like buttons. As a design example Bharatula *et al.* (2004) (Fig. 21.12) shows an autonomous sensor, consisting of a light sensor, a microphone, an accelerometer, a microprocessor and a RF transceiver. A solar cell powers the system even for continuous indoor operation. The two holes allow this system to be sewn in clothes like a normal button.

21.6.2 Accessories

The fusion of the mobile phone, personal digital assistant (PDA) and even the MP3 player into 'smart phones' offers an interface between the personal communication environment and public services. But today's 'smart phones'

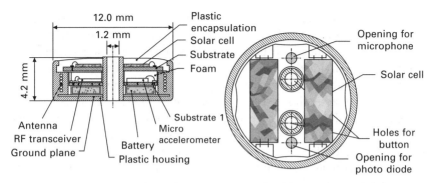

21.12 Design of an autonomous 'sensor button': cross-section and top view.

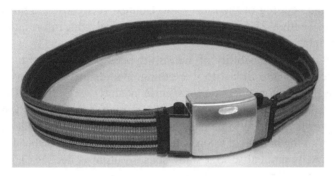

21.13 ETH-QBIC – a mobile computer (Xscale CPU, 256 MB SRAM, USB, RS-232, VGA, Bluetooth) integrated in a belt buckle; the belt houses the flexible batteries and interface connectors.

require manual handling and focusing on the interface. Stripped of bulky IO interfaces and large batteries, mobile computing and communication modules are small enough to be easily carried in a purse or be part of carry-on accessories such as a key chain or a belt buckle as depicted in Fig. 21.13 (Amft *et al.*, 2004).

21.7 Applications

21.7.1 Wearable back manager

Back pain is often caused by unhealthy behaviour. Personal circumstances and activities can increase the risk of musculoskeletal disorders and accidents. Many occupations are characterized by monotonic body postures, lack of body movements or high stress. European studies reveal that every third employee suffers from back pain and every second complains about exhausting and painful body postures during work. Consequences are often chronic pain, inability to work and long and expensive medical treatment. Prevention of back pain would be much more effective than therapy, but for most people it is difficult to change their behaviour. A personal wearable back manager could help an individual to adopt a healthier life-style in daily life. Using our approach of context-aware textiles the garment could monitor body posture, movements and activities, stress levels and other physiological data.

Figure 21.14 shows the concept of a wearable back manager with distributed sensors that are connected over a textile network. Combining different types of sensors enables detection of situations that are critical for the back. One example of a critical situation is lifting a heavy weight incorrectly. Bending the back instead of the knees can damage the spinal column. In order to know if the situation is critical it is necessary to detect not only the posture of the back but also the lifting movements of the arms, the weight of the object and the bending of the knee joint. Context recognition algorithms as described in Section 21.5 can be used to extract relevant information from the sensor data.

We evaluated the feasibility of measuring back postures with elongation sensors integrated into tight-fitting clothing. We identified regions of high elongations that are characteristic for specific postures (Fig. 21.15). However, existing textile elongation sensors are not suitable for this application as the back manager is intended to be used also in situations with low activity as, for example, sitting in front of a computer. During such activities the elongation does not vary dynamically, so a static sensor output is needed. The textile

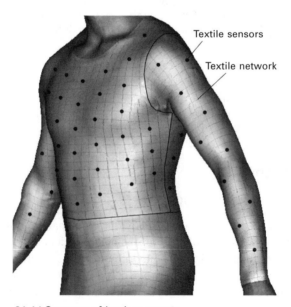

21.14 Concept of back manager.

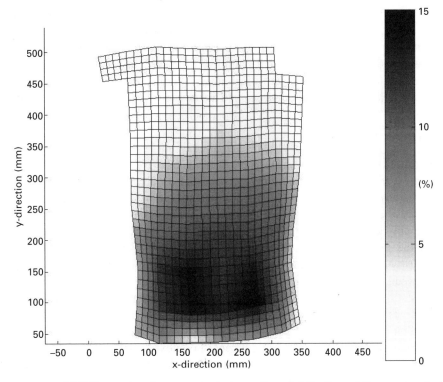

21.15 Example of a measured elongation distribution of the garment on the back.

elongation sensors have to be optimized to fulfil this requirement. Further sensors like pressure sensors, accelerometers, gyroscopes and magnetic field sensors can provide information about the movements of the extremities and pressure changes (e.g. in the shoes). Additional data can come from the objects, for example, by labelling heavy weights with RFID-tags.

21.7.2 Wearable heart manager

The EU-funded project MyHeart (MyHeart, 2004) aims at the reduction of cardio-vascular diseases using wearable health assistants. Cardio-vascular diseases cause roughly 45% of all deaths in Europe; 4 million deaths in Europe, 1.5 million of them in the EU every year. More than 20% of all European citizens suffer from chronic cardio-vascular diseases. Assuming that (only) 5% of the EU population, namely, 19 million persons, will use the MyHeart wearable health assistant, about 60,000 deaths per year caused by myocardial infarctions and strokes could be avoided, saving costs in the range of 12 billion Euros per year.

The project MyHeart focuses on five application fields: improving physical activities, nutrition and dieting, sleep and relaxation phases, stress prevention and early diagnosis and prediction of acute events. The MyHeart wearable health assistant comprises several 'intelligent' clothes with embedded sensors. The data communication with the family doctor, hospital or medical care centre enables an individually matched acknowledgement and health control.

21.8 Outlook

In this chapter we described how the combination of electronic textiles with context recognition technology and miniaturized wearable computers enables a wearable health assistant. Our approach is practicable in terms of modularity and also costs and will allow electronic clothing to become a mass product, one day being affordable for everyone. This trend will have a strong impact on the fashion business. It is not just a chance to strengthen the textile industry by innovation and new market potentials, it also requires a convergence between textile and the electronics industry. That means textile companies have to learn the rules for producing and marketing high-tech products, whereas the electronics companies have to understand the importance of fashion trends.

First implementations of our concept of context-aware textiles in the area of posture training and heart monitoring have been described. Further applications are foreseeable such as prevention of obesity or stress-related illnesses, assisted living for elderly or disabled people, as well as work assistance (e.g. for high-risk environments or remote working). Going one step further from wearable health assistants to even more general-purpose personal assistants will be the next challenge. Such a personal assistant not only monitors the health but also recognizes the needs of the user and provides automatic and active support like a personal servant or friend.

21.9 Acknowledgement

The authors would like to thank Jan Meyer and Corinne Mattmann for contributing their research results.

21.10 References

Amft O, Lauffer M, Ossevoort S, Macaluso F, Lukowicz P, Tröster G (2004), Design of the QBIC wearable computing platform. *Proc. 15th IEEE Application-specific systems, architectures and processors ASAP 2004.*

Bharatula N B, Ossevoort S, Stäger M, Tröster G (2004), Towards wearable autonomous microsystems. *Pervasive 2004*, Springer; 2004: 225–237.

Bickerton M (2003), Effects of fiber interaction on conductivity, within a knitted fabric stretch sensor. In *Proceedings of the IEE Eurowearable 2003*, pages 67–72, Birmingham, UK, September 4–5 2003.

Cottet D, Grzyb J, Kirstein T, Tröster G (2003), Electrical Characterization of Textile Transmission Lines, *IEEE Transactions on Advanced Packaging*, Vol. 26, No. 2, May 2003, pages 182–190.

Farringdon J, Moore A J, Tilbury N, Church J, Biemond P D (1999), Wearable sensor badge & sensor jacket for context awareness. In *The Third International Symposium on Wearable Computers*, Digest of Papers, pages 107–113, San Francisco, California, October 18–19 1999.

Gouaux F, Simon-Chautemps, Fayn J, Arzi M, Assanelli D *et al.* (2002), Ambient Intelligence and Pervasive Computing for the Monitoring of Citizens at Cardiac Risk: New Solutions form the EPI-MEDICS Project. *Computers in Cardiology*; 29; 2002: 289–292.

Kirstein T, Lawrence M, Tröster G (2003), Functional Electrical Stimulation (FES) with Smart Textile Electrodes, *Proc. Wearable Systems for e-Health Workshop*, Pisa Italy, 11–14 December 2003.

Kirstein T (2004), Medical Applications of Electronic Clothing, *Medical Device Technology*, Vol.15, 06/2004.

Kirstein T, Cottet D, Grzyb J, Tröster G (2005), *Wearable Computing Systems – Electronic Textiles, Wearable Electronics and Photonics*, edited by X. Tao, Woodhead Publishing Ltd., 2005.

Klemm M, Locher I, Tröster G (2004), A novel circularly polarized textile antenna for wearable applications, in *Proc. 34th European Microwave Week*, pp. 137–140, October 2004.

Korhonen I, Pärkkä J, van Gils M.(2003), Health Monitoring in the Home of the Future. *IEEE Eng. Medicine and Biology Mag.* May/June 2003: 66–73.

Küng C (2005), Wearable elongation sensors for human posture analysis, diploma thesis, ETH Zürich, 2005.

Locher I, Kirstein T, Tröster G (2004), Routing methods adapted to e-textiles, in *Proc. 37th International Symposium on Microelectronics* (IMAPS 2004), November 2004.

Locher I, Kirstein T, Tröster G (2005), From Smart Textiles to Wearable Systems, *mst news*, No. 2/05, April 2005, pages 12–13.

Lukowicz P, Junker H, Stäger M, von Büren T, Tröster G (2002) WearNET: A Distributed Multi-Sensor System for Context Aware Wearables, in *Proc. of the UbiComp2002*, Springer, 2002: pages 361–370.

Lukowicz P, Kirstein T, Tröster G (2004), Wearable Systems for Healthcare Applications, *Methods of Information in Medicine*, Vol. 43, 03/2004.

Marculescu D, *et al.* (2003), Electronic Textiles: A Platform for Pervasive Computing, in *Proceedings of the IEEE*, Vol. 91, No.12, Dec. 2003.

Massey P (2001), Mobile phone fabric antennas integrated within clothing, in *Proceedings of the 11th IEE Conference on Antennas and Propagation* (IEE Conf. Publ. No. 480), vol. 1, pp. 344–347, April 2001.

MyHeart (2004) Fighting cardio-vascular diseases by preventive lifestyle & early diagnosis. www.extra.research.philips.com/euprojects/myheart.

Oh K W, Park H J, Kim S H (2003), Stretchable conductive fabric for electrotherapy, *Journal of Applied Polymer Science*, 88: 1225–1229, 2003.

Pacelli M, *et al.* (2001), Sensing threads and fabrics for monitoring body kinematic and vital signs, in *Proceedings of Fibers and textiles for the future*, Tampere, Finland, 2001.

Piccard R, Vyzas E, Healey J (2001), Toward machine emotional intelligence: analysis of affective physiological state. *IEEE Trans. Pattern Anal. Mach. Intelligence*, vol 32, Oct 2001: 829–837.

Salonen P, Hurme H (2003), A novel fabric WLAN antenna for wearable applications, in *Proceedings of the IEEE Antennas and Propagation Society International Symposium*, vol. 2, pp. 700–703, June 2003.

Salonen P, Keskilammi M, Kivikoski M (2000), Single-feed dual-band planar inverted-f antenna with u-shaped slot, *IEEE Transaction on Antennas and Propagation*, vol. 48, pp. 1262–1264, August 2000.

Salonen P, Rahmat-Samii Y, Hurme H, Kivikoski M (2004), Dual-band wearable textile antenna, in *Proceedings of the IEEE Antennas and Propagation Society International Symposium*, vol. 1, pp. 463–466, June 2004.

Schultze C (2003), New technology for textile based monitoring of periodic physiological activity, in *Proc. Wearable Systems for eHealth*, Lucca, Italy, Dec. 11–14, 2003.

Scilingo E P, Lorussi F, Mazzoldi A, De Rossi D (2003), Strain-sensing fabrics for wearable kinaesthetic-like systems. *IEEE Sensors Journal*, 3(4), August 2003.

Tanaka M, Jae-Hyeuk J (2003), Wearable microstrip antenna, in *Proceedings of the IEEE Antennas and Propagation Society International Symposium*, vol. 2, pp. 704–707, June 2003.

Tognetti A, Lorussi F, Bartalesi R, Quaglini S, Tesconi M, Zupone G, De Rossi D (2005), Wearable kinesthetic system for capturing and classifying upper limb gesture in post-stroke rehabilitation. *Journal of NeuroEngineering and Rehabilitation*, 2(8), 2005.

Weiser M (1991), The computer for the 21st century. *Scientific American* 265, No. 3, September 1991: 94–104. Available from: URL: http://www.ubiq.com/hypertext/weiser/SciAmDraft3.html

Wijesiriwardana R, Dias T, Mukhopadhyay S (2003), Resistive Fibre-meshed transducers, in *Proceedings of the Seventh IEEE International Symposium on Wearable Computers (ISWC'03)*, pages 200–209, October 21–23 2003.

22
Intelligent garments in prehospital emergency care

N LINTU, M MATTILA and O HÄNNINEN,
University of Kuopio, Finland

22.1 Introduction

Prehospital emergency care is an essential part of the chain of survival in cases of trauma or acutely exacerbated diseases. An emergency is a more or less urgent and dangerous situation, because of weakening of some or many vital functions. Prehospital means the period before arrival at hospital. Prehospital emergency care includes all diagnostic and therapeutic procedures carried out by the ambulance (emergency care) team. This team consists of a paramedic and emergency care technician. In some organisations also an emergency physician or nurse can be involved. Appropriate medical control is crucial to guarantee high-quality prehospital care (Holroyd *et al.*, 1986). The organisation of emergency services differs greatly in different countries, and even in different areas of a country (Suserud *et al.*, 1998). For this reason the facts included in this chapter cannot apply to each national organisation. In order to understand the specific features of applications of smart textiles in prehospital care some of its basic characteristics will be described.

The working circumstances in prehospital emergency care are difficult and the tasks are demanding. The ambulance team gets its actual 'mission' by alarm centre personnel, the initial information on the particular 'case' is based on the alarm call and compiled mostly by someone who is inexperienced in health care. Each case is individual and diagnosis can be initially obscure on arrival at the patient. What has happened, what is the situation, how has it developed and what should be done, are logical questions in this situation. Based on patients described symptoms and signs, a preliminary diagnosis and estimation of degree of urgency are completed, but only on some certainty level. Clinical assessment includes patient's observations, measurements and palpations. However, these manual operations, based on human sensation, cannot detect deviations from the norm of all vital functions. Human senses can be clarified by patient monitors, which greatly supplement the clinical picture, and also opens new areas for applications of smart textiles and intelligent garments.

22.2 Different cases and situations

Emergency care services provide care and transportation for a range of different categories of cases. There is a scale from extremely urgent, through urgent to non-urgent missions. There is also a great variation in the disease or trauma that has caused the emergency and of which vital functions deviate from normal and by how much. It will be necessary to estimate these deviations from the normal range. If possible, it includes the stabilisation of vital functions and hopefully returning them to a normal level before transportation.

22.3 Circumstances

The location and circumstances of incidents varies greatly. In most cases they occur at home, but often also outdoors in different circumstances. The weather can be very cold, rainy and windy. Also during transportation the patient is often exposed to outdoor conditions. A smart protection against variable conditions is thus a challenge for emergency care services.

22.4 Vital functions

It is important that every organ in the human body has continuously optimal oscillating homeodynamics under the continuous neuronal and hormonal control. These are functions that are essential and vital for wellbeing. With the aid of intelligent garments it is possible to monitor different vital functions, such as ECG, circulation, respiration, EMG and skin temperatures.

22.4.1 Consciousness

Central and peripheral nervous systems monitor and control the body's functions and adapt them to surrounding ambience and its variations. They protect the body against external/internal influences and dangers. A fully conscious person can greatly facilitate the evaluation of a situation and survey its possible reasons by expressing sensations of the body and explaining their development. His/her important protective reflexes are active. Decrease of consciousness level means always decrease of body security linearly with the loss of perception. Then the danger of occlusion of airways increases as well as the probability of aspiration of gastric contents to bronchi.

Degrees of consciousness can be scaled based on eye opening, verbal and motor response (Glasgow coma scale). The opening of eyes can be spontaneous, a response to voice or to pain, or it is absent. The verbal response can be orientated, confused, inappropriate, and incomprehensible or absent. There is practically no instrumental method to monitor the level of consciousness.

22.4.2 Circulation

It is essential that every organ receives continuously enough oxygen and nutrients from blood circulation. Equally important is the removal of waste products including carbon dioxide. This situation requires normal blood volume, cardiac output and optimal circulatory distribution. The heart pumps blood to tissues through a vascular system consisting of arteries, capillaries and veins. Every contraction of the left ventricle creates a systolic pressure wave in great arteries, and the elasticity of arterial walls maintains the diastolic pressure in circulation. Blood pressure varies greatly between individuals and depends on physical activity (Thomas *et al.*, 2005). Blood pressure is necessary to create blood flow to tissues by overcoming the peripheral vascular resistance. Tissue circulation is controlled by dilation and constriction of local small arterioles. The heart acts according to the rhythm dictated by the sinus node. The impulse proceeds through atria to ventricles. Heart rate adapts to circulatory needs in accordance with the stroke volume which is the volume of blood pumped by one beat.

22.4.3 Respiration

Respiration requires rhythmic movements of the thoracic cage. This is necessary for the exchange of alveolar air and interchange of oxygen and carbon dioxide between alveolar air and blood in pulmonary capillaries. As a result arterial blood haemoglobin will be nearly completely saturated with oxygen. Haemoglobin in the red blood cells carries oxygen to tissues delivering part of the bound oxygen in the capillary to adjacent tissues. Alveolar ventilation also removes carbon dioxide from the body that is produced in tissues.

22.4.4 Body temperature

Thermal balance means that the temperature of the body is optimal for organ function. Core temperature, i.e., the temperature in the central inner parts of the body, is maintained within a narrow range, as human beings are homeothermic. In contrary to the core temperature, the temperature in the peripheral parts of the body, such as in the extremities, varies greatly in accordance with the ambient temperature.

22.5 Monitoring of vital functions

With our own senses we can estimate only roughly if vital functions are performing adequately or if they are deviating from the normal level. In some trauma cases the casualty is stained by blood and mud. Moreover, stress can cause a paleness of the face that can complicate visual estimation of the casualty's state. We can greatly augment our understanding of the actual situation of vital functions by recording parameters electrically, which represent cardiac function, circulation, oxygenation, ventilation and thermal balance (Konstantas *et al.*, 2004, Lymberis, 2004). Patient monitoring is based on probes and wires that can measure electric voltage or current, pressure or light flow. The necessary number and location of electrodes (probes) depends on which phenomenon (parameter) is recorded and how exact data are needed.

22.5.1 Electrocardiogram (ECG)

ECG measures voltage differences created by electric discharges in the heart during each cycle. From different leads of the electrocardiogram one can estimate the impulse propagation, heart rhythm, vascular resistance and possible lesions in the heart muscle as well as their location and extension. Myocardial infarction and cardiac arrhythmias can be detected with ECG. This registered ECG represents electric function of the heart and has no direct correlation with the blood pumping action of the heart. The most impressive pathological evidence is given by a difference in present ECG compared with an earlier recorded one. The exact location of ECG electrodes is a prerequisite for the utilisation of the method in morphologic diagnostics. The probes should have a firm skin attachment and should not be allowed to move at all.

22.5.2 Pulsation

Pulsation is a quite informative circulatory parameter in different situations. Pulsation is recorded at the wrist (radial pulsation), at the inguinal channel (femoral pulsation) or at the neck (carotid pulsation). Pulse rate can be counted by pressing lightly with a finger on an artery but this manual method is laborious. As well as the rate in beats per minute, the strength (strong, weak or absent) and rhythm (regular or not) can be detected.

22.5.3 Pulse oximetry

Pulse oximetry is a highly respected measurement among patient monitoring methods (Nuhr *et al.*, 2004). Oxygenation is such an essential and vulnerable function, that a non-invasive method providing reliable information on oxygenation status is more than welcome. At present the measurement is based on the differences in light absorbance between oxyhaemoglobin and reduced haemoglobin. They are measured with two different emitted wavelength lights (Sinex, 1999). The probe is located usually on a finger, light emitter and detector on opposite sides. A specific algorithm calculates oxygen saturation and pulse rate (Barker, 2002, Gehring *et al.*, 2002, Tobin *et al.*, 2002). Moreover, a visible pulse wave can be reproduced on the screen. The amplitude of the pulse wave mirrors circulation changes at the measuring site (finger) and gives information on the peripheral circulation.

In the interpretation of saturation values one must master the oxygen dissociation curve characteristics. As the normal saturation is 95–98%, 92% means an imminent hypoxaemia and at the 90% level hypoxaemia is already real. Changes in oxygenation are fast in many emergency cases, and active treatment of hypoxia is both essential and effective. Pulse wave amplitude is a relative parameter, which strongly reacts to several different stimuli.

22.5.4 Measurements of body temperature

The measurement of core temperature from rectum, tympanic membrane or oesophagus is technically easy. Skin temperature measurements are a totally different entity. Both individual data and their differences serve as valuable material in the interpretation and conclusions. Temperatures and their textile-integrated measuring sensors are intrinsic components of intelligent garments for prehospital emergency care.

22.6 Selection of monitoring methods

In prehospital conditions all actions should be undertaken easily and quickly because of shortage of time and manpower (Birk and Henriksen, 2002). The selection of methods and parameters in patient monitoring starts from the real needs and benefit/effort ratio of the parameter. In each case it must be considered, which of the vital functions are absolutely necessary to evaluate and which parameters are the most informative for the diagnosis and status estimation. This type of intelligently and individually tailored patient monitoring would greatly increase the benefit/effort ratio. At the same time it reduces the flow of information which is also a very important aspect in emergency care.

22.7 Interpretation of monitored parameters

The interpretation of recorded data requires clinical experience, knowledge and good familiarity with the monitoring method and physiological background of the function that it measures. Very seldom the recording expresses directly what the diagnosis is and the actual patient situation. There is a need to interpret symptoms, signs and monitored parameters together and thus obtain a diagnosis by utilizing all relevant means. Correct diagnosis and treatment is the main purpose of prehospital care.

22.8 Telemedicine

Therapeutic decisions require deep expertise in the interpretation of findings. There is clearly a need for teleconsultation to have an expert's advice on line

(Bhatikar *et al.*, 2002). Telemedicine application for prehospital emergency care includes transmission of all recorded data to an expert, who can reconstruct the situation, give the correct diagnosis and send treatment and action advice in real time (Anantharaman and Swee Han, 2001). The essential contents of a patient's chart are anamnesis, status, measurement and recording data, all in structured digital form. Thus an interpretation is possible and reliable for an experienced emergency physician to make therapeutic decisions that are transmitted in digital form as a reply to consultation.

The best solution for the organisation of telemedicine services in prehospital emergency care would be one national expert centre, which would always be ready to add high-quality expertise to emergency care both for ambulance and health centre personnel. In the evaluation of a situation correctly monitored and recorded parameters and their trends are valuable because they can be transmitted to experts' computer screens. The effects of therapeutic procedures can also be followed in real time (Gallego *et al.*, 2005).

Presently, the teleconsultation in emergency care in Finland is limited to suspected myocardial infarctions. Twelve lead ECG is transmitted from the field to the nearest central hospital for evaluation and decision of starting thrombolysis is received back from the hospital before patient's transportation. This practice of immediate action has had a positive effect on the infarction outcome. In ECG transmission usually telefax is used.

22.9 Negative effects of transportation on vital parameters

The aim of prehospital emergency care is to stabilise a patient's state to a safe level, before the start of transportation, if it is possible. In an apartment building, carrying the patient on a stretcher from the upper floors is a physically demanding task both for paramedics and the patient. Carrying the patient in the head-down or up position can seriously influence his/her blood circulation. Elevators, on the other hand, are often so small that the patient must be carried in a sitting position, which can be fatal for the patient. During carrying and transportation the paramedic needs to supervise the situation by continuous patient monitoring, and intelligent clothing would be a valuable resource to detect the dangerous development of a situation in real time.

If the patient is well stabilised at the scene there is no need to drive him/her to hospital with maximal speed. Angular acceleration, acceleration and deceleration of the ambulance vehicle all have harmful effects on circulation (Sagawa and Inooka, 2002). These physical factors can also provoke nausea and vomiting, which can lead to dangerous aspiration of gastric contents into bronchi.

22.10 Patient chart

Presently anamnesis, symptoms and signs, observations and measurements are written on patient charts in analogue form. This form contains also notes on times of dispatch, of arrival at the scene and start of transportation and arrival at hospital. It also includes remarks on therapeutic procedures and medication. It is difficult, however, to include information of continuous monitoring in this written form. Its contents are not in transmittable form. A digital patient chart would be a radical improvement for the chain of information (Meislin *et al.*, 1999). All data would be in digital form as a database in the network, simultaneously available in real time for all involved in this particular patient's care. In this networking, a selected abstract of patient's case history could also be augmented before arrival at the scene. These facts could decisively improve diagnostics in prehospital care. The monitoring system based on intelligent clothing would offer valuable assistance to prehospital care.

22.11 Data security

When patient data and case history are handled, data security and privacy have always seriously to be taken into consideration. There are very strict regulations concerning patient data secrecy, which should be known and obeyed. These secrecy rules should not, however, prevent the patient getting optimal treatment, especially in emergencies. It is also important that only authorised health care personnel have direct access to patient data. If the paramedic asks for advice from a national expert centre, patient identity can be omitted so the interchange of data is fully legal and acceptable.

Because decisions on treatment are done based on values and trends of monitored parameters, they must be highly reliable and real. The treatment decisions also include great responsibility because they concern human health and possibly even life. All artefacts are harmful and potentially dangerous. The chain of information includes many points where the registered data can be corrupted. One of the critical points in this chain is the interface between the human body and the probe, just in the area of smart textile applications. Motion easily causes disconnections in tight contact as well as variations in the mobile net intensity. On the other hand, movement may be one characteristic of the emergency care.

22.12 Day surgery

Post-operative care has fundamentally changed in recent years. Earlier, patients were under close supervision after surgical operations in hospital for days, whereas nowadays some patients are returned home only a couple of hours after completion of surgery. Also new anaesthesia methods and anaesthetics have made earlier discharge from hospital possible. The adaptation of vital functions to so short post-operative observation is demanding. The most critical periods are the transportation home and the first twenty-four hours there. Patients' safety would require established reliable monitoring services after day surgery.

These monitoring services would benefit from a smart garment, which would include sensors, connectors, collector and emitter as a patient interface. The smart collector would continuously record and save trends as well as analyse monitored parameters to detect all impending deviations from normal. These functions require a large amount of artificial intelligence because many factors should be simultaneously noticed. The emitter unit would send the data to a call centre for further interpretation in case a parameter or its trend seems to be suspicious. At this level artificial intelligence inevitably requires support from top medical expertise. The most important is the appropriate response of emergency medical services to every alarming situation (Bhatikar *et al.*, 2002). This post-operative monitoring service is a special short-term application of home monitoring services for elderly ill patients (Dittmar *et al.*, 2004, Prentza *et al.*, 2004). In this way an intelligent garment can enhance the safety and wellbeing of the users and support health care.

22.13 Protective covering

One example of the multidisciplinary approach in the development of better patient protection in health care has been the Ergovaate-special clothing for health and social sector-process at the University of Kuopio during 2001–2003. Its aims have been to incorporate technology into clothing for use in emergency care and first aid by developing a protective covering for injured casualties with integrated patient monitoring and data networking. A new concept for emergency care was developed (Mattila *et al.*, 2003). The concept consisted of three different parts: (i) protection of the patient with rescue covering, (ii) monitoring the vital functions of the patient and (iii) collecting the patient data to a digital patient chart and wireless transmission. The developed and tested prototype of the rescue covering is currently in everyday use in, e.g., rescue services and helicopters in Finland (www.telespro.fi). A demonstration prototype of the patient monitoring and data transmission system was also developed. The protective covering is easy to use in typically demanding emergency circumstances. It does not hamper emergency care procedures. It is possible to touch, examine and treat separate parts the body by exposing only highly the relevant part to the effects of weather.

The selected special materials of the covering allow maintenance after each use. It is easy to remove contaminated blood and other body fluids and to clean the covering without damaging its protective and operating characteristics. This requires a high-class service system. When a casualty is transported to hospital in the protective covering, a replacement covering should be provided for the ambulance staff. This is the only way to guarantee uninterrupted preparedness for prehospital protection against hostile conditions, such as cold and rain. It is also essential that the effective cold protection continues in the hospital during the first hours of stabilisation.

22.14 An integrated monitoring of vital functions

Integrated patient monitoring, data collection and data transmission systems are separate additional modules, which are most important and beneficial in multiple casualty traffic accidents, but they are useful also in single case emergencies. The final goal is a smart integration of probes and cables within the protective textile. A digital patient card would offer obvious advantages in the evaluation of the state of vital functions and their development over hand-written notes. In multiple casualty situations digital patient charts serve as a guideline for setting casualties in priority order to emergency care procedures and transportation. Digital information is not site limited, and is simultaneously available in the same form to all those who understand and need it for decision making. The principal implement in the utilisation of digitalised information could be a Palm PC.

The follow-up of location (accident scene, emergency care tent, local hospital, trauma centre) of casualties in the chain of rescue should be totally automatic, utilising wireless phone networks or special tags and detectors.

22.15 Mobile isolation

Some severe infectious diseases require efficient isolation to protect health care personnel and to prevent infection dissemination and a serious epidemic. The isolation need covers the time from the first suspicion of possible infection, through verification to full recovery. Because care of these infections demands special expertise, they are usually nationally concentrated on special units. As a consequence, suspected cases are often transported long distances and in a critical condition. This requires special preparedness of emergency care both to simultaneously care and isolate, and protect emergency care staff. A seamless isolation of infected patients is the most effective way to prevent the distribution of the problem. Efficient preventive and disinfecting measures are necessary during the whole chain of transportation from home to final isolation room in a central hospital. This includes public stairways, transportation vehicles and long hospital corridors.

The contamination risk concerns both the emergency care personnel and interior of the vehicle and all pathways the patient is carried through. It is difficult to clean the whole area and impossible to disinfect it. It is much easier is to isolate the patient within an individual containment (Hänninen *et al.*, 2003). Such containment should be clean and ready to use and provide good isolation. It should integrate the opportunities for monitoring and care and be 'portable' to fit in the vehicles and elevators. The protective clothing should be impermeable to microbes and washable at high temperature without losing its benefits. This offers a challenge to materials.

22.16 Optimal smart solution for prehospital emergency care

Because time is the critical factor in sudden emergency situations all the components of the monitoring chain should be instantaneously applicable. As a matter of fact, all the time spent on attaching separate sensors, connecting separate cables, looking at separate screens and trying to get connections for real time devices means delay in diagnosis and appropriate treatment. This point alone provides an opportunity for an intelligent garment to provide all essential components in one entity (Anliker *et al.*, 2004) (see Fig. 22.1).

22.16.1 Sensors

Many of the monitoring sensors can be integrated into textiles, e.g., a shirt or a thorax-surrounding belt so that a firm touch with the body at appropriate

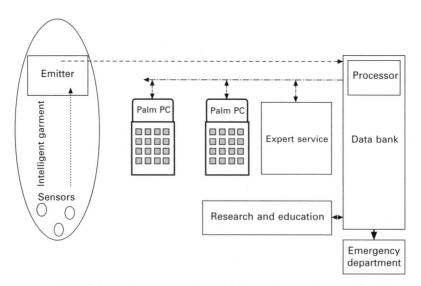

22.1 Schematic presentation of information pathways. The links make patient data available to all participants in medical rescue with the aid of sensors in the intelligent garment. Expert service is connected on line to the accident scene and databank. The databank serves current problem solving and research as well as educational needs.

points can be achieved. Subsequently, the time of probe application would be greatly decreased. The differences in thorax size would provide inbuilt difficulties. All important parameters cannot be recorded from the thorax area, but need some other specific points of recording, e.g., for blood pressure, skin temperatures and end tidal carbon dioxide.

22.16.2 Cables

Different cables and wires that connect probes to collector units disturb measurements and they can become loose and be a source for faults in measured data. Textile integrated electrodes and wireless connections can solve the present problem of lead agglomeration.

22.16.3 Patient chart

Treatment decisions are based on different components of information such as actual anamnesis, case history, status, observations, monitored parameters, procedures and medication. This information should be in easily readable structured form, preferably in graphic phenotype. This entity is called a digital patient chart (electronic patient chart) and it represents a modern high-tech information tool as a completion to an intelligent garment. The same real-time information can be seen on the screen of the Palm PC of the paramedic at the scene, in the emergency department of the responsive hospital and in the experts' telemedicine office. If all digital charts are collected in one national archive, it could serve as a comprehensive source for research, development and education. So far this reliable documentation of different cases in prehospital care has been lacking.

22.16.4 Teleconsultation

The cases in emergency situations are often complicated and the need for expert advice for diagnostics and optimal treatment is urgently needed (Soysal *et al.*, 2005). This requires a standby advice service system, instantly ready to give appropriate medical advice on a wide range of problems. In a small country a national expert centre would be a justified solution to guarantee a high quality of teleconsultation round-the-clock seven days a week.

22.17 Conclusions

The presently available sensor and information transfer technologies make the follow up of the vital functions possible. Nevertheless, only a minimal part of the technically possible applications is in routine use in prehospital emergency care at present. Prehospital emergency care opens distinctive challenges for the application of modern technology and design in extremely demanding conditions. The applied monitoring technique can be decisive for survival through correct diagnosis to appropriate therapy enabling also teleconsultation on line. A partial solution could be an intelligent garment, which would provide integrated probes, cables and wireless connection to Palm PC (mobile PC screen) in one entity (Barnard and Shea, 2004). The adoption of the potentially valuable technologies takes time because of the difficult and multidimensional working conditions.

There is a need for efficient services from alarm to response, whenever intelligent garments are used in home care for elderly sick patients and post-operative monitoring after day surgery. Smart applications for emergency care require multidisciplinary cooperation of top experts in emergency care, in smart textile solutions and monitoring/information technology because end products and organisations should be usable in difficult field conditions.

22.18 References

Anliker U, Ward J A, Lukowicz P, Troster G, Dolveck F, Baer M, Keita F, Schenker E B, Catarsi F, Coluccini L, Belardinelli A, Shklarski D, Alon M, Hirt E, Schmid R and Vuskovic M (2004), 'AMON: a wearable multiparameter medical monitoring and alert system', *IEEE Trans Inf Technol Biomed*, 8 (4), 415–27.

Anantharaman V and Swee Han L (2001), 'Hospital and emergency ambulance link: using IT to enhance emergency pre-hospital care' *Int J Med Inform*, 61 (2–3), 147–61.

Barker S J (2002), Motion-resistant pulse oximetry: a comparison of new and old models', *Anesth Analg*, 95 (4), 967–72.

Barnard Rand Shea J T (2004), 'How wearable technologies will impact the future of health care', *Stud Health Technol Inform*, 108, 49–55.

Bhatikar S R, Mahajan R L and DeGroff C (2002), 'A novel paradigm for telemedicine using the personal bio-monitor', *Biomed Sci Instrum*, 38, 59–70.

Birk H O and Henriksen L O (2002), 'Prehospital interventions: on-scene-time and ambulance-technicians' experience', *Prehospital Disaster Med.* 17 (3), 167–9.

Dittmar A, Axisa F, Delhomme G and Gehin C (2004), 'New concepts and technologies in home care and ambulatory monitoring', *Stud Health Technol Inform*, 108, 9–35.

Gallego J R, Hernandez-Solana A, Canales M, Lafuente J, Valdovinos A and Fernandez-Navajas J (2005), 'Performance analysis of multiplexed medical data transmission for mobile emergency care over the UMTS channel', *IEEE Trans Inf Technol Biomed*, 9, (1), 13–22.

Gehring H, Hornberger C, Matz H, Konecny E and Schmucker P (2002), 'The effects of motion artifact and low perfusion on the performance of a new generation of pulse oximeters in volunteers undergoing hypoxemia', *Respir Care*, 47, (1), 48–60.

Holroyd B R, Knopp R and Kallsen G (1986), 'Medical control. Quality assurance in prehospital care'. *JAMA*, 256 (8), 1027–31.

Hänninen O, Lintu N, Holopainen J, Seppälä S, Mattila M A K (2003) 'Preparedness needed to isolate suspected severe infection patients during transportation' in NBC 2003 symposium on *Nuclear, biological and chemical threats – a crisis management challenge,* K. Laihia (ed.), U Jyväskylä Research Reports 98. 154–7.

Konstantas D, van Halteren A, Bults R, Wac K, Widya I, Dokovsky N, Koprinkov G, Jones V and Herzog R (2004), 'Mobile patient monitoring: the MobiHealth system', *Stud Health Technol Inform*, 103, 307–14.

Lymberis A (2004), 'Research and development of smart wearable health applications: the challenge ahead', *Stud Health Technol Inform*, 108 (1) 55–61.

Mattila M A K, Lintu N, Holopainen J, Seppälä S and Hänninen O (2003), 'Smart protection of trauma patients against weather influences in prehospital care', *Acta Anaesth Scand,* 47, Suppl 116, 57.

Meislin H W, Spaite D W, Conroy C, Detwiler M and Valenzuela T D (1999), 'Development of an electronic emergency medical services patient care record', *Prehosp Emerg Care*, 3 (1), 54–9.

Nuhr M, Hoerauf K, Joldzo A, Frickey N, Barker R, Gorove L, Puskas T and Kober A (2004), 'Forehead SpO2 monitoring compared to finger SpO2 recording in emergency transport', *Anaesthesia*, 59 (4), 390–3.

Prentza A, Angelidis P, Leondaridis L and Koutsouris D (2004), 'Cost-effective health services for interactive continuous monitoring of vital signs parameters – the e-Vital concept', *Stud Health Technol Inform*, 103, 355–61.

Sagawa K and Inooka H (2002), 'Ride quality evaluation of an actively-controlled stretcher for an ambulance', *Proc Inst Mech Eng,* 216 (4), 247–56.

Sinex J E (1999), 'Pulse oximetry: principles and limitations', *Am J Emerg Med*, 17 (1), 59–67.

Soysal S, Karcioglu O, Topacoglu H, Yenal S, Koparan H and Yaman O (2005), 'Evaluation of prehospital emergency care in the field and during the ambulance drive to the hospital', *Adv Ther*, 22 (1), 44–8.

Suserud B O, Wallman-C:son K A and Haljamae H (1998), 'Assessment of the quality improvement of prehospital emergency care in Sweden', *Eur J Emerg Med*, 5 (4), 407–14.

Thomas S H, Winsor G, Pang P, Wedel S K and Parry B (2005), 'Near-continuous, noninvasive blood pressure monitoring in the out-of-hospital setting', *Prehosp Emerg Care*, 9 (1), 68–72.

Tobin R M, Pologe J A and Batchelder P B (2002), 'A characterization of motion affecting pulse oximetry in 350 patients', *Anesth Analg*, 94, (1 Suppl), S54–61.

23
Intelligent textiles for children

C HERTLEER and L VAN LANGENHOVE,
Ghent University, Belgium and R PUERS,
Katholieke Universiteit Leuven, Belgium

23.1 Introduction

During the late 1990s the textile sector entered a new era, the one of intelligent or smart textiles. Due to the emergence of new materials, functional textiles were elevated to sensors and actuators, which are among others, required building blocks of smart textiles. Throughout the years however, the definition of smart textiles has altered. Nowadays they are more frequently defined as system concepts merging textile materials and electronics, resulting in a textile with extensive capabilities. Although smart textiles are applicable in a wide range of areas, a great deal of the research focuses on their use in garments. It is obvious that textiles used for clothing are the ideal interface between the human body and external technologies. Moreover, several of these technologies are aimed at facilitating our life. Implementing them into a garment can therefore considerably increase our level of comfort and safety in a non-obtrusive way. This new genre of clothing has been entitled 'smart garments'. They can be applied in areas such as healthcare, protection, sports and leisure.

When used in a medical environment they are often referred to as smart biomedical garments. This clothing typically has monitoring and processing capabilities for bio-physiological signals. Furthermore, advances in telecommunication technologies have led to the development of stand-alone garments. It is clear that a successful merging process between new textile materials and wearable microelectronics enables these developments.

Children especially can benefit from this evolution because today's monitoring methods are not always child-friendly. Generally, children being monitored are prevented from moving freely because wires connect the sensors to the related instruments. By embedding sensors, interconnections, antennas and electronics in the garment, a child-friendly stand-alone suit can be obtained. In addition, the use of embedded textile components guarantees washability (and thus reuse) of the suit.

Based on these ideas, a Flemish project was set up in the late 1990s to develop a smart suit for infants, named the Intellitex suit. A consortium was founded with three partners involved: the Department of Textiles of Ghent University, the Electronics Department MICAS of Katholieke Universiteit Leuven and the Paediatrics Department of Ghent University Hospital. The four-year project was funded by IWT (Belgium), which is acknowledged for its support. The realisation of the Intellitex suit will be presented here.

23.2 State of the art

Smart textiles belong to a fast-evolving research area. New prototypes of smart biomedical garments are unremittingly being presented. It started in the late 1990s when Georgia Tech (USA) introduced the Wearable Motherboard™.[2] This smart shirt was developed for the ambulatory monitoring of soldiers in combat situations. It is a single-piece circularly woven undergarment onto which sensors for monitoring biosignals can be plugged. In addition to monitoring vital signs, the shirt can also detect bullet penetration. The garment itself consists of a grid of optical and electroconductive wires, acting as a 'data bus' through which data coming from the sensors is transmitted to a processing unit. The applied weaving process had to be adapted in order not to have discontinuities in this wiring system. This 'textile motherboard' can be tailored to each individual and provides a platform for a suite of sensors.

Another smart garment WEALTHY was developed by SMARTEX (Italy).[3] It is a wearable monitoring system that fully exploits the possibilities of textiles. Strain fabric sensors, piezoresistive yarns, fabric electrodes and electroconductive interconnections are all knitted into a garment allowing the recording of vital signs such as heart and respiration rate, electrocardiogram, activity pattern and temperature. The sensitive garment is provided with a portable electronic unit that processes and transmits the acquired data.

The Lifeshirt™ by Vivometrics Inc. (USA) is a Lycra vest whose core sensor system is based on inductance plethysmography. The sensor is a sinusoidally arranged electrical wire embedded in a stretchable shirt. In addition, state of the art conventional sensors are used to measure respiration, electrocardiogram, posture and activity.[4] The complete LifeShirt™ system is composed of three parts, a garment, a data recorder and PC based analysis software.

A last biomedical garment that will be mentioned here is a sensorised T-shirt developed within the French project VTAM (Vêtement de Télé-Assistance Médicale Nomade).[5] The T-shirt is equipped with four dry ECG (electrocardiogram) electrodes, a breath rate sensor, a shock/fall detector and two temperature sensors. Sinusoid-like conductors integrated in a textile belt monitor respiration rate, whereas electronic monitoring of the three-component acceleration of the body enables the shock/fall identification. A miniature GSM/GPRS module for signal precomputing and transmission together with a power supply are kept on a belt around the T-shirt.

The enumerated smart biomedical garments are just a limited selection of initiatives in progress. Developments particularly aimed at children, however, are fewer. The Mamagoose pyjama, developed by the Belgian company Verhaert is an example. The baby suit is marketed as a prevention tool for SIDS (Sudden Infant Death Syndrome). Conventional heartbeat and respiration sensors collect data from the infant and an alarm is sent out in case of potential danger.[6] Also the previously described Lifeshirt™ system is available in a paediatric size for children from five years old.

The above-mentioned state of the art smart garments differ in their exploitation of the textile material. Only a few of them use it as a sensing device while many still rely on conventional sensors. In more applications textile material is utilised for interconnections, being woven or knitted into the garment. Nonetheless all quoted research efforts express the feasibility and the great potential of garments to be used as wearable monitoring systems.

23.3 The intellitex suit

The main aim of the research carried out by the Flemish Consortium was to explore new textile materials and microelectronics to combine them in a smart biomedical garment for long-term, continuous monitoring of children in hospital. Babies having to be monitored as protection against SIDS are only one example in which the garment could be applied. The Intellitex suit distinguishes from the Mamagoose pyjama in the more extensive use of textile material.

Children who have to stay in hospital for a while will generally experience this as unpleasant. This is partly caused by the fact that their body is covered with different kinds of electrodes and wiring connects them to monitors. Furthermore, monitors using conventional sensor technology often cause skin problems. The monitoring systems currently in use are generally not really 'child-friendly'. Recent advances in smart textiles and telecommunication can remedy these shortcomings and provide the small patients with sufficient freedom of movement. The possibility of integrating sensors, data processing units, storage and transmission circuitry and interconnections into clothing means patients will be provided with more comfort, mobility and privacy. Moreover, children will experience wearing smart biomedical clothing as normal, without even noticing they are being monitored.

From the beginning of the project, it was decided to focus mainly on the development of a textile sensor-based system for long-term continuous monitoring of heart and respiration rate, named the Intellitex suit. In contrast to other projects we resolved not to miniaturise and integrate conventional sensors but to exploit the capabilities of textile material as such. Therefore,

Intelligent textiles for children 437

we adopted electroconductive textiles, particularly stainless steel yarns and structures. These yarns are currently being knitted, woven and embroidered, hence using existing textile technology to apply them into sensorised garments. In doing so the aim of the development of smart textiles was met by exploiting the potential of the textile material such as giving it sensing capacities. Moreover, as textile materials are used to manufacture other required components, integration will increase, resulting in a washable and patient-friendly garment.

23.3.1 Respiration measurements

A textile sensor for measuring respiration rate was developed.

Respiration rate is measured in various ways, such as sensing pressure, detecting CO_2/O_2 concentration in inhaled and exhaled air, applying body plethysmography (inductance, impedance, capacitance, etc.) and using strain gauges. In the framework of this research, textile-based strain gauges were chosen. These are knitted structures consisting of elastic and electroconductive yarns (stainless steel yarns by Bekintex). Knitted structures were chosen on the one hand because of their inherently elastic properties and on the other hand because undergarments are mainly knitted fabrics. By placing the structure as a coil around the abdomen or thorax, a variation in resistance caused by breathing is obtained. Therefore the elastic component is important, allowing the structure to adapt itself to the moving upper body and detecting circumference changes.

Several electroconductive, elastic textile structures were successively tested in a controlled environment to evaluate their ability for long-term use and possible property changes after washing.[7] Based on preliminary research, which proved anisotropy in resistance change for knitted structures, they were stretched in the wale direction, as shown in Fig. 23.1. Resistance changes in course direction are considerably less pronounced because the current flows through one thread uninterruptedly and this conducting path is not changed upon stretching. In wale direction however, the contact points between the successive rows of loops play a prevailing role in the resistance changes. Upon stretching, an increase in contact points results in a decrease of the overall resistance of the structure.

To determine these resistance changes, a dynamic resistance measurement set-up was developed whose core instrument was a yarn tensile tester. It was programmed to perform a continuous cyclic elongation simulating breathing movement. The structure being tested is circularly knitted, consisting of elasthane and Bekinox stainless steel yarn as the electroconductive component. Figure 23.2 shows the resistance change upon elongation of the structure up to 40% and relaxing back to 10%. During the first eight seconds the resistance stays constant, followed by a decrease as the structure is elongated.

438   Intelligent textiles and clothing

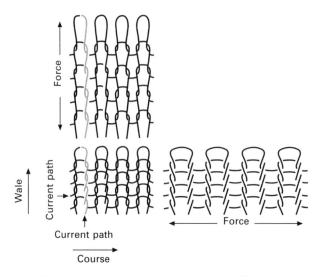

23.1 Conducting path in course and wale direction upon stretching.

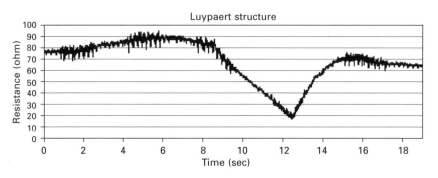

23.2 Change of resistance during one cycle of cyclic elongation.

Subsequently the resistance increases as the structure is relaxed to 10% elongation.

This cyclic elongation was repeated for a period of 50 minutes to study the evolution of the signal. Figure 23.3 shows the signal at defined times: after 25 seconds and after 5, 25 and 50 minutes. These graphs show a clear drift of the resistance. The signals' amplitude is slowly decreasing after each deformation cycle but the cyclic motion of elongating and relaxing can still be distinguished.

The same series of tests (50 minutes of cyclic elongation) were performed after washing the structure successively 5, 10 and 25 times in a domestic washing machine. An overview of the results is given in Fig. 23.4. This graph represents the evolution of the relative signal amplitude $\Delta R/R_{max}$ as a function of time, where ΔR is $R_{max} - R_{min}$ for each cycle. The graph

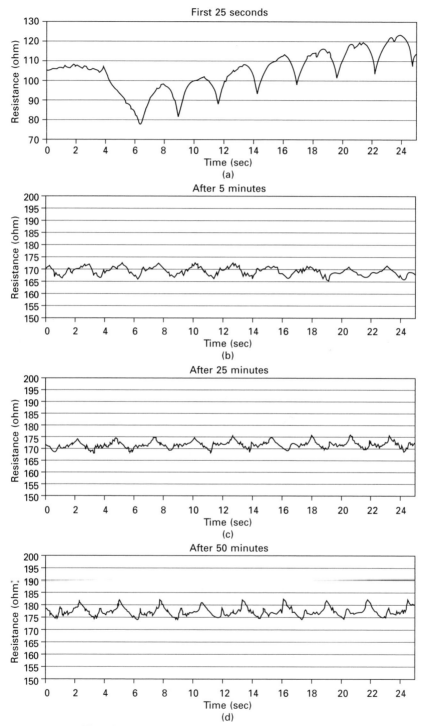

23.3 Electrical resistance change during 50 minutes of cyclic elongation.

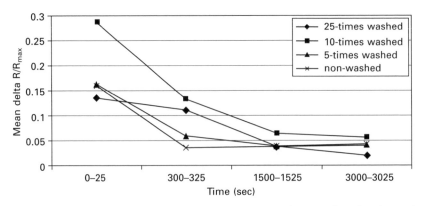

23.4 Relative signal amplitude during 50 minutes of cyclic elongation at selected time periods for non-washed, 5-, 10-, 25-times washed structure.

shows that there is a tendency for the relative signal amplitude of non-washed and washed materials to equalise over time without damaging the reliability of the sensor; however, the influence of washing on the overall resistance of the structure has to be taken into account. It can be concluded that textile structures containing electroconductive material, are useful as strain gauges when they are carefully engineered and characterised.

23.3.2 Electrocardiogram (ECG) measurements

In order to measure the ECG, textrodes were developed. They are a textile structure constituted of stainless steel yarns (by Bekintex) which may be used in direct contact with the skin. The choice of stainless steel was led by the following properties:

- it is a good conductor
- the fibres have a good touch
- it has a low toxicity to living tissue
- it can be processed as a textile material.

Conventional electrodes are always used in combination with electrogel to establish a good conductive contact with the skin, consequently improving the output signal. However, many patients experience some discomfort since electrogel may cause skin irritation and softening. These inconveniences impose restrictions on the use of this kind of electrode for long-term monitoring. Using textrodes can overcome these limitations because the textile material is in direct contact with the skin hence making electrogel unnecessary.

Because of the intrinsic weakness of the potential of the heart measurable at the skin (1.5–3 mV), a close contact between electrodes and skin is of major importance. Therefore elasticity of the garment is a highly required property, improving close fitting of the suit around the thorax. As far as textrodes are concerned, the influence of sweat is experienced as a benefit since it is an electrolyte thus improving the conductivity of the electrical signal towards the electrode.

In recording the ECG and other bioelectric events, the input impedance of the recording system has to be many times greater than the impedance of the electrode/bioelectric system. If this is not the case, not only a loss of amplitude of the bioelectric signal will arise, but also a distortion of the waveform.[8] Moreover, low electrode/skin impedance considerably improves the quality of the measured signal as the noise signal is less amplified. Since the textile electrodes differ to a large extent from the conventional electrodes, the electrode/skin impedance had to be determined. This was done in a frequency domain of 5 to 100 Hz. As was expected, the textile-electrode/skin impedance had an order of magnitude of $1.5 \text{ M}\Omega \text{ cm}^2$, which is much higher than for conventional gel electrodes, where the impedance typically is $10 \text{ k}\Omega \text{ cm}^2$ in the same frequency range.[9] In addition to determining the electrode/skin impedance, a study was carried out on a number of electroconductive textile structures since stainless steel electrodes are available in three structures: woven, non-woven and knitted (Fig. 23.5).

An electrochemical cell was developed to enable a quality evaluation of these different structures when used as electrodes on the skin. The study is extensively discussed in the book *Analytical Electrochemistry in Textiles*;[10] however one topic will be highlighted here. The three textile structures were exposed to four concentrations of artificial sweat (10^{-3}, 10^{-2}, 10^{-1} and 0.5 mol/l) for 24 hours and then analysed to evaluate a possible change in their functioning. The results are summarised in Table 23.1.

It is concluded that the knitted and the woven structure show no significant changes in electrical behaviour as a function of time. The non-woven electrode, however, does reveal a change as for the higher electrolyte concentrations the impedance increases considerably. This might be due to corrosion of the structure or adsorption of species at the fibre surface. The phenomenon is more obvious with very high electrolyte concentrations, which means that the influence is limited. Compared to the knitted and woven electrodes, the non-woven one has a more open structure and a much larger surface area owing to individualising of fibres, which makes it more sensitive to chemical and mechanical interaction. This should be taken into consideration when non-woven electrodes are applied as sensing devices in biomedical clothing. In relation to the sensation of comfort, non-woven fabrics will perform worse since protruding metal fibres will irritate the skin more easily.

In order to have a good elasticity and accordingly an improved contact with the skin, we decided on knitting a sensor integrated belt (Fig. 23.6). Interconnections between the sensors and electroconductive fasteners were knitted in as well. At this stage the electronics were connected to the belt

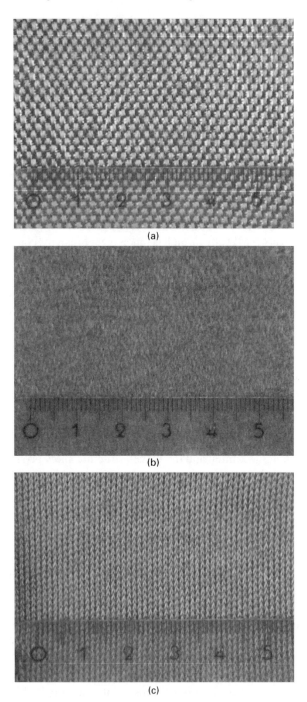

23.5 Stainless steel electrodes in three structures (a) woven, (b) non-woven and (c) knitted.

Table 23.1 Electrical resistance measured in an electrochemical cell containing knitted, woven or non-woven stainless steel electrodes and different concentrations of artificial sweat. Measurements are taken at different exposure times of the electrodes to artificial sweat

Electrolyte concentration (mol/l)	10^{-3}			10^{-2}			10^{-1}			0.5		
Exposure time (hours)	t = 0	t = 12	t = 24	t = 0	t = 12	t = 24	t = 0	t = 12	t = 24	t = 0	t = 12	t = 24
Knitted	31 610	31 530	31 510	3985	3980	3980	316	316	315	26.1	26.8	27.8
Woven	31 510	31 540	31 495	3989	3980	3982	319	316	316	25.9	27.1	27.2
Non-woven	31 520	31 510	31 540	3981	3978	3.95	315	330	351	25.9	35.0	60.3

23.6 Belt with integrated sensors and interconnections.

through the fasteners. The belt has a double layer knitted structure; the actual sensing material (stainless steel) is only present at determined positions on the inside of the belt, while Viloft/Co^{TM} is used for the other parts and the outside.

To measure the ECG, a three-electrode configuration was used.[11] Two electrodes were placed on a horizontal line on the thorax, while a third one, acting as a reference (named the 'right leg drive'), was placed on the lower part of the abdomen (not integrated into the belt). In order to assess their performance, the signal originating from a conventional electrode (gel electrodes by the company 3M) and the textile electrodes were recorded at the same time. The results of these measurements are shown in Fig. 23.7. Despite the Textrodes generating more noise, the figures show the accuracy of the signal.

The measurements shown Fig. 23.7 were carried out in a laboratory. Later on, clinical tests were performed in an operating room at University Hospital Ghent, Belgium. Anaesthetised children between two and four years old were simultaneously monitored with the textile and the conventional electrodes. The test focused exclusively on the performance of the Textrodes. Because of the ease of handling, a prototype in the form of a belt (as described above) was chosen for the initial clinical tests. The belt can more easily be put on and removed using a Velcro fastener in the front. It also leaves sufficient space on the body to attach other conventional sensors. However, the aim was finally to manufacture a complete baby suit.

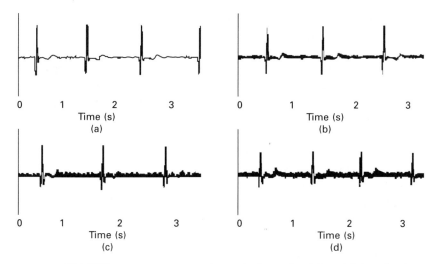

23.7 ECG measurements using gel electrodes (a) and (b) and using textrodes (c) and (d).

Clinical tests revealed the following deficiencies:

- The interconnections between the electrodes and the electroconductive fasteners cause problems after repeatedly opening and closing the belt. Metal fibres stick out of the belt and make contact with the surface of the textile sensor. This causes the signal to be interrupted.
- From the tests, it clearly appeared that a good skin-electrode contact has a very great influence on the quality of the output signal (for ECG measurements).
- The Velcro fastener appeared to be unreliable, coming off too frequently during measuring. A fixed fastening system seems to be indispensable.
- The electrical circuit had to be adjusted because the textile electrodes' signal causes the amplifiers to go into saturation.

23.3.3 Wireless communication and energy transmission

In order to extend the autonomy of the textile-based system, an existing inductive link was modified using electroconductive textile material.[12] The implementation of this inductive link has a dual function, enabling wireless bi-directional data transmission on the one hand and power transfer on the other. The data transmission downlink from a base station to the baby is useful, for instance, to change parameters, to determine minimum and maximum respiration rate or to adapt the measuring algorithm to the needs of the patient, while the uplink from the baby to the receiver sends out the measured data. Inductive powering avoids the use of batteries, reducing the volume and enabling continuous measurements. The developed system samples the

ECG signal with a frequency of 300 Hz, while simultaneously a power transfer of 50 mW is transmitted from the base station to the suit.

An inductive link requires two coils: a primary coil (the base station) and a secondary coil, operating within a maximum coil separation distance of 6 cm. Since the Intellitex suit is meant for babies, the base station of the inductive link can be hosted in the mattress of the cradle or the bed, while the secondary coil is integrated into the garment. This enables both functions to be realised while the baby is lying in bed. The most suited technique to integrate the coil into the baby suit is embroidery, allowing stitching a very flexible electroconductive yarn with high precision and accuracy on a textile carrier. The embroidering technique was fully exploited to compromise conflicting requirements; as many turns as possible that do not touch each other. Touching turns would create a short circuit while more turns would produce a superior coil. The useful area on the baby suit, however, also restricts the diameter. The ZSK embroidery machine, equipped with a special W-head, succeeded in manufacturing a 20-turns coil with an external diameter of 10 cm.

23.3.4 Intellitex suit: final prototype

Based on the conclusions drawn from the clinical tests, a subsequent prototype was manufactured (Fig. 23.8). This garment hosts more components than the former belt such as the electronic circuit and a wireless power and bi-directional data transmission link are included. In this prototype the electronic circuit is the only non-textile component but special attention has been given to its miniaturisation. To prevent damage through washing, these components have to be removed during maintenance of the prototype garment.

23.3.5 Conclusion

The Intellitex suit is an example of a smart biomedical garment exploiting the potential of electroconductive textile materials. It offers a solution to the disadvantages of conventional techniques. As an alternative to conventional electrodes, both knitted and woven stainless steel electrodes (textrodes) revealed promising results. The main advantage of dry textrodes is the non-irritating and integrating qualities. A disadvantage is the poor skin/electrode contact which is translated into higher demands on the electronic circuit. Therefore the analogue front-end had to be redesigned and optimised. These electronics were assembled on a flexible printed circuit which can be integrated in the baby suit. A set of clinical tests has on the one hand uncovered some practical deficiencies but on the other hand proven the usability of textile sensors. An improved prototype was manufactured. The result is a child-friendly garment that allows small patients to be monitored in the best possible conditions.

Intelligent textiles for children 447

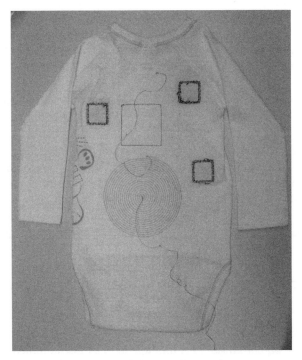

23.8 Final prototype of the Intellitex suit.

This work demonstrates that textile materials themselves have a strong potential to be used as sensor elements, interconnections and transmission links in smart biomedical garments. Despite the low quality of the textile sensors, the use of electronics with high requirements provides reliable monitoring. Major benefits are improved patient comfort and reusability of the sensors. Washable packaging of the electronics and durable interconnections remain major challenges to be tackled, not only for the specific system presented here, but for all wearable electronics and intelligent textiles developments. There is still a long way to go to obtain reliable commercial smart biomedical garments.

23.4 Future trends

Not only in this study but likewise in many others the feasibility of smart textiles has been demonstrated and the prototypes proved to be beneficial. The manufacture of truly wearable smart biomedical garments has taken a cautious start, but there is still a long way to go. Slowly these new textile-based products are finding their way into society. Possible problems however should be recognised and overcome by continuously searching for improved materials and technologies. Treating these systems in the way we treat our daily garments is very demanding and therefore a huge challenge. Consequently, evaluation of long-term behaviour, durability and system performance after repeated laundering should invariably be included in the research tasks. Therefore fundamental research has to support the use of these textile materials sufficiently.

Children suffering from diseases such as diabetes, hypoglycaemia and cystic fibrosis could considerably improve their quality of life with the help of smart biomedical garments. Being monitored in a non-obstructive way is of utmost importance to them, not only from a medical point of view but also psychologically. Since they do not notice they are being monitored, it will have a more relaxing effect which will improve the level of comfort. Not being able to pull off the wires of the system will also reassure parents.

In the future garments will detect risks and report them, resulting in an appropriate reaction and thus prevention. In case of hazard, the textile will support, protect and inform. Additionally, during rehabilitation, the textile will deal with the administering of medication through the skin as well as, e.g., performing physiotherapy. The textile will likewise follow up the recovery process and adjust the applied treatment. When exploiting the many advantages of textiles, such as large contact area with the skin, permanent availability and comfort, much physiological data will be collected in a non-invasive way through the garment fabric. This might lead to a more complete clinical picture and consequently new medical insights, enabling the formulation of a patient's medical profile on the basis of which anomalies can be detected. All this is not yet within reach but current multidisciplinary technological developments progressively head towards that direction. Hence, as technology is gradually provided, smart textile systems for children will become available; adapted and accepted as a second active skin and enhancing their level of comfort and well-being.

23.5   Acknowledgements

This research project was supported by IWT, in the framework of the STWW programme, contract no. 000160. The authors would like to acknowledge Bekintex and ZSK.

23.6   References

1. Gopalsamy C., Park S., Rajamanickam R., and Jayaraman S., 'The wearable motherboard: the first generation of adaptive and responsive textile structures (ARTS) for medical applications', *J. Virtual Reality*, vol. 4, pp. 152–168, 1999.
2. Park S., and Jayaraman S., 'e-Health and Quality of Life: The Role of the Wearable Motherboard', *Wearable eHealth Systems for Personalised Health Management*, Lymberis A., De Rossi D., ISBN 1 58603 449 9, p 239–252.
3. Paradiso R., Loriga G., and Taccini N., 'Wearable System for Vital Signs Monitoring',

     Wearable eHealth Systems for Personalised Health Management, Lymberis A., De Rossi D., ISBN 1 58603 449 9, p 253–259.
4. Grossman P., 'The Lifeshirt: A multi-function ambulatory system monitoring health, disease and medical intervention in the real world', *Wearable eHealth Systems for Personalised Health Management*, Lymberis A., De Rossi D., ISBN 1 58603 449 9, p 133–141.
5. Weber J.-L., Blanc D., Dittmar A., Comet B., Corroy C., Noury N., Baghai R., Vayasse S., and Blinowska A., 'Telemonitoring of vital parameters with newly designed biomedical clothing VTAM', *Proc. of International Workshop-New Generation of Wearable Systems for e-Health: Towards a Revolution of Citizens' Health and Life Style Management?*, Lucca, Italy, pp. 169–174, 2003.
6. Lanfer B., *The development and investigation of electroconductive textile strain sensors for use in smart clothing,* E-Team Master Thesis, Academic Year 2004–2005.
7. Verhaert, Mamagoose pyjama, Information available online at: www.verhaert.com/pdfs/Verhaert%20mamagoose.pdf
8. Swanson, D.K., and Webster J.G., 'A Model for Skin-Electrode Impedance', in Miller H.A. and Harrison D.C., (eds), *Biomedical Electrode Technology*, Academic Press, 1974.
9. Geddes L.A., and Baker L.E., 'The relationship between input impedance and electrode area in recording the ECG', *Med. & biol. Engng.* Vol. 4, pp. 439–450.
10. Westbroek P., Priniotakis G., and Kiekens P., *Analytical Electrochemistry in Textiles,* Woodhead Publishing, Cambridge ISBN 1 85573 919 4.
11. Neuman M.R., 'Biopotential Amplifiers', in: Webster J.G., (ed.), '*Medical Instrumentation – Application and Design,* John Wiley & Sons, 1998, pp. 233–286.
12. Catrysse M., Hermans B., Puers R., 'An inductive powering system with integrated bidirectional datatransmission', *Proc. Eurosens.* (2003) 843–846.

24
Wearable biofeedback systems

B J MUNRO, University of Wollongong, Commonwealth Scientific and Industrial Research Organisation (CSIRO) Textile and Fibre Technology, Australia and J R STEELE, T E CAMPBELL and G G WALLACE, University of Wollongong, Australia

24.1 Introduction

Biofeedback systems are becoming increasingly popular due to their non-invasive nature and the beneficial effects they provide both as rehabilitative devices and training tools. However, biofeedback systems have typically encompassed large, rigid electronic devices best used when an individual is stationary or required only to move slowly through a small range of motion. Integration of inherently conducting polymer coatings onto textiles has created new non-rigid biofeedback options in the form of textile sensors. When combined with conventional but wearable electronics, these unique textile sensors can be integrated directly into existing clothing and equipment without changing the material properties or functions of these items and without interfering with normal human motion. The result is truly wearable systems capable of providing immediate biofeedback to the wearer on a set of pre-determined conditions. This innovative technology has been used to develop the Intelligent Knee Sleeve, a wearable biofeedback system able to provide immediate, individualised and objective feedback to the wearer about knee flexion angle. The Intelligent Knee Sleeve is currently used in landing training programs to assist athletes to learn how to land correctly to prevent injury as well as in rehabilitation programs. This chapter describes the concept, design and use of wearable biofeedback systems, focusing on the development of textile sensors and their integration with functional electronics to specifically monitor joint motion using the Intelligent Knee Sleeve. Other applications of this technology for rehabilitation, technique modification and injury prevention are also highlighted, as well as proposed advancements of this technology to ensure its optimal use in today's society.

24.2 Is there a need for biofeedback technology?

Biofeedback training is the use of electronic instruments for subliminal learning to change the body's response in order to improve performance or health, such as during rehabilitation following injury or illness. The field of biofeedback training is burgeoning as it is safe, non-invasive and pain-free and its efficacy and long-term benefits are well documented in the medical literature (Middaugh and Pawlick, 2002; Chiotakakou-Faliakou *et al.*, 1998; Lake, 2001; Barthel *et al.*, 1998; Parekh *et al.*, 2003; Wiener *et al.*, 2000). Often used to alter brain activity, blood pressure, chronic pain, muscle tension, heart rate and other bodily functions normally beyond voluntary control, biofeedback can be provided in various ways, including visual (e.g. a flashing light), auditory (e.g. an audible tone) or tactile (e.g. vibration) forms of feedback. For example, muscle biofeedback involves placing electrodes on the skin overlying muscles of interest and recording the electrical signal generated by these muscles. Feedback pertaining to muscle activity, such as an audible tone to indicate when muscle intensity has reached a desired level, can then be provided to individuals for a wide variety of applications, such as teaching muscle relaxation for the treatment of headaches (Cinciripini, 1982), preventing or treating occupational overuse syndromes (Thomas and Vaidya, 1993), eliciting muscle contraction during strengthening exercises (Storheim *et al.*, 2002), treating incontinence (Jundt *et al.*, 2002) or during rehabilitation following a stroke (Moreland *et al.*, 1998) or surgery (Draper, 1990). As the role of biofeedback is expanding across a broad range of fields such as sport, recreation, rehabilitation, performance of daily living activities and industry, advancements are vital to ensure biofeedback technology can accurately monitor human performance and provide reliable and valid feedback about select aspects of this performance to the user during dynamic motion.

24.3 Are there problems with current biofeedback devices?

The original systems used to provide biofeedback were typically immovable systems of massive size due to the feedback modality and the need for data processing, memory and power supply. These devices were restricted to applications where individuals were stationary or performing slow movements through a restricted range of motion in the vicinity of the biofeedback system. More recently, portable devices have been designed to provide feedback during human motion, such as heart rate monitors and some muscle biofeedback machines. However, these devices are predominantly made of relatively bulky, rigid components that can interfere with, or alter, the wearer's natural motion during dynamic tasks and rarely provide real-time feedback, particularly of joint motion. Therefore, the challenge confronting those involved in monitoring human function is the ability to design truly wearable systems, systems that are conceived as 'unobtrusive as clothing' (Engin *et al.*, 2005, p. 174), although capable of sustained real-time data processing during dynamic forms of activity.

24.4 Can we provide biofeedback for joint motion?

For practitioners involved in directly evaluating and modifying human motion, the ability to monitor the biomechanics that characterise human performance in the field, such as joint motion, and then to 'feed' this information back to the wearer in real time so the performer may modify their motion during the actual skill to achieve the desired outcome, has been elusive. Currently, coaches, trainers and medical personnel typically 'eyeball' the performance of their athletes and/or patients and 'guess' whether the correct motion is being used before providing verbal feedback to change their movement patterns. Alternatively, they can use highly sophisticated apparatus for biomechanical analysis of human performance, such as optoelectronic motion analysis systems. However, this equipment is often extremely costly and requires both significant expertise to use and extensive data processing (e.g. manual analysis of video images) before meaningful information can be relayed back to the performer. Furthermore, although this information can be relayed using both visual and verbal forms of feedback, it is usually not received in real time. Alternatively, to receive real-time feedback, performers may be required to view a monitor whilst performing a task, restricting the movements that can be performed. Other devices, particularly those that are attached to the user's body, such as electrogoniometers, can be used to provide feedback about joint motion in real time. These devices also have the benefit of being able to provide the feedback through audible or tactile forms during the performance of dynamic activities. Despite this benefit, electrogoniometric-like devices are often inappropriate as they have rigid components that do not conform to the user's body shape, thereby interfering with their natural motion during dynamic movements and, potentially posing a safety hazard to the user. Furthermore, care must be taken when providing tactile biofeedback as these feedback forms may induce muscular response, potentially leading to injury.

Advances in textile technology have seen the emergence of electronic textiles, which can overcome the limitations traditionally encountered with rigid biofeedback devices. For example, conductive coatings can be used to transform fibres, yarns or textiles into electrically conductive materials through electroless plating; evaporative deposition; sputtering; chemical, electrochemical or admicellar polymerisation with a conductive polymer (Lekpittaya et al., 2004); and filling or loading fibres and carbonising (Meoli and May-Plumlee, 2002), without significantly altering existing textile properties. These conductive textiles have strain-gauge-like properties which, in preference to conventional, non-textile strain gauges, are wearable, have a wide dynamic range (De Rossi et al., 1999a; Fletcher, 1996) and are relatively inexpensive. Electrically conductive metals such as ferrous alloys, nickel, stainless steel, titanium, aluminium, copper and carbon can also be applied to textiles to produce highly conductive and wearable sensors. However, these metals are brittle and may potentially damage fibres within a knitted textile structure (Kaynak *et al.*, 2002). Specific coating procedures must be developed to overcome problems with adhesion and corrosion resistance.

In comparison, textiles coated with conductive polymers, namely polypyrrole, polyaniline or polyterthiophene, have excellent adhesive and non-corrosive properties (Kuhn, 1993) and are less expensive than metallated textiles. When combined with parallel advances in the miniaturisation of conventional electronics and computing, and with a concomitant decrease in size and power consumption, these unique textile sensors are ideal as novel, non-rigid wearable systems capable of providing direct biofeedback with respect to joint and segmental motion. That is, they can be integrated directly into existing clothing and equipment without changing the material properties or functions of these items and without interfering with normal human motion. Therefore, these wearable textile biofeedback systems are considered superior to traditional biomonitoring systems as they contain minimal rigid components, conform to body shape, are extremely light, do not impede human performance, are safe to be worn during physical activity and are of low cost. Furthermore, due to their simplicity in design, these devices have multiple applications in sport, recreation, industry and when performing activities of daily living.

Independent of the application, wearable biofeedback systems require the optimal integration of three components, which are attached to a base garment, namely the textile sensor, functional electronics and the interconnections between the sensor and electronics (see Fig. 24.1). These components will be discussed as isolated components in the following sections and then how they are integrated to form the Intelligent Knee Sleeve.

24.5 The development of a functioning wearable textile sensor

Electronic sensors are devices that detect and then transform measured quantities of physical phenomena into an electrical signal (Engin *et al.*, 2005; Gniotek and Krucinska, 2004). The ideal electronic sensor to monitor human motion is a wearable textile strain gauge that has a wide linear dynamic range with appropriate sensitivity, a fast response time (low time constant) enabling use over the frequency range of interest in biomechanical measurements and an initial resistance compatible with the electronics to be used. The sensor should also generate minimal resistive force upon stretching over the frequency range of interest and have little or no hysteresis in either the electrical or mechanical response. Ideally, these wearable sensors should also be robust against external physical, electrical and electromagnetic disturbances as well as impervious to sweat, moisture, washing processes, temperature, mechanical impacts, repeated bending, compression and light

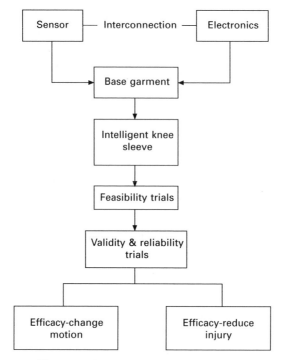

24.1 The steps involved in the development and assessment of a wearable biofeedback system.

Table 24.1 Properties of the ideal textile sensor for use in a wearable biofeedback system

Properties of the ideal wearable sensor
Appropriate textile substrate
Ability to control initial resistance (coating conductivity)
Appropriate sensitivity
Large linear dynamic range
Small response time
Minimal resistive forces and limited hysteresis
Robust and immune to environmental effects
Compatible with simple electronics

(Gniotek and Krucinska, 2004; Engin *et al*., 2005). In addition, it is vital that wearable sensors are designed for low power consumption. Therefore, conducting polymer-coated textiles offer several attractive features with regard to their use as wearable electronic sensors to monitor human motion. The ideal properties of a wearable sensor, as summarised in Table 24.1, with respect to the performance of polypyrrole-coated nylon lycra, are discussed in the following sections.

24.5.1 Textile substrate

Nylon lycra has proven to be an excellent textile substrate for wearable strain gauge sensors when compared to other commercially available textiles (Campbell *et al.*, 2003). This composite textile substrate exhibits very low resistive forces such that when worn, body movement is not impeded. Nylon lycra is also easily coated with conducting polymers using the Kuhn (1993) method involving *in-situ* polymerisation on the surface of the textile (see Fig. 24.2). Use of this *in-situ* polymerisation technique readily obtains thin, uniform, adherent coatings (see Fig. 24.3) as each of the individual fibres is separately coated with no fibre bonding, such that there is no resulting deterioration in the mechanical properties or the handle of the fabric.

24.5.2 Control of initial resistance

Controlling the initial resistance of the coated textile is important for electronics compatibility and to set the baseline signal for the sensor. Several factors influence the initial resistance of the coated fabric, including the composition and structure (knit) of the base textile, textile pre-treatment, the concentration of monomer/oxidant/dopant used for coating (see Fig. 24.2), polymerisation time and polymerisation temperature, as reported previously (Wu, 2004).

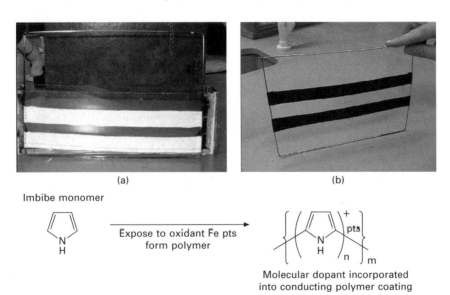

24.2 The *in-situ* polymerisation process used to coat nylon lycra strips (2 cm × 25 cm) with polypyrrole. The uncoated strips (a) were stretched horizontally across a wire rack, then stood vertically in the polymerisation solution for coating before being removed for drying and washing (b).

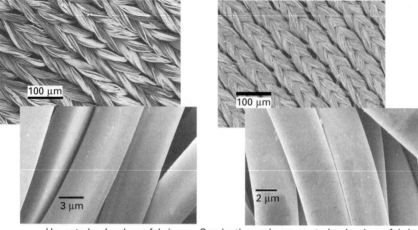

Uncoated nylon lycra fabric
(a)

Conducting polymer-coated nylon lycra fabric
(b)

24.3 Scanning electron microscope (SEM) photographs of uncoated nylon lycra (a) and nylon lycra coated with polypyrrole (b) displaying the uniformity of the polymer coating.

24.5.3 Sensitivity

Sensor sensitivity is another important factor for monitoring joint motion. If highly sensitive, then sensors for wearable biofeedback systems will undergo significant changes in conductivity, that is, have a high gauge factor*, corresponding to a small degree of movement. Conventional silicon and metal strain gauges record gauge factors between 100–170 and 0.3–4, respectively. Gauge factors reported for polypyrrole-coated textiles are of the order of –12 (De Rossi *et al.*, 1999a). The negative gauge factor reflects a decrease in resistance when the textile is strained. By altering the coating procedures and textile substrate, the sensitivity of the sensor can be tuned for different applications. The gauge factors reported to date are adequate for monitoring joint movement within the accuracy limits of a standard goniometer (Gogia *et al.*, 1987) although further research work aims to increase the gauge factor of these sensors to further improve the accuracy of biofeedback during joint motion.

---

* The gauge factor of the strain gauge is defined as the ratio of the fractional change in resistance to the fractional change in length Cobbold, R. S. C. (1974) In *Biomedical Engineering and Health Systems* (ed., Milsum, J. H.) John Wiley & Sons, New York, USA.

24.5.4 Linear dynamic range

The range over which a strain gauge is linear is vital to ensure repeatability and provide feedback over the required range of joint motion. However, this range is affected by the uniformity of polymer coating across the surface of the textile, textile composition, textile structure and stability of the polymer on the textile. Compared to the limited linear dynamic ranges displayed by other strain gauge types, polypyrrole-coated nylon lycra consistently displays a large linear dynamic range between 10–70% strain, as displayed in Fig. 24.4 (Campbell *et al.*, 2003). The large linear dynamic range displayed by these sensors corresponds to the greatest degree of joint motion required during human motion. However, the sensor must be prestrained a minimum of 10% to ensure joint motion falls within the reported linear range.

24.5.5 Response time

The response time of the textile sensor is required to be as small as possible to ensure an instantaneous response during rapid human movements. De Rossi *et al.* (1999b) noted that polypyrrole-coated lycra fabrics displayed a fast response time when cycled at 1 Hz, which makes them conducive to monitoring human motion. However, this response did not achieve steady state for several minutes (see Section 24.5.7). The response time of the polypyrrole-coated nylon lycra sensors has been shown to be adequate in providing valid and reliable biofeedback during rapid and repetitive human movements such as jumping (Campbell *et al.*, 2003; Munro *et al.*, 2002). The time to steady state did not affect the performance of these sensors under human testing conditions, such that they were proven valid and reliable (see Section 24.9.2).

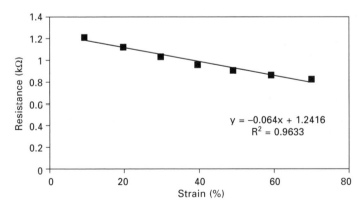

24.4 The calibration curve of a polypyrrole-coated nylon lycra sensor displaying a large linear dynamic range between 10–70% strain when being cycled at 1 Hz for an average of 15 cycles at each strain value.

24.5.6 Resistive force and hysteresis

An ideal wearable textile strain gauge would exhibit consistently identical and minimal forces on both the stretch and relaxation portions of a cycle, at a given frequency. That is, when dynamically cycling the textiles between varying strain limits, both resistive force and mechanical hysteresis should be minimal. However, similar to other properties discussed previously, the mechanical and electrical hysteresis of the textile sensor are dependent upon textile composition, textile structure (knit) and coating procedures. At commencement of cycling, polypyrrole-coated nylon lycra sensors display an initial decrease in resistive force until they plateau after approximately ten cycles, regardless of frequency. Once the plateau is reached there is no further change in mechanical or electrical properties even after 60 minutes of continuous cycling at 1 Hz as displayed in Fig. 24.5 (De Rossi *et al.*, 1999b, Campbell and Wallace, unpublished data). Therefore, to ensure consistent feedback during human motion, the polypyrrole-coated nylon lycra sensors should be conditioned before use. The force hysteresis generated by the polypyrrole-coated nylon lycra was 0.15 N over the strain range 10–70%. Similarly, the hysteretic electrical response of these sensors approximated 0.02 kΩ (Campbell *et al.*, 2003). It is postulated that these hysteretic effects are due to fibre movement within the textile and the reorganisation of the polymer contacts upon elongation (Campbell and Wallace, unpublished data). Regardless of the cause, the hysteresis responses provide further support for ensuring the sensor is prestrained a minimum of 10% when applied to the supporting garment due to the initial non-linear resistance variation recorded with applied strain (see Section 24.5.4).

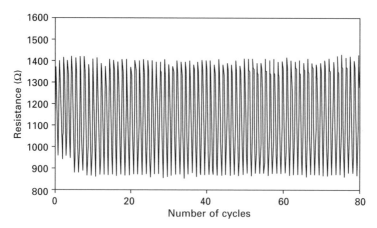

24.5 The electrical resistance-time responses of a polypyrrole-coated nylon lycra sensor being cycled between 0–40% at 1 Hz for 80 continuous cycles.

24.5.7 Robustness and environmental effects

The ideal sensor to be used to monitor human motion will be subjected to varying environmental effects and human demands. Consequently, to ensure that the results being received throughout a biofeedback session are not adversely affected by environmental factors, wearable sensors must display minimal mechanical and electrical creep and be stable both on the benchtop and when in use, irrespective of changing environmental conditions. That is, if the textile length or electrical response differs markedly during cycling, feedback could be provided at varying joint angles, negating the effectiveness of the wearable biofeedback system. For polypyrrole-coated nylon lycra there was minimal change in textile length and permanent mechanical textile degradation was recorded only when the textiles were strained beyond their yield point, which varied depending on the amount of lycra within the nylon lycra substrate. With prolonged elongation, the electrical resistance exhibited a slow decline, reflecting electrical creep (De Rossi *et al.*, 1999b). The cause of this creep is multifactorial and, although postulated to be due to the textile substrate, is still under investigation. However, electrical creep did not affect the performance of the Intelligent Knee Sleeve as it is evident only during prolonged elongation and not during fast repetitive movements (see Section 24.9.2). Nonetheless, care must be taken to ascertain the effect of creep and stress relaxation within textiles for other applications, particularly those that may require prolonged sensor elongation.

The difficulty in evaluating many of these factors remains with the stability of the conducting polymer coating under different environmental conditions. Currently, polypyrrole-coated nylon lycra sensors are able to be stored within air-tight plastic bags with minimal change in initial electrical resistance after six months (Campbell and Wallace, unpublished data). In addition, these sensors have been shown to be stable when being used on the Intelligent Knee Sleeve by athletes participating in biofeedback training and being stored in air-tight plastic bags between sessions, provided total usage time is limited to 60 minutes. However, these sensors are still affected by sweat, washing processes and distinct changes in temperature and/or humidity. To overcome these potential limitations whilst the technology is still advancing, disposable sensors, validated for 60 minutes usage time, are currently used in wearable biofeedback systems. However, it is envisaged that with further research, the stability of these sensors for use in the field will improve leading to more robust wearable biofeedback systems.

24.5.8 Compatibility with simple electronics

The simplicity in design and function of these wearable electronic sensors is such that they require minimal power consumption. As the battery is typically the most bulky part of the electronics (see Section 24.6), the associated electronics circuitry to be integrated with these sensors is small and light, facilitating a wearable biofeedback system that can be worn in the field. However, the concept of disposable sensors requires additional functionality within the electronics design and innovative interconnects (see Section 24.7) for operation.

24.6 Functional electronics

Electronics are required to add both power and functionality to wearable biofeedback systems. However, their necessary inclusion in wearable biofeedback systems is problematic, as the battery, electronics boards and feedback modality are rarely flexible and may at times be bulky. In addition, their inclusion, due mainly to the battery, increases the mass of the electronics and may affect the operation of the wearable biofeedback system by affecting the placement of the garment on the joint of interest. Alternative energy sources and flexible motherboards are being investigated in an attempt to make these systems truly wearable (Engin *et al.*, 2005). However, in the current environment, time must be invested when determining the functional specification for a wearable biofeedback device as greater processing or communication requirements typically require greater power supply, greater mass and greater production cost (Engin *et al.*, 2005; Gniotek and Krucinska, 2004). Some important aspects to consider when devising the functional specification for compatible electronics within a wearable biofeedback system have been included in Table 24.2.

24.7 Interconnections

A fundamental component of wearable biofeedback systems is the interconnections between the textile sensor itself and the electronics. That is, the interconnections require optimal pressure and/or good adhesion for optimal conductivity. Interconnections must also be sufficiently robust, a need that is heightened in a system where the sensors are disposable. This problem should be overcome with the use of conductive interconnects that aim to increase the size of the electrode rather than being dependent upon the pressure. However, with movement of the sensor around a joint, the contact between the sensor and the electronics often provides problems as, while they are strong in the shear direction, they are weak in compression and the bending force applied to the sensor encompasses both shear and compression. In addition, conductive fibres are often not able to withstand the constant wearing stresses applied to the biofeedback garment and therefore do not maintain an adequate conductive connection. As inappropriate interconnections can radically affect the performance of the wearable biofeedback system, as well

Table 24.2 An example of some functional specification details for manufacture of electronics components for wearable biofeedback devices

Requirement	Definition/example
Function	What will the system do? System will monitor joint motion during dynamic activity and provide an audible tone when a preset threshold is exceeded. Tone will be audible only within +/– 10% of set threshold.
Features	What features will the system provide for ease of use? System will include one-button push control for self-calibration, sensor replacement alarm, low battery indication and automatic sleep after system has been inactive for two minutes.
Sensor interface	How will the sensor be attached and what are the specifications of the sensor? The sensor part will be easily replaced. The sensor resistance shall be of the order of 10–300 kΩ.
Performance/ calibration	How will the unit operate and under what environmental conditions? For a given use and situation, the angle at which the unit provides the audible tone will be repeatable within 5% over the full environmental and physical operating conditions (10–40 °C; 95% humidity). The unit will be splash proof and self-powered with an operating time of 200 hours. The end-user will be responsible for calibration.
Maintenance	Will the unit be serviceable? No maintenance other than sensor replacement is required; the battery is not replaceable.
Physical/ environmental	What is the manageable size of the unit? The electronics unit will be as small as possible (30 × 30 × 5 mm) and will weigh less than 10 g.
Data handling	What does the unit need to do with the data? No data will be stored within the unit although the unit will be compatible with telemetry devices to transfer data from a field environment to a stationary computer. Future developments of the unit may include data logging, wireless control, collection of multiple signals and processing of information with respect to joint angle.
Cost/quantity/ timing	What is the cost of the system? The system will be designed to be of low cost. The quantity of units and receipt of units can be variable.
Regulatory conform	What regulatory bodies do you need to conform to, what are the conformity procedures and issues and how long will the process take? The unit will conform to the requirements of regulating authorities in countries of interest.

as the ability of the user to easily remove and replace the sensor as required, metal press-studs have been used as they are accepted in clothing products and form a robust interconnection with the electronics. However, in addition to the electronics, they introduce a rigid component into the system and therefore research will continue to alleviate this problem in designing a truly wearable and non-rigid biofeedback system.

24.8 The Intelligent Knee Sleeve: a wearable biofeedback device in action

One example of a wearable biofeedback system using the technology described above is the Intelligent Knee Sleeve*. The Intelligent Knee Sleeve is a lightweight fabric sleeve worn around the knee, which incorporates a disposable polymer-coated textile sensor that is placed over the patella (kneecap; see Fig. 24.6). The textile sensor, integrated into an appropriate electronic circuit (3 V), acts as a textile strain gauge with a wide dynamic range whereby as the sensor is stretched when the wearer bends their knee, resistance within the textile sensor strip changes. At a predetermined threshold resistance based on knee flexion angle, which can be varied, an audible tone is emitted to alert the wearer that the desired knee flexion angle has been achieved. The

24.6 The Intelligent Knee Sleeve. The base sleeve is placed around the knee with the textile sensor placed over the kneecap (patella) and the electronics placed on the side of the sleeve hidden by a fabric pocket. Photo courtesy of CSIRO Textile and Fibre Technology.

---

* Wallace GG, Steele J, Innes P, Spinks G, and Zhou D. *Sensors in fabrics with audio feedback: A training tool for enhanced performance and rehabilitation*. International Patent Application WO 03/01 4684, August, 2001.

Intelligent Knee Sleeve is a unique example of a wearable biofeedback system that can provide immediate, individualised, and objective biofeedback to the wearer with respect to knee joint motion during dynamic tasks performed in the field. It therefore increases the objectivity, frequency, and speed of feedback provided to individuals about their landing technique to ensure they are reinforcing the correct technique throughout training sessions. Consequently, the effectiveness of landing training programs is improved, with the aim of reducing the high incidence of non-contact ACL injuries in sports.

24.9 Why is the Intelligent Knee Sleeve needed?

The human knee joint has a high susceptibility to injury due to its incongruent structure and the high forces imposed on the joint, particularly during dynamic activities such as landing. Of all the knee ligaments, the anterior cruciate ligament (ACL), one of two ligaments crossing within the knee joint, is the most frequently injured (Johnson, 1983). When the native ACL is ruptured, the knee joint is predisposed to episodes of giving way, further risk of damage to the cartilage discs within the knee joint, loss of proprioception via damage to mechanoreceptors in the joint and ligament itself, recurrent pain, and likely degeneration of the knee joint because of excessive laxity and persistent instability (Acierno et al., 1995). Although ACL reconstructive surgery can be a viable treatment option, it is preferable to prevent these potentially debilitating non-contact ACL injuries from happening in the first place, as ACL reconstruction often results in losses in joint range of motion, muscle strength, and control (Draper, 1990). As nearly one-quarter of non-contact ACL ruptures are caused by poor landing technique (Noyes et al., 1983), a device such as the Intelligent Knee Sleeve, designed to assist in teaching correct landing technique, is urgently required.

Most non-contact ACL injuries occur when landing from a jump with the knee flexed less than 30° (Cochrane et al., 2001). When landing using this extended knee posture, contraction of the quadriceps muscles on the anterior thigh, which are activated to prevent the lower limb from 'collapsing', unfortunately also increases ACL strain (Torzilli et al., 1994; Draganich and Vahey, 1990). Furthermore, with the knee extended, the hamstring muscles, on the posterior thigh, are inefficient in providing sufficient posterior tibial draw to counteract the quadriceps-induced anterior tibial translation, due to their inefficient line of action (Pandy and Shelbourne, 1997). For this reason, it is strongly advocated that individuals should bend their knees when landing to enable the hamstring muscles to more effectively protect against high ACL strain (Cochrane et al., 2001). Flexing the knees throughout the landing action can also 'cushion' the forces generated at foot-ground contact, thereby reducing the jarring effects of landing, as well as lower an individual's centre of gravity, in turn, enhancing their stability (Steele and

Milburn, 1987a,b). Therefore, to reduce the potential for non-contact ACL injury, it is advocated that individuals should land with a relatively high knee flexion angle combined with a large range or amplitude of joint motion over which to dissipate the energy in muscles (Mizrahi and Susak, 1982a,b; Cochrane *et al.*, 2001).

The benefits of landing programs in reducing ACL injuries are readily acknowledged by the implementation of landing training programs in sports such as Australian Rules Football (Seward *et al.*, 1999), soccer, volleyball, and basketball (Hewett *et al.*, 1996, 1999; Caraffa *et al.*, 1996). However, participants in such programs currently have no method to ensure they are bending their knees sufficiently during training. That is, no field-based method currently exists that can provide immediate feedback to individuals with respect to knee flexion angle during dynamic landing tasks. Therefore, a wearable joint biofeedback device, such as the Intelligent Knee Sleeve, fills this current void.

24.9.1 Proof of concept: feasibility of using the Intelligent Knee Sleeve in the field

To establish whether using the Intelligent Knee Sleeve as an instantaneous biofeedback device to teach correct landing mechanics was feasible, members of an elite Australian Rules football team completed three standard landing training sessions per week using the device to assist them to bend their knees appropriately when performing landing movements. The results of this field-based feasibility study were positive in that the base sleeve was found to be comfortable, remaining in the same position on the lower limb with the sensor over the kneecap during performance of all activities, and the feedback challenged the players during performance of landing activities. Furthermore, although minimal time was available to educate the players on independent use of the sleeve, the sleeve users had little difficulty in learning its correct operation and welcomed the biofeedback device and training program as they were novel, challenging and informative.

24.9.2 The Intelligent Knee Sleeve: how valid and reliable is the audio feedback?

Although proven to be a feasible biofeedback device, it is imperative to establish validity and reproducibility of the feedback provided by the device if it is to be effective in improving the user's landing technique. To establish validity and reliability of the Intelligent Knee Sleeve, 12 subjects (mean age 26.1 ± 3.2 years) involved in sports requiring landing movements and with no history of knee joint disease or trauma performed ten trials of four landing movements while wearing the Intelligent Knee Sleeve on their dominant limb. The audible feedback tone was set to be emitted at two programmed knee flexion angles during the four movements, which included: a shallow hop (25° of knee flexion); a deep hop (45°); a 30 cm step down (45°); and a 30 cm step down followed by a rebound movement (45°). The total time to collect the data per subject replicated the typical time of a landing training session (30 minutes). The two knee flexion angles at which the audible tone was emitted were programmed manually using a goniometer and goniometric measurements reconfirmed the knee angle at which the audible tone was emitted at the completion of all trials of each movement.

During the trials each subject's landing action was characterised by collecting the ground reaction forces generated at landing using a force platform (1000 Hz) and monitoring their motion using an optoelectronic motion analysis system (200 Hz). Data from the knee sleeve were also sampled (1000 Hz) to determine the onset of the audible tone. Paired t-tests revealed that the knee sleeve was valid, in that the programmed angle (set using a goniometer) was equal to the angle at which the audible feedback tone was actually emitted (calculated using the motion analysis system). The knee sleeve audible tone also proved to be highly reliable with intra-class correlations calculated from the knee angle at audio onset ranging from $R_1 = 0.903$ to 0.988. Based on these results, the knee sleeve was deemed to be a valid and reliable device to provide information about knee flexion angle during dynamic landing movements.

24.9.3 Does feedback provided by the Intelligent Knee Sleeve improve knee flexion during landing?

To ascertain whether the biofeedback provided by the Intelligent Knee Sleeve was effective in assisting athletes to learn to flex their knees more during dynamic landing movements, a pilot trial was completed involving 37 subjects (mean age 23.6 ± 4.0 years), all of whom were involved in sports requiring landings and who had no history of knee joint disease or trauma. Each subject performed a series of landing movements, whereby they landed on their dominant limb with their foot centrally located on a force platform whilst catching a football, before and after a six-week training program. During each testing session ground reaction force data were collected (1000 Hz) using a force platform whilst kinematic data characterising landing technique were collected (200 Hz) using an optoelectronic motion analysis system. At the completion of initial testing, subjects, matched for age, height, body mass, injury history and playing ability, were divided into three groups:

1. subjects who participated in a landing training program and who received audible feedback from the Intelligent Knee Sleeve during this training – 'knee sleeve trained';

2. subjects who participated in a landing training program wearing the Intelligent Knee Sleeve but without receiving any audible feedback during training – 'placebo trained'
3. subjects who did not participate in the landing training program – 'control'.

Subjects in the knee-sleeve-trained and placebo-trained groups then participated in a six-week landing training program to learn correct landing mechanics, completing three 30 minute training sessions per week.

When comparing each subject's landing technique displayed pre- and post-training, it was noted that the control and placebo trained groups displayed, on average, less knee flexion, as evidenced by percentage change data, during the post-intervention session at the time of initial foot ground contact during landing (control = –11%; placebo = –23%); at the time of the peak resultant force (control = –8%; placebo = –4%) and the maximum knee flexion angle (control = –4%; placebo = –6%). In contrast, the knee-sleeve-trained subjects displayed the desired increases in knee flexion post-intervention at initial foot-ground contact (+14%), peak resultant force (+1%) and maximum knee flexion angle (+7%). In fact, although not statistically significant, the increase in maximum knee flexion angle from pre- to post-intervention displayed by the knee-sleeve-trained group was on average 8°, an increase that could be considered functionally relevant. Interestingly, although both the placebo and knee-sleeve-trained groups participated in the same intensive six-week landing training program, only those subjects who received immediate feedback from the Intelligent Knee Sleeve during training achieved positive changes in their knee flexion angle. As the subject numbers and statistical power in this pilot investigation were low, further investigation is warranted to confirm the trends displayed in the present investigation whereby participating in a landing training program using the Intelligent Knee Sleeve to provide audible feedback with respect to knee flexion angle assisted in teaching athletes to bend their knees more during landing after decelerating abruptly when catching a ball (Munro and Steele, 2005).

24.9.4 The Intelligent Knee Sleeve as a rehabilitation tool

Apart from the landing training application described previously in this chapter, the Intelligent Knee Sleeve can also be incorporated into rehabilitation programs to ensure patients perform their rehabilitation exercises properly, moving their limbs through the desired range of motion. For example, the Intelligent Knee Sleeve system can be used to assist patients in rehabilitation following ACL reconstructive surgery. Following arthroscopic surgical reconstruction of the injured ACL, most patients, particularly those who are involved in running or jumping sports, notice an asymmetric loss of knee flexion and extension (Millett et al., 2001). Therefore, the primary goals of postoperative

ACL reconstructive surgery are to regain full knee range of motion and to recover muscle strength and control (Draper, 1990) with this re-education process usually taking between four to seven months of physiotherapy. The Intelligent Knee Sleeve can assist ACL reconstructed patients to learn how to move their knee through a desirable range of motion throughout typical rehabilitation exercises and, in turn, promote the recovery of knee range of motion and function. The Intelligent Knee Sleeve may therefore increase the effectiveness of rehabilitation for ACL reconstructed patients, perhaps simultaneously promoting the recovery of knee muscle strength and neuromuscular control. Furthermore, when coupled with data logging capabilities, the Intelligent Knee Sleeve would provide the ability to evaluate rehabilitation routines, providing new knowledge to rehabilitation specialists on the effectiveness of these routines.

Trials are currently in progress to determine the efficacy of the Intelligent Knee Sleeve as a biofeedback device in enhancing post-knee-replacement surgery rehabilitation. As it is suitable for use by people of a wide age range, both independently and supervised at home or in the clinic, and on most joints of the human body, wearable biofeedback systems have a great many applications in a broad spectrum of activities, ranging from technique training to enhanced performance or for injury prevention as well as technique monitoring during rehabilitation.

24.10 Other applications of wearable biofeedback technology

Wearable biofeedback systems that monitor joint motion have broad application and can be used to provide immediate feedback pertaining to movement of most major segments of the human body. For example, placing sensors on a glove may be able to improve a golfer's putting technique, and thereby handicap, by providing feedback when there is excessive wrist motion which may detract from an optimal performance. However, regardless of the application, care is required to ensure the base garment does not provide support for the joint, is comfortable for the wearer and remains in the same position on the limb to ensure reliability in the feedback being received. Therefore, the base garment provides additional challenges to the designers of wearable technology, particularly those that provide biofeedback to the user during human motion.

24.11 Future directions

As wearable sensing technologies reach the prototype stage and undergo field trials there is no doubt that the need for specific improvements will be identified. However, even now it is obvious that parallel developments in the following areas will facilitate the emergence of new wearable technologies to monitor human motion.

24.11.1 Integration of electronic materials into textiles

Recent studies have shown that some electronic polymers (sulfonated polyanilines) act as highly effective dye molecules (Wu, 2004) being readily incorporated into host textiles of differing composition and structure and imparting electronic properties with no effect on textile handle. Undoubtedly future studies will result in the synthesis of improved (higher conductivity) electronic polymers tailored to match textile substrates at the molecular level.

24.11.2 Electronic fibres

An alternative to the dying approach to impart electronic properties is to weave electronic fibres through host textile structures. At present, there are only limited fibres, predominantly based on the organic conductor polyaniline, with handle akin to wearable textiles. Pomfret *et al.* (1998) have achieved conductivities in excess of 1,000 Scm^{-1} with a Young's modulus in the order of 40 MPa for fibres wet spun from a solution of polyaniline and therefore improvements to impart these electronic properties to fibres continue to be made in laboratories around the world.

24.11.3 Novel power sources and storage devices

The power sources and storage devices we currently utilise were not conceived, designed or fabricated with wearable systems in mind. While in the short term, off-the-shelf components will be used, truly wearable energy conversion and storage devices deserve a revolution in thought. The need, at least, is being recognised with projects in wearable solar energy conversion (www.natick.army.mil) and wearable batteries (www.uow.edu.au/science/research/ipri/innovations.html), such as polymer-based fibre batteries (Wang *et al.*, 2005), being initiated.

24.11.4 Electronics and interconnects

It is also true that the evolution of conventional electronics was not influenced by the 'wearables' community. However, continued miniaturisation may make electronics so unobtrusive that the lack of natural compatibility disappears as a major issue or concern. While miniaturisation may well provide seamless integration and compatibility, it will undoubtedly exacerbate the need for more innovative and effective interconnection systems to the outside world. Even now, when using conventional electronics, the interconnection between the soft(er) world of textiles and the hard(er) world of electronics is a challenge. Innovative approaches, undoubtedly involving new intermediaries and new 'contact materials', are required.

24.11.5 Biomechanical applications

Research leading to the development of robust wearable systems that operate independently of the environment, are machine washable, sweat resistant and, in some cases, biocompatible, is required. Furthermore, improved data processing, downloading and information transfer capabilities from multiple sensors are needed to ensure wearable biofeedback systems can withstand the rigours of physical activity, under all dynamic conditions likely to be encountered in activities of daily living, work, and recreation.

24.12 References

Acierno, S. P., D'Ambrosia, C., Solomonow, M., Baratta, R. V. and D'Ambrosia, R. D. (1995) *Orthopaedics,* **18**, 1101–1107.

Barthel, H. R., Miller, L. S., Deardorff, W. W. and Portenier, R. (1998) *Journal of Hand Therapy,* **11**, 191–199.

Campbell, T. E. and Wallace, G. (unpublished data).

Campbell, T. E., Munro, B. J., Steele, J. R. and Wallace, G. G. (2003) in *Proceedings of the International Society of Biomechanics XIXth Congress* (eds, Milburn, P. D., Wilson, B. and Yanai, T.) Dunedin, New Zealand, p. 48.

Caraffa, A., Cerulli, G., Projetti, M., Aisa, G. and Rizzo, A. (1996) *Knee Surgery, Sports Traumatology, Arthroscopy,* **4**, 19–21.

Chiotakakou-Faliakou, E., Kamm, M. A., Roy, A. J., Storrie, J. B. and Turner, I. C. (1998) *Gut,* **42**, 517–521.

Cinciripini, P. M. (1982) *Perceptual and Motor Skills,* **54**, 895–898.

Cobbold, R. S. C. (1974) In *Biomedical Engineering and Health Systems* (ed., Milsum, J. H.) John Wiley & Sons, New York, USA.

Cochrane, J., Lloyd, D., Buttfield, A., Seward, H. and McGivern, J. (2001) *Analysis of Anterior Cruciate Ligament Injuries in Australian Rules Football,* Department of Human Movement & Exercise Science, University of Western Australia, Perth. pp. 16.

De Rossi, D., Della Santa, A. and Mazzoldi, A. (1999a) *Materials in Science and Engineering,* C, 31–35.

De Rossi, D., Mazzoldi, A., Lorussi, F. and Paradiso, R. (1999b) in *International Society for Optical Engineering (SPIE) Conference: Smart Structures and Materials 2000: Sensory Phenomena and Measurement Instrumentation for Smart Structures and Materials,* Vol. 3986 (eds, Claus, R. O. and Spillman, W. B.), pp. 2–10.

Draganich, L. F. and Vahey, J. W. (1990) *Journal of Orthopaedic Research,* **8**, 57–63.

Draper, V. (1990) *Physical Therapy,* **70**, 11–17.

Engin, M., Demirel, A., Engin, E. Z. and Fedakar, M. (2005) *Measurement,* **37**, 173–188.

Fletcher, R. (1996) *IBM Systems Journal,* **35**, 630–638.

Gniotek, K. and Krucinska, I. (2004) *Fibres and Textiles in Eastern Europe,* **12**, 13–16.

Gogia, P. P., Braatz, J. H., Rose, S. J. and Norton, B. J. (1987) *Physical Therapy,* **67**, 192–195.

Hewett, T. E., Stroupe, A. L., Nance, T. A. and Noyes, F. R. (1996) *American Journal of Sports Medicine*, **24**, 765–773.
Hewett, T. E., Lindenfeld, T. N., Riccobene, J. V. and Noyes, F. R. (1999) *American Journal of Sports Medicine*, **27**, 699–706.
Johnson, R. J. (1983) *Clinical Orthopaedics and Related Research*, **172**, 14–18.
Jundt, K., Peschers, U. M. and Dimpfl, T. (2002) *European Journal of Obstetrics, Gynecology, & Reproductive Biology*, **105**, 181–185.
Kaynak, A., Wang, L., Hurren, C. and Wang, X. (2002) *Fibres and Polymers*, **3**, 24–30.
Kuhn, H. H. (1993) in *Characterization and Application of Polypyrrole-Coated Textiles* (ed., Aldissi, M.) Kluwer Academic Publishers, Dordrecht, pp. 25.
Lake, A. E. R. (2001) *Medical Clinics of North America*, **85**, 1055–1075.
Lekpittaya, P., Yanumet, N., Grady, B. P. and O'Rear, E. A. (2004) *Journal of Applied Polymer Science*, **92**, 2629–2636.
Meoli, D. and May-Plumlee, T. (2002) *Journal of Textile and Apparel, Technology and Management*, **2**, 1–12.
Middaugh, S. J. and Pawlick, K. (2002) *Applied Psychophysiology & Biofeedback*, **27**, 185–202.
Millett, P. J., Wickiewicz, T. L. and Warren, R. F. (2001) *American Journal of Sports Medicine*, **29**, 664–675.
Mizrahi, J. and Susak, Z. (1982a) *Engineering in Medicine*, **11**, 141–147.
Mizrahi, J. and Susak, Z. (1982b) *Journal of Biomedical Engineering*, **104**, 63–65.
Moreland, J. D., Thomson, M. A. and Fuoco, A. R. (1998) *Archives of Physical Medicine & Rehabilitation*, **79**, 134–140.
Munro, B. J. and Steele, J. R. (2005) *Journal of Science and Medicine in Sports*, **8**, 169.
Munro, B. J., Steele, J. R. and Wallace, G. (2002) *Journal of Science & Medicine in Sport*, **5**, 10.
Noyes, F. R., Mooar, P. A. and Matthews, D. S. (1983) *Journal of Bone & Joint Surgery*, **65A**, 154–162.
Pandy, M. G. and Shelbourne, K. B. (1997) *Journal of Biomechanics*, **30**, 1015–1024.
Parekh, A. R., Feng, M. I., Kirages, D., Bremner, H., Kaswick, J. and Aboseif, S. (2003) *Journal of Urology*, **170**, 130–133.
Pomfret, S. J., Adams, P. N., Comfort, N. P. and Monkman, A. P. (1998) *Advanced Materials*, **10**, 1351.
Seward, H., Wrigley, T., Waddington, G. and Lacey, N. (1999) in *Fifth IOC World Congress on Sport Sciences* (ed., Sports Medicine Australia) Sydney, NSW, pp. 17.
Steele, J. R. and Milburn, P. D. (1987a) *Journal of Human Movement Studies*, **13**, 399–410.
Steele, J. R. and Milburn, P. D. (1987b) *Australian Journal of Science and Medicine in Sport*, **19**, 23–27.
Storheim, K., Bo, K., Pederstad, O. and Jahnsen, R. (2002) *Physiotherapy Research International*, **7**, 239–249.
Thomas, R. E. and Vaidya, S. C. (1993) *Ergonomics*, **36**, 353–361.
Torzilli, P. A., Deng, X. and Warren, R. F. (1994) *American Journal of Sports Medicine*, **22**, 105–112.
Wang, J., Too, C. O. and Wallace, G. G. (2005) *Journal of Power Sources*, **150**, 223–228.
Wiener, J. S., Scales, M. T., Hampton, J., King, L. R., Surwit, R. and Edwards, C. L. (2000) *Journal of Urology*, **164**, 786–790.
Wu, J. (2004) Synthesis, characterisation and applications of conducting polymer coated textiles (PhD thesis) Wollongong: Department of Chemistry University of Wollongong, pp. 235.

25
Applications for woven electrical fabrics

S SWALLOW and A P THOMPSON,
Intelligent Textiles Limited, UK

25.1 Smart fabric technologies

25.1.1 What is a smart fabric?

In the absence thus far of any established taxonomy, this section will seek a framework of definitions and categories, to better understand and classify the member technologies. Firstly, where can a distinction be drawn between technical fabrics and smart fabrics? For example, do smart fabrics as a class include the many proprietary fabrics that use a laminated polytetrafluoroethylene (PTFE) membrane to render them waterproof yet breathable? Subjectively, instinct is to exclude such breathable textiles, but the reasoning is not immediately explicit. They seem to be clever, but not necessarily 'smart'.

The Larousse technical dictionary defines smart or intelligent materials as those that 'respond to an external stimulus in a specific, controlled way', listing as exemplars photochromic, piezoelectric and liquid crystal materials. To create a definition it is possible to borrow a key concept here, of a material that responds to a stimulus. To put it another way, the material has a number of distinct states, with differing physical properties in these states. Furthermore, the material will transition repeatedly between these states with a given external stimulus.

In the case of a photochromic material, such as the coatings used in light-reactive sunglasses, the stimulus is light energy and the physical states are high and low transparency (or strictly, polarisation angle of transmitted light). A liquid crystal material, of the type that composes laptop computer screens, also moves between high and low transparency, but the stimulus is electrical energy. In a piezoelectric material used, for example, to generate the alarm bleeps of a digital wristwatch, the stimulus is again electrical current, and the states are an expansion and contraction of the material. It is this vibration that causes the bleeping. By this borrowed distinction, then, the breathable waterproof PTFE membrane fabric is not a smart fabric. It may appear at first inspection to behave in two distinct manners, remaining permeable when presented with water in vapour form, whilst forming an impenetrable barrier to liquid water. However, the fabric itself always occupies the same state.

The microscopic pores in the PTFE membrane remain the same whether presented with liquid water droplets, which are simply too large to pass through the pores, or water vapour, in which the water molecules are separate and thus small enough to pass through. It is the water molecules that behave differently according to their physical state, rather than the fabric exhibiting any change. In contrast, a smart fabric might actually alter the size of its pores, and thus occupy distinct high and low states for its permeability or thermal insulation. The external stimulus might be humidity, temperature or even a controlling electrical current. This latter point illustrates a further useful subdivision within smart fabrics, which is explored in the following section.

25.2 Active and passive smart fabrics

Consider the stimulus that might spur a variable-pore-size fabric into action. A fabric that responds to either humidity or temperature as a stimulus would have what can be termed a passive response. The fabric would react and undergo its state change according directly to some environmental factor. The earlier example of photochromic coatings used in light-reactive sunglasses is another case of a passive smart material.

Hopefully, the state change proves useful, in both manner and direction of change. In this example case, a rise in humidity or temperature would advantageously trigger an increase in pore size. This in turn should serve to better ventilate the fabric or lessen the warmth afforded by it. This manner and direction of state change would thus perform a useful homeostatic function. There would be no need for intervention or control, and the smart fabric would quietly and automatically go about its work, attempting to regulate the humidity or temperature within a garment, for instance. Incidentally, the physical mechanism through which this might be accomplished could be some hygroscopic (moisture absorbing) polymer or a shape memory material. It is possible to imagine a woven or knitted fabric structure that alternately tightens or loosens, as its constituent hygroscopic or shape memory yarns change shape.

The alternative to a passive smart fabric is an active one. Both terms are borrowed from electrical engineering, where 'active' is used to denote a device where there is an input of energy. By implication, this required energy input is used to control or drive the device, and in the vast majority of cases this energy input is electrical. The archetypical active and passive devices in electronic engineering are filters. Whilst a passive filter can only ever reduce unwanted frequencies, an active filter uses an amplifier to add further energy, and can instead boost the desired frequencies. Turning once more to the example smart fabric with the variable pores, the required stimulus might be an input of electrical energy. An electric current might serve, say, to heat a shape memory metallic yarn to a transition temperature that effects its change in dimension. This variant of the fabric would thus show an active response.

As noted earlier, the vast majority of active smart fabrics tend to be electrical in nature. Whilst this is not in itself a definitive feature, electrical current is usually the most convenient means of imparting an energy input, and one which can be very competently manipulated. One notable exception is the use of light in fibre-optic based fabrics, which can be active, but not electrical. Whilst 'electrical' or 'conductive' smart fabrics might easily constitute their own subdivision within another taxonomy, it is important to note that they are not interchangeable with 'active' smart fabrics as a category. Similarly, passive smart fabrics usually appear to rely upon chemistry or physics, rather than electrical effects. Once again, however, this correlation is neither absolute nor definitive.

25.2.1 Sensor and effector smart fabrics

Another useful distinction within smart fabrics is whether they are used as sensors or effectors. Both are electrical engineering definitions for types of transducer. Transducers convert some physical variable into an electrical signal, or vice versa. Sensors are used in one direction, converting a variable such as light, sound or temperature into an electrical signal. Effectors are used in the other direction, converting electrical energy into some physical manifestation.

Piezoelectric materials are also very commonly used as both sensors and effectors, sometimes simultaneously. In medical ultrasound machines, a piezoelectric crystal generates pulses of ultrasound, the reflections from which are sensed an instant later by the same crystal, working alternately as effector and sensor. Note that this arrangement can also be termed 'active' sensing, as energy is introduced by the pulse into the sensed medium. Effectors, by definition, are active. An input of electrical current is required for conversion into another form of energy. Sensors can be both passive and active. Note the parallels between this and the 'pulse-echo' system for medical ultrasound, above.

In the passive smart fabrics of the previous section, the sensing and effecting tend to be intrinsically coupled together. Our aforementioned variable porosity fabric can be seen to both sense and react simultaneously – the change of state that accompanies a stimulus is the same change of state that effects the variation in pore size. The same is true in photochromic sunglasses, where the response to sensed light and the desired effected change are one and the same. The taxonomy is illustrated in Fig. 25.1. Table 25.1 summarises the relevant regions R to Z of our informal taxonomy. Intelligent Textiles Limited weave electrical smart fabrics, regions T to X, which will be discussed in the following sections.

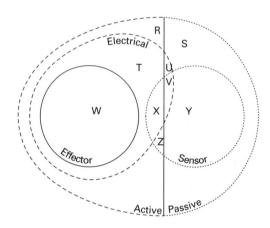

25.1 Taxonomy.

Table 25.1 Summarises the relevant regions R to Z of our informal taxonomy

Region	Description	Examples
R	Active, non-electrical, non-transducer	
S	Passive, non-electrical, non-transducer	Thermal phase-change materials (PCMs); photochromic fabrics
T	Active, electrical, non-transducer	Computational fabrics; photovoltaic fabrics; thermoelectric generator fabrics
U	Passive, electrical, non-transducer	Resistor, capacitor, inductor components; power and communication backplanes
V	Passive electrical sensor	Mechanical switch and keypad fabrics; pressure sensitive fabrics (some); capacitive sensor fabrics (some); temperature measurement fabrics; light sensitive fabrics
W	Active electrical effector	Heating fabrics; electroluminescent and LED illuminating and display fabrics; Peltier thermoelectric fabrics
X	Active electrical sensor	Pressure-sensitive fabrics (some); 2D and 3D positional pressure-sensitive fabrics; capacitive sensor fabrics (some)
Y	Passive, non-electrical sensor	Chemical indicator fabrics (thermochromic, moisture, acidity)
Z	Active, non-electrical sensor	Fibre-optic strain measurement fabrics

25.3 Electrical smart fabrics

25.3.1 Woven electrical fabrics at Intelligent Textiles Limited

Intelligent Textiles Limited has operated a woven electrical fabrics program since 1998. From the platform of the earliest technology, a pressure sensitive fabric named Detect™, they have since developed a versatile toolkit of raw materials, weave structures and circuit components from which they can fabricate a broad range of electrical smart fabrics. These fabrics can be variously active, passive, sensing and effecting in their function. Using this toolkit, it is possible to construct fabrics that incorporate multiple conductive yarns, introduced during the weaving process and integral to the very structure of the cloth. These conductors are mutually interconnected in a fashion that is controlled by the woven structure. Instead of weaving checked or tartan patterns in different coloured yarns, colours can be swapped for conductors to weave electrical circuits.

The elemental circuit components that can be fabricated at present are all passive, namely resistors, capacitors and inductors, plus some electromechanical structures such as switches. Note that the term 'passive' is used here in the specific sense of denoting electrical components that have no ability to amplify a signal, and by that implication, contain no semiconductor material (or strictly, no transistors). The resistor, capacitor and inductor components are themselves composed of tightly dimensioned geometric networks of conductive elements. Each element is a segment of a conductive yarn. At their simplest, networks can comprise a handful of such elements, or use recursive and combinatorial structures to quickly rise into thousands of elements. The woven conductive structures also provide the equivalent of conductive tracks, as found on conventional printed circuit boards (PCBs). Thus, multiple components can be positioned arbitrarily within a given piece of fabric, and route signals and power between them. Effectively, hard PCBs can be translated into pliable woven fabric. In fact, given a sufficiently large area of fabric, entire circuits of resistors, capacitors and inductors can be fabricated. That said, the density of components is dramatically less than conventional circuit boards, and all but the simplest circuits can demand many metres of fabric.

Fortunately, many useful electrical devices are very simple circuits, or even single components of a particular design. For instance, a large variety of conventional sensors are based upon little more than single resistors, whose construction and composition are so chosen as to respond to a stimulus with a change in resistance. Light, temperature and mechanical strain are all routinely measured with such specialised resistors, although sound transducers tend to be inductive or capacitive in nature.

It is interesting to compare the fabrication scale and density of these woven electrical components against the historical development of mainstream electronics. The original electrical pioneers machined large parts in glass and metal in order to make their seminal electrical components. In the 1960s and 1970s, machined metal and blown glass gave way to plastic packages and etched semiconductor materials. Components shrank, from centimetres in length, to millimetres, to the current micro- and nanometre scales

The change in medium from wafers of semiconductor to woven fabric, signals a return to the centimetre- and millimetre-sized components of the early pioneers. Similarly, the complexity and functionality of the resulting woven components has regressed to electrical first principles, but reapplying these first principles to a new fabrication medium is a fertile area of development.

25.3.2 Passive, non-transducer fabrics

As introduced in the preceding subsection, the most basic electrical devices that can be constructed as woven textiles are passive components – resistors, capacitors and inductors. Although these basic components all find application in various guises as sensors and effectors, they can also conceivably be used purely as components within larger circuits, and not necessarily as transducers. It is entirely possible, for instance, for a piece of fabric with these components to constitute a complete passive audio filter circuit.

Resistors of specific value can be fabricated in arbitrary shapes and sizes within a stretch of fabric. A resistor's value is generally independent of the area of fabric it covers, which is achieved by using large networks of parallel and series conductive elements. The resistors' parallel nature means they possess great redundancy, and so their function and value is fault tolerant, degrading gracefully if the fabric is partially damaged. The resistance values can range from a few ohms to many megohms, and when manufactured on automated power looms their tolerance on desired resistance is typically within 5%.

The area of a resistor is usually only a few square centimetres for signal wattages. For greater power dissipation, the resistor can be scaled up in area. Every square metre of fabric will comfortably dissipate several hundred watts of power. The configuration of the fabric resistor, as a very thin, broad sheet, is ideal for efficient heat dissipation. This efficiency at dissipating heat is one factor that makes the fabric resistor particularly useful as a heating element. This is a mode of use which constitutes an active effector, rather than a passive component.

The same is true of the capacitor and inductor components that can be fabricated with the same techniques. They both find applications as various types of sensor and effector, but when one considers them not as transducers but as replacements for conventional PCB mounted devices, they are dreadfully cumbersome and costly. Again, these two remaining types of passive component

Applications for woven electrical fabrics 477

can be embodied as networks of contiguous conductive elements within the fabric structure. Unlike resistors, however, the construction of capacitors and inductors relies to a greater extent upon the specific geometry of the conductive elements. By forming the elements into various loops, spirals and meanders within the fabric, it is possible to duplicate some of the geometric features of conventional capacitive and inductive devices. However, attaining specific values, from within an adequately broad range, is far more problematic than for woven resistors.

Another passive, electrical and non-transducer application for these woven fabrics is to provide wiring looms or circuit backplanes for conventional electrical. For example, a conductive fabric backplane might incorporate the 'wiring' for distributing power or audio signals between a number of portable devices worn upon the body. Telephones, music players, computers and their requisite headphones and microphones may all be interconnected in this manner, in an arrangement termed the personal area network, or PAN. Another attractive prospect is to consolidate and centralise the power supplies for these devices, reducing weight. A similar application lies in utilising the textile upholstery in vehicle interiors to additionally replace the conventional wiring looms. The conductors in a fabric wiring loom can even be designed to terminate in broadened areas for use as electrodes, when incorporated into garments for biomonitoring.

25.3.3 Passive sensor fabrics

As mentioned earlier, many common sensor types are based upon little more than a single resistor, whose arrangement and chemical composition are so chosen as to change the resistance according to a stimulus. Given that it is possible to produce resistor components with the woven technique toolkit, a number of these sensors can be translated into fabric. One physical variable that is routinely measured in this way is temperature. Thermistors are specialised resistors, usually based on semiconductor materials, that show a pronounced 'resistance temperature coefficient', or change in value according to temperature.

Alternatively, pure metallic conductors are often used, in which case the resistors in question are often named 'resistance temperature devices', or RTDs. The resistance temperature coefficient of pure metals is much smaller than that of the semiconductors developed specifically for use in thermistors, but is still adequate for reliable measurement when metallic yarns are used. Making RTD measurements from the fabric resistors described here can typically be achieved with around 5% accuracy, within the operating temperature range of the fabric. Advantageously, the same resistor that might constitute a fabric heating element can also be used as the sensor for the resistance temperature measurement. The fabric resistor alternates between effector and sensor, much like a piezoelectric ultrasound probe. Other physical variables that can be sensed with these passive resistors, in various configurations, include moisture, galvanic skin resistance and mechanical strain.

Conventional capacitive and inductive sensors are primarily applied to non-contact proximity detection. That is, determining the distance of nearby objects via their disturbance of a capacitor or inductor's electrostatic or electromagnetic field, respectively. Fabric capacitors and inductors can be used in this manner, albeit currently with greatly reduced acuity than their conventional counterparts. It has proved possible, at least, to perform 'event detection', in which the sensor is not required to measure the distance of an object, but merely that it is present nearby or absent.

A wholly different approach can also be taken for sensing using electrical fabrics. The fabric 'Detect'™, by Intelligent Textiles Limited, can sense physical contact or pressure. The sensing is achieved with millimetre-sized, mechanical switches in a regular array across the surface of the fabric. The switches are created by controlling the local weave structures around the conductive yarns. A 10 cm square piece of fabric, say, may typically comprise several hundred such electrical switches. It may help to imagine a very broad computer keyboard, with hundreds of keys, miniaturised and flattened into fabric form. Detect is a second-generation technology, which offers touch sensitivity in a single, conventionally woven sheet. To all other practical intents, including magnitude of manufacturing costs, Detect is an ordinary piece of fabric. It can also be cut, sewn and handled in a conventional manner.

The fabric can be used in both active and passive modes. When used in a passive sensor fashion, as a simple switch, Detect can replace a variety of input devices such as pushbuttons, keypads, limit switches and keyboards. Its function is that of a momentary contact, normally open push button. Because it is a mechanical switch, with real moving contacts, Detect is a direct replacement for conventional pushbuttons, keys and microswitches. It requires no additional signal conditioning or processing, beyond the debouncing that might be afforded any conventional switch, which is a prime advantage of its passive nature. Detect fabric is designed and manufactured bespoke, to suit a specific application, some of the general specifications that can be achieved are as follows. Operating pressure for the switches can be as low as approximately 5 kiloPascals, but a pressure of around 50 kiloPascals is more typically used in switchgear applications, which is a comfortable, light force to apply with a fingertip. Operating pressure can be predetermined at time of manufacture.

The minimum pitch of the switches is approximately 2 mm by 5 mm, although 2.5 mm is often used for compatibility with conventional electrical components, connectors and PCBs. This results in a possible switch density of over 100,000 switches per square metre, which overcapacity is often used to introduce redundancy and improve robustness. The switches' open circuit resistance has been measured at over 40 MΩ, as befits a switch with genuine mechanical separation between contacts. The closed circuit resistance varies according to specific design, but it is most cost-efficient to limit this specification to around 1 kΩ, if acceptable. Operational lifespan of the switches is typically in excess of 1,000,000 operations, and the abrasion resistance of the fabric is similarly good, perhaps surprisingly so. According to the nature of the base fibre and the fabric weight, the Martindale abrasion resistance has variously been measured at between 20,000 and 40,000 cycles.

Detect is routinely woven with base fibres of cotton, polyester, wool and nylon, whilst a typical finished weight is *circa* 250 grams per square metre. Obviously, the base fibre also affects fabric washability. Woollen base fibres, whilst very resistant to abrasion and affording excellent elastic recovery, suffer poor washability. A Detect fabric woven from polyester base yarns, conversely, can be reliably washed over 30 times on a 40 °C cycle.

25.3.4 Active sensor fabrics

In addition to using Detect as a passive switch, it can also be used as an active sensor. This can be achieved by energising the switch array with a low reference voltage, and then making an analogue measurement from one set of switch contacts. Comparing the analogue sensed voltage with the reference voltage then allows more information to be gleaned from the fabric about the nature of the physical contact. In one fabric configuration, this analogue measure can be used to determine the location of a physical contact. This can be a one-dimensional measurement, whereby the fabric senses a single distance measurement between the point of contact and another fixed point. This configuration proves useful for emulating slider-style controls or keys arranged on a continuum, like piano keys.

If the same switch configuration and measurement are duplicated twice within a piece of fabric, ideally measuring one horizontal and one vertical distance, the fabric can be interrogated for an absolute two-dimensional position. This same measurement system is used in resistive touchscreens for palmtop computers. The fabric can also take relative measurements, with a small alteration in interface software, which is then analogous to the tracker pads found on many laptop computers. Whether the distance measurements are treated as continuous, as in touchscreen and tracker pad styles of interaction, or the measurements are thresholded and quantised to provide an array of discrete keys or buttons, this two-dimensional positional technique finds a great many useful applications.

The primary advantage of these analogue techniques is their reduction in interconnections compared to the simple passive switch sensing described earlier. Taking the example of implementing a qwerty keyboard with perhaps one hundred keys, for instance, a two-dimensional positional sensing fabric will require only two analogue signal lines, plus two energising inputs. This means that only four signal tracks need be routed through the fabric, and the off-fabric connector remains small, inexpensive and unobtrusive. Conversely, treating each key as a separate passive switch would require over one hundred signal lines, one for each key. Even treating the keyboard as a row-column multiplexed matrix, a standard keyboard technique, would require some twenty signal lines.

The major disadvantage of the analogue technique is its requirement for fairly complex interface circuitry. Not only must this circuitry comprise analogue-to-digital conversion (ADC) hardware and the microprocessor means to interpret the results, but also the signal conditioning, shielding and filtering to remove the noise that inevitably affects analogue systems. Pragmatically, each application case must be assessed on its own merits, and the best compromise solution selected. There is, at least, a spectrum of solutions of increasing complexity from which to select. Using similar interface circuitry, but altering the fabric configuration slightly, allows Detect to sense the area of a contact, rather than its position. Again, an analogue measurement is taken, but in this case it is proportional to the number of switches within the array that are closed. Simultaneously taking more than one such area measurement allows a picture to be composed of the shape of a contacted area. With sufficient separate measurements, a pixel by pixel pressure map can even be determined. These techniques are particularly useful in upholstery applications, for monitoring posture or classifying the seat occupant, the latter to better inform airbag deployment in cars.

Rather than making many area measurements separated in space, a different monitoring system can alternatively make many area measurements separated through time, from a single transducer. Inspecting the manner in which an area of contact changes through time can provide information about the velocity, impulse and dynamic pressure of a contact or impact. The Detect fabric, then, demonstrates great versatility as a contacting sensor device, with a variety of fabric configurations and a variety of means for interrogating these sensors. For non-contacting sensing with fabrics, the capacitive and inductive sensors described in the previous subsection provide a possible solution. These types of sensors are most often used in an active fashion, energised with an alternating current to establish their electrical fields.

25.3.5 Effector fabrics

By far the simplest electrical effector fabric is the resistance heating element. Any conductive material that carries an electrical current will generate heat, usually as an unwanted and inefficient by-product. The resistance heating element, such as those in electric kettles and ovens, makes virtue of this phenomenon, and is the only 100% efficient effector. In previous sections, the use of woven resistor components has been discussed, including their ability to be repeatedly fabricated to an arbitrary resistance value, size and shape, as well as their efficiency at dissipating heat. All of these factors result in woven electrical fabrics being excellent in flexible heating applications. In fact, Intelligent Textiles Limited manufacture a specific format of low-voltage woven heater under the name 'Heat'™ fabric. Like Detect, Heat fabric appears to be an ordinary woven fabric. It is thin, flexible, drapeable and breathable. It can be cut and sewn in a similar manner to conventional fabrics, so is easy to incorporate into the manufacturing of apparel or soft furnishings.

The base material for Heat fabric can again be chosen from cotton, polyester or wool, with typical fabric weight around 250 grams per square metre. Composition affects washability, once more. Heat fabric can be designed to survive upwards of 30 washes, but in many of the common upholstery applications, this is not a requirement. Upholstery weight Heat fabric, with a coarse micron woollen base fibre, is less washable but has excellent abrasion qualities. Martindale results for Heat are generally superior to Detect, and woollen Heat variants have achieved over 100,000 cycles.

Turning to Heat fabric's electrical specifications, elements can be manufactured to within 5% of a specific resistance, within the range of five ohms to perhaps a megohm. Heat fabric can comfortably dissipate over 100 watts per square metre (named the fabric's 'energy density'), up to an operating temperature of around 100 °C. The flexibility in specifying a resistance for the heating element gives great freedom to design a heating system of a given wattage, working at a particular voltage. This becomes very useful in battery-powered portable applications where, for example, a battery can be selected on grounds of its capacity or package size, rather than its rated voltage. A further advantage of Heat fabric is its ability to provide a non-homogeneous distribution of heat output across its surface. That is, the combination of resistive elements and wiring loom within the same fabric allows us to place heaters wherever they are most effective or efficient within that piece of fabric. For example, one component produced is for a heated glove lining, where power is conducted from a wrist mounted battery to heating elements in each fingertip. This serves to minimise wasted power, concentrating the heat distribution where it is needed most. Again, this factor is particularly relevant in battery-powered applications, where energy capacity is the bounding factor for these products' effectiveness.

It seems usually to be a surprise outside of electrical circles just how power-hungry electrical heating can be, compared to the majority of electrical applications. Whilst a digital watch might consume only microwatts, and talking on a mobile phone perhaps some few hundred milliwatts, perhaps five or ten watts is consumed just to heat the fingertips to any significant degree. This means that battery-powered heated products usually suffer from large or expensive batteries, short lifespan or low heat output, or perhaps all four at once. It is particularly useful, then, that Heat fabric can also be used as a temperature sensor, as described earlier as a passive sensor property. A product that can monitor and regulate its own temperature, effecting climate control, can prevent excessive heating and thus further conserve energy. Heat fabric constitutes the largest proportion of Intelligent Textiles Limited's work into effector fabrics, not least because the marketplace for flexible heating is relatively mature, and the technology is straightforward and easily appreciated by the consumer. Despite this, the techniques involved in creating Detect and Heat fabrics show promise for application to other effector types in future.

One very attractive effector fabric would be capable of graphical display. Whether this is based upon electroluminescence, electrophoretic materials, liquid crystals or light emitting diodes as an enabling technology, the pixels will each still need power and control, and thus require a conductive fabric backplane. Another promising fabric effector technology lies in the application of inductor components. Broad, flat fabric components lend themselves well to incorporating antennae and induction coils. Such devices may be concerned with radio signal transmission and reception, or for induced contactless power transmission.

25.3.6 Active non-transducer fabrics

This category of electrical smart fabrics is perhaps the strangest assortment. In the archetypical electronic engineering diagram of a computer or information system, the three building blocks are 'input' (that is, sensors), 'output' (effectors), and between them, the inscrutable 'processing'. Processing here means computation, which equates to logic circuitry and memory, which in turn means transistors and semiconductors. The remaining major class of electrical device which has yet to be represented is prime movers. Prime movers are the source of electrical energy, be they primary or secondary cells (batteries, and rechargeable batteries), photovoltaic solar cells, electromechanical, piezoelectric or thermoelectric generators.

The majority of these prime movers also rely upon semiconductor technology becoming available in some fabric-compatible form. It would seem that semiconductor yarn or fabric is the current frontier in electrical smart fabrics. Their advent is eagerly anticipated, for they promise to bring, amongst many other things, all of the components for a computer that can unfurl like a handkerchief – power source, input devices, memory, processor and display.

25.4 Products and applications

25.4.1 Overview

Textiles can be warm, soft, lightweight, tactile and embracing in contrast to technological devices which by their very nature can be cold, hard, heavy, smooth and certainly unsympathetic to human contact. It is clear why people choose to surround themselves with textile products that are soft to the touch, thermally insulative and breathable. It can also explain why modern technology is often perceived from the outset as being difficult, uncomfortable and alienating?

In ubiquitous computing's seminal 1991 paper, Weiser (1991) states,

> We are trying to conceive a new way of thinking about computers in the world, one that takes into account the natural human environment and allows the computers themselves to vanish into the background.

If computers, and technology in general, are to become as ubiquitous as fabric, it must embed itself successfully into textile products, if only because so much of people's everyday environment is already swathed in fabric. This collision of engineering and textile design disciplines is spawning a new class of soft products that can be startlingly different from their technological predecessors yet as familiar as an old chair or favourite coat.

Having defined and explored the various electrical smart fabric technologies that are of interest at Intelligent Textiles Limited, the following sections examine their applications and products. The discussion is subdivided into the traditional large vertical markets for textile products, namely apparel, furnishings and interior fabrics, healthcare and vehicle interiors.

25.4.2 Apparel applications

Thus far, the majority of take-up on smart fabrics in the apparel sector has been with sports and leisure wear manufacturers. Perhaps this is because they are the branch of apparel that is most comfortable and familiar with technical fabrics, and have a consumer base that demands technical function and advancement. Biometric applications are an obvious starting point, for monitoring both sporting technique and physiological performance. The latter can make use of conductive fabric sensors incorporated unobtrusively into garments, measuring heart rate with integral skin contact electrodes or temperature with passive RTD sensors. Monitoring of sporting technique can be well served by contacting pressure sensors such as Detect fabric. In shoe insoles, pressure mapping fabric can measure plantar pressures, gait or activity levels. Similar technologies can give information about grip technique in bat, club and racquet sports.

Heated garments have an obvious appeal for outdoor activities. The skiing and snowboarding marketplace is clearly very large and affluent, but is still outweighed by the sheer number of smaller niche applications for heated wear, from motorcyclists and scuba divers to pylon riggers and cold-store workers. The limitations of battery technology in these mobile products, as discussed earlier, restrict the scale of electrical heating that can conveniently be applied. At present, we believe that our Heat fabric's ability to concentrate power output and conserve battery life enables genuinely useful products to be manufactured, that can strike a reasonable compromise between cost, bulk, heat output and lifespan. Our glove liner, for example, has a battery lifespan of around four hours when used with a high-capacity lithium polymer rechargeable cell. We also produce a boot liner component, with similar specifications. Heating more of the body than these most vulnerable extremities remains problematic with current battery technology. That said, their pace of development has quickened noticeably, and fuel cell or supercapacitor technology may soon become a realistic alternative.

Incidentally, the technological Philosopher's Stone for this sector of the market is the smart fabric that can cool as well as heat. Phase-change materials are intriguing, but of limited capacity. Far more enticing is the prospect of an actively powered thermoelectric fabric, which can pump heat away from the body as an electric current. The sport and leisure companies are also the most prevalent early adopters of wearable computing concepts. Many such brands currently list high-end garments and accessories that have inbuilt controls for MP3 music players. Accordingly, many smart fabric companies produce components or systems for incorporation into these garments.

As these portable products, such as music players, mobile phones and palmtop computers, become ever smaller, they suffer from the limited space available for their interfaces, as well as being cold, hard, heavy and smooth. For the typical case of a mobile phone, Cochrane *et al.* list the limitations to size as being batteries, keyboard size and display size, in that order (Cochrane *et al.*, 1997). On the second of these limitations, they state: '... multi-function displays and buttons all add to users' operating difficulty. Buttons with four or five functions are not uncommon in the race to reduce size and price.' Reconfiguring these devices to use fabric controls allows their interfaces to be spread over much larger surface areas. The larger controls can be integrated into clothing to eliminate bulk and encumbrance, and even help to make interaction more natural and less intrusive to use, requiring less visual focus. Interactions become gestural, rather than fine manipulation tasks. On the practical side, wearable interfaces cannot be dropped or misplaced, and are always ready for use.

25.4.3 Furnishings and interiors applications

Consider the television remote control. It is a typical small, cold, hard, heavy and smooth technological device, at odds with its soft environment of furnishings, people and pets. Its small size compromises the number, labelling, accessibility and affordance of its control keys. Being cold and hard makes it uncomfortable to sit upon. Being heavy and smooth allows the device to slip efficiently between cushions and become lost. All of these problems can be addressed by a soft device solution using Detect fabric. A remote control built into the upholstery of a chair or a cushion is not small. Its switches can be twentyfold greater in size than a conventional remote control. Nor is the fabric controller heavy, as its size increase has been achieved with little increase in mass or indeed material volume. The keypad has simply been reorganised into a lower profile and lower density package.

The fabric remote control is not hard. It and other soft fabric devices can be at least as robust in general use as the equivalent hard devices. Imagine selecting a material to withstand repeated hammer blows. A typical mass production polymer, such as polystyrene or ABS, will split or shatter. Now imagine trying to destroy a piece of fabric in the same way. This illustrates the distinction between strength (which tends in general to correlate with a material's hardness, stiffness and density) and toughness. Fabric is not strong, or stiff, but can be very tough. Nor is the fabric remote control cold, or smooth. Its fibrous nature imparts it with warmth to the touch and favourable tactile qualities, in addition to arresting its attempts to slip around and disappear.

Whilst this remote control is in itself an interesting and illuminating example of soft product design, it is perhaps more notable for its implications to ubiquitous computing. This type of touch sensor interface can be incorporated into furniture, bedding or carpeting. Detect fabric, as a low-profile, low-complexity, low-cost interface technology can be used to invisibly sensitise all manner of surfaces around us. Expanding upon the example of remote controls incorporated into upholstery and soft furnishings, these can range from simple keypads, to qwerty keyboards, to mouse control tracker pads for the navigation of integrated computerised entertainment systems.

All manner of household textile objects, such as seating, carpets, bedding and towels can be heated, without intrusive heater cables. Heated curtains and carpets might oust ugly radiators and vents, freeing valuable areas of wall in our ever-more compact urban living spaces. Carpets can be sensitised to react to the passage of people. This sensing can be used as a component of intelligent building control, controlling lighting and ventilation according to a room's occupancy. The carpet might monitor the elderly or vulnerable in their homes, discriminating the telltale footprint of a collapsed occupant, lying prone. In public environments, pedestrian traffic could be monitored to aid evacuation, detect intruders or unattended packages, or just glean information for building planning or marketing purposes.

As the technology pundit Bill Buxton has remarked (Harris *et al.*, 1998), 'When people are asked to draw a computer, about 80 percent of the time they draw the I/O devices [interface devices, such as keyboards and monitors].' With these smart fabric transducers now able to replace many of these interface devices, the technology can now begin to disappear, literally, into the very fabric of our surroundings.

25.4.4 Automotive and transport applications

A client from a major European car manufacturer once informed us that their most well-appointed car interior features an extraordinary five hundred individual buttons. In order to service this plethora of switchgear, an additional 14 km of wiring loom is also required. Managing to incorporate this many controls, without resorting to a 'flight deck' aesthetic, has become a grave design issue. Electrical smart fabrics like Detect can serve to rationalise many of these controls and their wiring into the existing textile upholstery components, door panels and headlining, thus greatly reducing the component count.

By way of example, Fig. 25.2 illustrates a conventional qwerty computer keyboard, exploded into its hundred or so constituent components. The corresponding Detect fabric keyboard, with a similar number of discrete pushbuttons, streamlines this into a single component. As well as reducing cost, complexity and component count, upholstered controls can also reduce weight. This is particularly relevant in the aerospace industry, where another client claimed that every extra kilogram carried by a commercial airliner costs over $100,000 per annum in additional fuel.

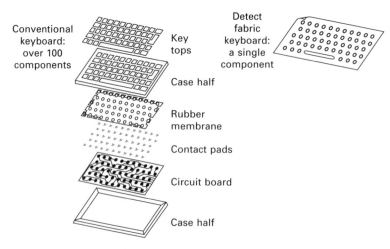

25.2 A conventional qwerty computer keyboard, exploded into its hundred or so constituent components.

Within the seats themselves, the most obvious candidate technology is Heat fabric. Within the same fabric panels, or as an underlay to leather upholstery, Detect switches can discriminate between adult and child occupants, and thus inform intelligent safety systems whether to deploy airbags and seatbelt tensioners in an accident, and with what force. Eventually, one might envisage a wholly textile car interior, where the fabric components form a continuous backplane and medium for the majority of the switchgear, interior illumination and even displays for instruments and entertainment devices.

25.4.5 Healthcare applications

As in sporting apparel applications, a major healthcare application for electrical smart fabrics is biomonitoring. The sensing of a patient's physiological variables might take place through their bedding, seating or clothing, but in any event could include electrodes for electrocardiography, RTDs for skin temperature, some configuration of strain or pressure sensors for respiration, and even give warning that moisture has been detected.

In bedding, pressure mapping fabrics might also monitor for the areas of increased load and prolonged inactivity that result in the formation of pressure ulcers. Pressure ulceration poses serious problems for bedridden or paralysed patients, and exacts a devastating cost, both financial and in quality of life. All of these techniques are easily transferable from ward beds to operating tables, to wheelchairs and neonatal cots. In cots, particularly, much of this may also find consumer product applications. The same pressure sensitivity in seats or mats can help monitor posture or technique during rehabilitation and occupational therapy.

Turning to medical dressings, sensors intrinsic to the fabric of surgical gauze, bandages or splinting materials may allow the examination of wounds without their continued re-exposure and re-dressing. Resistive moisture sensors may detect bleeding, or give a measure of alkalinity, indicative of infection. Stabilised electrical heating may be applied to relieve pain or accelerate the healing of wounds. Conversely, phlebology dressings that are designed to slow healing, and thus reduce scarring, by constricting a wound might be monitored to ensure they apply the correct pressure. Healthcare is a market sector where technology is in particularly intimate contact with our skins, or even underneath it. The imperative to make technology soft and sympathetic is at its strongest here.

25.5 References and bibliography

1. Cochrane, P., Payne, R. and MacDonald, B. (1997) From Kirk to Picard – a vision of mobility. *Personal Technologies*, Springer-Verlag, Vol. 1, April 1997, 6–10.

2. Harris, M., Sinclair, M., Freeman, W., Buxton, W., Lucente, M., Ishii, H. (1998) Interfaces for Humans: Natural Interaction, Tangible Data and Beyond... (panel). In *Conference Abstracts and Applications of SIGGRAPH '98*. July 1998. Orlando, Florida, USA: ACM Press, 200–202.
3. Lee, S. (2005) *Fashioning the Future: Tomorrow's Wardrobe*, Thames and Hudson.
4. Newman, C. and Wolinsky, C. (2003) Dream weavers, *National Geographic Magazine*, January, 50–73.
5. Swallow, S.S. and Thompson, A.P. (2001) Sensory fabric for ubiquitous interfaces, *International Journal of Human–Computer Interaction* (IJHCI), 13(2), 147–159.
6. Weiser, M. (1991) The computer for the 21st century, *Scientific American*, September, 94–104.

Index

abrasion testing 272
accelerometers 386
accessories, electronic 401, 414–15
accuracy 372
acid treatments 290–1
acoustic sensor arrays 276
acrylic fibres 44, 45
activation energy 161
active control 403
active diffusion 150
active electrical effector fabrics 474, 480–2
active electical non-transducer fabrics 474, 482
active electrical sensor fabrics 474, 479–80
active insulation 26
active smart fabrics 472–4
active wear 26–8
    outdoor 51, 484
actuators 1–2
    intelligent textiles as in medical applications 382, 388–90
  MSM 95
  Nitinol 136
Adaptamat Ltd 95
adhesive bonding 250
aeronautical applications 125, 135
aerospace applications 135, 486
aesthetic degradation 171
aesthetic properties 7, 360, 66
aesthetics 165–89
    engineering textile and clothing aesthetics with SMMs 172–82
    interactive applications of shape changing smart textiles 182–4
    mood changing textiles 184–6

agitation, detection of states of 378–9
air-conditioning 55
alkyl hydrocarbons 23–4, 36–9
alloys, shape memory *see* shape memory alloys
alternative energy sources 206
Alzheimer's disease 378–9
Ambience 05 Conference 2–4
ambient intelligence 369–70
ambulatory electroencephalography (AEEG) 377
American Society for Testing and Materials (ASTM) tests 28–9
amorphous silicon solar cells 213
amorphous thin film transistors 257–8
amphiphilic block copolymers 385
animations 360, 365–6, 367–8
annealing 291, 296–8
anodisation 346
antennas 272–4, 349–50, 404–5
anterior cruciate ligament (ACL) 463–4
    rehabilitation following surgery 466–7
  see also Intelligent Knee Sleeve
antifoam agent 48
anti-scalding SMA valves 136
apparel *see* clothing
applications 2–4
applied research 11
Aracon fibre 363
Arcus 147
armbands
  GSR sensor 385
  Healthwear 401
armour 136–8, 139

489

attachment of electronic devices 252–6
attention deficit hyperactivity disorder (ADHD) 379
audio feedback 464–5
  see also Intelligent Knee Sleeve
audio output mood change trigger 186
austenite 86–7, 88, 89, 125, 167
austenite finish temperature 87
austenite start temperature 87
automotive applications
  electrical fabrics 486–7
  PCMs 55
  SMAs 135–6
  weight sensors for seats 275–6
awning, with solar cells 209

back-knitting 345, 350
back manager, wearable 415–17
backplanes 477
ball connectors 253
ballistics 136–8, 139
band theory 218, 219
bandwidths 94, 101
base treatments 290–1
BASF 361
basic research 11
Basofil fabrics 361, 362
Bayer MaterialScience 100
Bayesian decision theory 414
bedding 54
belt, sensor integrated 441–5
bending stiffness 272
Betalloy coupling 135
binder 47–8
biofeedback 232, 379
  need for biofeedback technology 450–1
  problems with current devices 451
  wearable biofeedback systems see wearable biofeedback systems
Biofoot/IBV 387, 388
biological micro-electromechanical systems (Bio-MEMS) 101
biomechanical applications 469
  monitoring biomechanical signals 385–7
biomechatronics 351–4
Bion II film 147
blankets
  electrically heated 54
  therapeutic 54–5
blinds 53–4, 183–4
block copolymers 110, 113–15, 116
blood-clot filters 135

blood pressure 423
  monitoring 376
Bluetooth antenna 404–5
body area network (BAN) 342, 369–70
body coverage 78
body temperature 22, 42, 423
  measurement of 385, 425
BodyMedia 401
Bragg sensors 388
braiding 345, 354
brass (copper-zinc) 86, 104, 125
brassieres 136
breakage of conductive yarns 241
breathable textiles 143–64, 471–2
  breathability and clothing comfort 144–5
  designing with nonporous films 150–1
  future trends 162–3
  merits of nonporous films for 148–50
  requirements for waterproof breathable textiles 145, 146
  types of 145–8
  water vapour permeability through shape memory polyurethane 152–62
breathing rate 384
Buehler, William J. 86, 104
bulk in-situ polymerisation 233

cables 431
  USB 271–2
calibration 372
camphor sulphonic acid (CSA) 236, 311
canopy, retractable 209–10
capacitance 267, 268
capacitively coupled electrical stimulation 389
capacitive sensors 476–7, 478, 479
carbon 219, 221, 222
  as a conductor 220
carbon black 325
  conductive polymer composite textile sensors 331–9
carbon filled silicon rubber sensor 411, 412, 413
carbon filled thermoplastic elastomer fibres 412–13
carbon nanotubes (CNTs) 221–2, 388–9
cardiomyopathy 375–6
carpets 485
carsolchromic materials 193
casual clothing 120
catheter guides 135
ceramics, shape memory 94

chain extender 111
chemical coating methods 214
chemical *in-situ* polymerisation 233–4
chemical metallised threads 348–50
chemical vapour deposition (CVD) 293–6
children 434–49
    future trends in smart textiles for 447–8
    Intellitex suit 436–47
    state of the art smart clothing 435–6
cholesteric liquid crystals 198
chromic materials 2, 3–4, 169, 193–205
    colour-changing inks 200–1
    electrochromic materials 193, 194, 201–3
    photochromic materials 193, 194–6, 471
    thermochromic materials 169, 193–4, 196–200
chronic diseases 371
circuit backplanes 477
circulation 423
Climaguard 146
clinical monitoring *see* monitoring
clothing
    ICPs 232
    requirements for shape changing textiles in 171–2
    SMAs 136
    solar textiles 215
    textiles containing PCMs 51–3
    woven electrical fabrics 483–4
    see also under different types of clothing
clothing theory 6–7
CLR-coated yarns/fibres 386
CMY colour weaving 343–4
coacervation 37–8
coating composition 48
coating processes 47–8, 214
    conductive fibres/yarns 310–11, 312
    solution coating 233
coatings, shape memory 168, 181–2
cold working 92–3
collaboration, multidisciplinary 12–15, 166
colour-changing inks 200–1
colour metals 220–1
commercially successful applications 3–4
comfort
    breathability and 144–5
    heat balance and thermophysiological 22
ComforTemp 24, 67
communication 372
    electronic textiles 403–5
    Intellitex suit 445–6
    medical applications 382, 390–3

communication between components 391
communication between user and textile 392
    data transmission 392–3
communication protocols 392
complementary explorations 374, 375–7
compliant mechanisms 351–4
compliant textile structures 346–7
components, communication between 391
Composite Technology Development (CTD) 100
composites
    ballistic 136–8, 139
    conductive polymer composites *see* conductive polymer composites
    SMP 176–8
computational fabrics 482
computer animations 360, 365–6, 367–8
computer keyboard 486
computerised jogging shoe 2
concept design process 8
Conductive Felt 247
conductive layers, for solar textiles 213–14
conductive materials 2, 217–38, 452–6
    application technologies 231–6
    carbon 221–2
    electrical conductivity 217–19
    future trends 236
    inherently conducting polymers 223–31
    ionic conductors 222–3
    metals 220–1
    see also conductive polymer fibres/yarns; conductive polymer composites; conductive polymer textile sensors; electronic textiles; micro system technology; polypyrrole-coated textiles
conductive plastics 219, 389
conductive polymer composites (CPCs) 325
    textile sensors 331–9
        effect of external parameters on electrical properties 336–8
        electrical properties under strain 336
        electrical properties without strain 334–5
        materials 331–2
        mechanical properties 332, 333–4
        testing methods 332–3
conductive polymer fibres/yarns 308–23, 468
    damage of power handling 317–20
    electrical circuits 246–9

electrical resistance 313–15
environmental effects on electrical properties 315–17
materials 311
morphological and electrical characterisation 312
morphological properties 312–13, 314, 318–20
preparation 309–11
prototypes 320, 321
solution preparation and coating 311–12
conductive polymer films 249
conductive polymer textile sensors 326–31
electromechanical properties 327
materials 327
mechanical sensor applications 327–9
sample preparation 327
temperature influence on electrical resistance 329–31
conductivity 217–19
conducting polymers compared with metals, semiconductors and insulators 246, 248
metals 218
plastics 219
polypyrrole-coated textiles 284–5
conductivity changes 286–9, 293, 294
conductivity gradient 245–6
congestive heart failure (CHF) 375–6
consciousness 422
constrained recovery 97
consumer needs model 7, 360, 366
context awareness 399, 400
context recognition technology 414
contextual environment of an artefact 11
cooling apparel 30, 31–2, 53
cooling effect 25, 40, 42–3, 63, 484
environmental step tests for PCM outdoor clothing 76–8
copolymers
block 110, 113–15, 116
polyacetylene 226
copper 220–1, 349
copper base alloys 133, 134
Cu-Al-Be 133, 134
Cu-Al-Ni 91, 92, 133, 134
Cu-Zn-Al 91, 92, 133, 134
fibres 403–4
stabilisation of polypyrrole-coated textile with copper ions 292
copper iodide 222
copper sulphide 222
copper wire-based electronic textile 347

Corpo Nove shirt 166, 169, 183
corrosion resistance 93
cotton 169–70
couplings 93, 135
CRG Industries 100
crossover point interconnections 250–1
crosstalk noise 242, 243, 269
Cryofit couplings 93, 135
'Curl test' 271
curtains 53–4, 196
customisation 372–3
cyanide 350

damping capability Q 131–2
damping effect 126, 130–2, 137
data security 427
data transmission 392–3, 445–6
day surgery 427–8
deep fryer 136
deformation sensors 388
dementia 378–9
denim 30
design
criteria 8
innovative concepts 165–6
MeMoGa project 5–18
participatory 13–15
'design for all' approach 14
design research 9–13
Detect fabric 475, 478–80, 483, 485, 487
keyboard 486
development work 11
devices, attachment of 252–6
diabetes 387
diagnosis/complementary explorations 374, 375–7
Diaplex membrane 98–9
dielectric elastomers 388–9
differential scanning calorimetry (DSC) 29, 364
digital patient charts 429, 431
diisocyanate 111
dimethyl formamide (DMF) 49
dip-coating 214
disconnects of conductors 251–2
disease management 373
dispersant 48
display 203, 351, 482
disposable hygiene products 117, 119
divers, dry suits for 53
DMDHEU (dimethyloldihydroxyethylenurea) 40
dodecylbenzene sulphonate (DBS) 290, 298–9

dodecylbenzene sulphonic acid (DBSA) 311
domestic textiles 53–4, 485–6
dopants 208, 246
  stabilisation of polypyrrole-coated textiles 290, 298–9
double comb structures 350
double memory effect 89–90, 126, 129, 130, 167, 179
Dream Cloth 163
dressings, medical 487
drift 217
drift velocity 217
DRIFTS orthosis project 380, 381
drug delivery 380, 390
dry cleaning 269–70
dry heat flux 27
dual loop antennas 273
durability 171, 372
dye-sensitisation for solar spectrum matching 215

E-glass fibres 212
e-Tags 391
effector electrical fabrics 474, 480–2
effector smart fabrics 473–4
eicosane 23, 37, 49
electric blanket 54
Electric Plaid 203
electrical circuits *see* electronic textiles
electrical conductivity *see* conductivity
electrical creep 459
electrical properties
  CPC textile sensors 332–8
    effect of external parameters 336–8
    under strain 336
    without strain 334–5
  electronic textile circuits 266–9
  PANI-coated conductive yarns 313–15
    damage of power handling 317–20
    environmental effects 315–17
electrical stimulation 389–90
electrically conductive plastics 219, 389
electroactive polymers (EAPs) 388–9
electrocardiogram (ECG) 342, 375–6, 383–4, 424
  Intellitex suit and measurement 440–5
electrochemical polymerisation 235
electrochemical surface modification technologies 346, 348–50
electrochromic materials 193, 194, 201–3
  applications 203
  definitions of electrochromism 202

electroconductive fluid 413
electrodeposition of paint 346, 348
electroencephalography (EEG) 377
electrogel 440
electrogoniometers 452
electrografting 346
electroluminescence devices 350–1
electromechanical properties 327–31
electronic textiles 239–82, 402–14, 452–3
  applications 272–6
  characterisation 266–72
    electrical properties 266–9
    mechanical properties 269–72
  copper wire-based 347
  development of textile-based circuits 240
  fabrication processes 240–6
  materials 246–65
    attachment of devices 252–6
    conductive fibres and films 246–9
    crossover point interconnects 250–1
    disconnects of conductors 251–2
    film or fibre-based transistors and integrated circuits 256–65
    non-conductive elements 249
  potential for the future 276–7
  System-on-Textile 409–12
  textiles for communication 403–5
  textiles for sensing 408–14
  textiles for signal pre-processing 405–8
  woven *see* woven electrical fabrics
electronics
  approaches to on-body electronics 401–2
  integration into textiles 362–4, 391, 468
    levels of integration 371–2, 393
  and wearable biofeedback systems 459–61, 468–9
electrons
  band theory 218
  Pauli exclusion principle 218
  valence 207, 218
electropolymerisation 346, 348
electrostatic spinning 235–6
electrostrictive materials 124
ElekTex 409
elongation corresponding to break point 333–4
elongation sensors 410–14
embedded microsystems 414, 415
embroidery 345–6
  fabrication of electrical circuits 241
  Intellitex suit 446

emeraldine base (EB) PANI 226–7, 228
emergency care *see* prehospital emergency care
encapsulation 363
   see also microencapsulation
energy bands 218, 219
energy sources 3–4
energy/power transmission 392–3, 445–6
energy supply 2, 363, 393, 468
environmental step change tests 69–71, 73–8
environmental temperature, changes in 64–5
Ergovaate 428–9
etching agent 245
ethyl-vinyl acetate (EVA)-HSBC composites 331–9
ethylene oxide-ethylene terephthalate (EOET) block copolymer 110, 114
evaporation 144–5
evaporative heat flux 27–8
evaporative resistance 69, 72, 73
event detection 478
EVOPRENE-HSBC composites 331–9
Excepor-U 147
exercise/rest tests 71–2, 78–80
explorations, medical 378, 379–81
expressiveness 7, 360, 366
extrinsic instability 287–9
extrusion
   ICPs 232–3
   SMPs 172–3

fabric insulation value 69, 72, 73
fabric structures 343–6
   and electrical circuits 242–4
   goal of application of compliant textile structures 346–7
   and solar textiles 212–13
fabrication
   processes for fabric-based circuits 240–6
   shape memory alloys 92–3
falls, prevention of 378
fashion 183
fault recovery 372
ferric metals 220
fibre Bragg grating (FBG) sensors 388
fibre material technology 9–13
fibre optic sensors 363–4
Fick's law 151
filler concentration (threshold point) 325, 333–5, 338
films
   conductive 249

optically active ICPs 231
shape memory polymer films 143–64
   thickness 151
fire sprinklers 136
fixing phase 108
flammability 65, 68–9, 73
FlecTron 247, 270
FlecTron N-Conductive 247
flexible circuit board 254
flexible screen 364, 392
floats 251
floor coverings 54
foam, PU 46, 47
foot pressure 387, 388
force application 388–9
force sensors 387
four-point probe technique 268–9
fractional free volume (FFV) 155–6, 161–2
France Telecom 203, 364, 392
functional electrical therapy 389–90
functionality 7, 360, 366
furnishings 53–4, 485–6

galvanic metal deposition 346, 348–50
galvanic skin response (GSR) 385
gate array switching grids 252
gauge factor 456
gelation 314
gels, shape memory 100
Georgia Tech Wearable Motherboard 383, 391, 435
glass fibres 212
glass plates 209
glass transition temperature 108–9, 168
   transition point for water vapour permeability through SMPU 152–6
glassy state 108–9, 168
Glauber's salt (sodium sulphate) 24, 30
gloves
   heated lining 481, 484
   sensing 275
   wearable biofeedback technology 467
glycerol stearate 390
gold 220–1, 349
gold-cadmium 86, 104, 125
Gore-Tex 147
Gorix 363
Grätzel cells 207
gyroscopes 386

H_2Off 147
handheld electronics 401
healing/medical treatment 374, 379–80

healthcare *see* medical applications
healthcare workwear 361, 364–7
HealthWear armband 401
heart manager, wearable 417–18
heart rate monitoring 375–6
heart rhythm 383–4
heat balance 22
Heat fabric 481–2, 484
heat flux 27–8, 56
heat stress 53
heat transfer
    heat losses in step change environmental manikin tests 70–1, 74–8
    through textiles containing PCM 56–7
    total heat transfer 144
heating effect 40, 42–3
    environmental step tests for PCM outdoor clothing 74–6, 77–8
heating fabrics 480–2, 485
heliochromic compounds 193
helmets, safety 53
Henry's law 151
heptadecane 23, 37
hexadecane 23, 37
high-density fabrics (HDF) 146–7
high structure carbon black (HSBC) 331–9
high-temperature workwear 361, 364–7
highly conductive ICPs 224–6
home area network (HAN) 392
human subject tests 68, 71–2, 78–80
humanities 10
humidity 337–8
humidity-activated shape-memory materials 119
humidity ageing 299–303
hydrated inorganic salts 24, 36, 39
hydrochromic materials 194
hydrophilic film
    laminates and coatings 147
    top coat over microporous coatings 147–8
hydrophilic groups 149
hysteresis 87–8, 458

IFM 203
images, 2D and 3D 365–6
impedance
    electrode/skin impedance and ECG measurement 441, 443
    evaluation for electronic textile circuits 266–9
impermeable fabrics 145
implants, medical 101

in-situ polymerisation 232, 233–5, 455
inductive link, wireless 445–6
inductive sensors 274, 476–7, 478, 479, 482
Infineon Technologies 2
inherently conducting polymers (ICPs) 219, 223–36, 324–5, 453, 468
    application technologies 231–6
    fibres for electric circuits 246–9
    highly conductive materials 224–6
    optically active ICPs 230–1
    polyaniline and its derivatives 226–7
    potential applications 326
    semiconductive polymers 227–30
inks, colour-changing 200–1
inorganic PCMs 24, 36, 39
inorganic thermochromism 199
insect wings 352
insulation effect 25, 52
    see also thermal insulation
insulators 246, 248
integrated circuits, film or fibre-based 256–65
integrated patient monitoring 429
intelligent building control 485
Intelligent Knee Sleeve 379, 392, 450, 459, 462–7
    feasibility of use in the field 464
    and knee flexion during landing 465–6
    as a rehabilitation tool 466–7
    validity and reliability of feedback 464–5
intelligent systems 1–2
    applications 2–4
intelligent textiles 104, 105–6, 359
    active and passive 472–4
    defining 471–2
    sensor and effector 473–4
    types of 124
Intelligent Textiles Ltd 474
    woven electrical fabrics 475–82
        applications 483–7
Intellitex suit 434–5, 436–47
    ECG measurements 440–5
    final prototype 446, 447
    respiration measurements 437–40
    wireless communication and energy transmission 445–6
Interactive Photonic Textiles 2
interconnections
    active electrical sensor fabrics 479–80
    crossover point interconnections 250–1
    wearable biofeedback systems 460–1, 468–9

interdigital structures 344, 350–1
interdisciplinarity 5, 9–12
interface circuitry 480
interface pressure 386–7
interfacial polymerisation 37–8, 48–9
interiors
    electrical fabric applications 485–6
    mood changing textiles 184–6
    PCMs 53–4
    SMMs 183–4
internal friction (damping effect) 126, 130–2, 137
interpenetrating network (IPN) 232, 235
intrinsic instability 286–7
intrinsically conductive polymers (ICPs) *see* inherently conducting polymers
inverted-F antennas 273–4, 404
ionic conductors 222–3
ionic polyelectrolytes 223
ionic polymer conductors 223
ionomers 223
Iowa Thin Technologies 209
isolation, mobile 429–30
isothermal tests 69

jacquard shedding system 242
joint motion
    biofeedback 452–3
    monitoring 385–6
    see also Intelligent Knee Sleeve

Kapton slit films 249, 264–5
Kevlar-covered conductive yarns 320, 321
keyboard 486
keyhole surgery 117–18
knee flexion angle 463–4, 465, 466
knee sleeve *see* Intelligent Knee Sleeve
knitted fabrics 212, 345, 350
    electroconductive 244, 410–11, 412
    micro system technology 352–4
    strain gauges in Intellitex suit 439–40
knitted stainless steel electrodes 441, 442, 443
knotting, smart sutures and 107, 117–18, 119
knowledge environment 11
Konarka Technologies 209

lamination 147
landing technique 463–6
large anion dopants 290, 298–9
latent heat 22–3, 34–5, 41
layers, number of 73–8

leucodyes 200–1
life balance factor (LBF) 400
Lifeshirt 274, 383, 384, 401, 435, 436
life style apparel 51
light-based elongation sensor 413–14
light-sensitive inks 201
line patterning 231
linear alkyl hydrocarbons 23–4, 36–9
linear dynamic range 457
linings 30–1, 32
liquid crystal themochromism 198
liquid crystals 471
    thermochromic inks 200
liquid-gas PCMs 23
load matching 210
long chain polyol 111
loop antennas 273

machining 93
magnetic anisotropy energy (MAE) 94
magnetic shape memory (MSM) materials 94–5, 101
magnetostrictive materials 124
Mamagoose pyjama 436
manikin tests 67–8, 69–71, 73–8
martensite 86–7, 88, 89, 125, 167
martensite finish temperature 87
martensite start temperature 87
martensitic transformation 86, 88–9, 125–6
mask 245
material-locked joints, miniaturised 352–4
    applications 354
    benefits 354
mature sciences 9–11
mechanical properties
    CPC textile sensors 332, 333–4
    fabric-based circuits 269–72
    polypyrrole-coated textiles 285–6
    shape memory alloys 91–2
mechanical switch fabrics 478–9
medical applications 369–98
    children *see* children
    diagnosis/complementary explorations 374, 375–7
    disease management 373
    electrical fabrics 487
    future trends 393–4
    healing/medical treatment 374, 379–80
    importance of intelligent textiles 370–3
    levels of integration of devices 371–2, 393
    modes of application 373–4
    PCMs 54–5

Index   497

prehospital emergency care *see* prehospital emergency care
remote monitoring 373–4
secondary prevention 374, 377–9
SMMs 101
   SMAs 134–5
   SMPs 118–19
technological solutions 380–93
   intelligent textiles as actuators 388–90
   intelligent textiles in communication 390–3
   intelligent textiles as sensors 382–8
user requirements 372–3, 381
wearable health assistants *see* wearable health assistants
workwear for health workers 119, 361, 364–7
see also monitoring
medical ultrasound 473
melamine-formaldehyde microcapsules 48–9
melt mixing process 309–10
melt-spun fibres 44–6
melting 92
metal salts
   as conductors 222–3
   stabilisation of polypyrrole-coated textiles 292
metals 246, 248
   conducting layer for solar textiles 214
   conductive coatings 456–7
   conductivity 218
   conductors 220–1
   wires 347, 348
Methods and Models for Intelligent Garment Design (MeMoGa) project 5–18
   background context 6–9
   interdisciplinarity 9–12
   multidisciplinary collaboration 12–15
   research and design procedure 8–9
micro-Brownian motion (thermal vibration) 98–9, 161–2
micro-electro-mechanical system (MEMS) applications 93, 101
micro-grippers 101
micro system technology 342–56
   compliant mechanisms in microengineering and biomechatronics 351–4
   copper wire textile electronic circuit technology 347
   embedded micro systems 414, 415
   galvanic modification of yarns 348–50

   light effects based on textiles with electrically conductive microstructures 350–1
   objective of application of compliant textile structures 346–7
   textile micro system technology 342–3
   textiles as inherent microstructures 343–6
microcapsule dye technology 196
microencapsulation 63
   colour-changing inks 200, 201
   drug delivery 390
   PCMs *see* microPCMs
microengineering 351–4
microfibres 146–7
microorganisms 148–9
microPCMs 23–4, 37–8, 44–50
   incorporation into fibres 44–6
   incorporation into fibrous structures 46–50, 51, 52
   lamination of PU foam containing microPCMs onto a fabric 46
microporous films 143, 147
   coatings with hydrophilic top coat 147–8
   compared with nonporous films 148–50
Microspirit 146
microstrip feedline 405
Microtex 147
miniaturisation 3–4
Mitsubishi Heavy Industries 98–9
MnemoScience 100, 118
mobile isolation 429–30
mobile phones 484
modular system concept 401–2
moisture permeability *see* water vapour permeability
moisture transfer 43–4
molecular rearrangement thermochromism 198–9
monitoring 275, 369–98, 487
   children *see* children
   physiological signals 375–7, 382–5, 400, 487
   recovery after day surgery 428
   remote 373–4
   vital functions 423–5
      integrated monitoring 429
      interpretation of data 425
      selection of monitoring methods 425
mood changing textiles 184–6
   technology 185–6
motorcyclists 30

498 Index

movement, patterns of 376–7
multicolour electrochromism 202
multidisciplinary collaborations 12–15, 166
multilayer garments 119
multipurpose textile-based sensors 324–41
  conductive polymer composite (CPC) textile sensors 331–9
  conductive polymer textile sensors 326–31
musical jacket 362–3
MyHeart project 383, 417–18

nanocrystalline thin film transistors 258, 259, 260
nanopolyacetylene blends 225
nanotechnology 3–4, 366
NASA 21, 34
natural sciences 10
needle wire localisers 135
net linings 30–1
network structures 109–10
Neutratherm 40
nickel 349
nickel-titanium (Ni-Ti) alloys (Nitinol) 86, 87, 88, 92, 104, 125, 132–4, 167
  effect of adding other elements 132–4
  fabrication 92–3
  properties 91, 92, 133
noble metals 220–1
Nomex 249
non-conductive threads 249
non-porous films 143–4
  benefits over porous films for breathable laminates 148–50
  designing breathable textiles with 150–1
non-uniform conductivity 245–6
non-woven fabrics 212–13
  TRFs for non-wovens with PCM 58–60
non-woven stainless steel electrodes 441, 442, 443
nonadecane 23, 37
Norsorex 110–11
Nuno Corporation 166
nylon 333
  coated with CPCs 332–9
nylon lycra, polypyrrole-coated 455–60

object of research 11–12
octadecane 23, 37, 48–9
office chair 54
Ohm's law 217
on-body electronics, approaches to 401

one-way shape memory effect 89, 126, 128–9
one-way stretching 230
optical data memory systems 195
optical fibres 363–4
  flexible screen 364, 392
optically active inherently conducting polymers 230–1
Optima (virtual working environment) 8
optoelectronic motion analysis systems 452
organic compound thermochromism 197–9
organic solar cells 213
organic thin film transistors 258–62
orthodontic wires 135
orthoses, wearable 380, 381
outdoor clothing
  active wear 51, 484
  PCMs in 63–81
    change in environmental temperature 64–5
    change in skin temperature 65–6
    environmental step change tests with a manikin 69–71, 73–8
    exercise/rest tests with human subjects 71–2, 78–80
    materials and garments 67–8
    measurement of textile properties 68–9, 72–3
  shape memory polymers 119
Outlast fabric 24, 25, 30–1, 32, 67
Outlast fibre 44, 45
Outlast Technologies 22, 30, 40, 67
output signal 267, 268
oxidation 286–7
oxidative polymerisation 228–9
oxygen 287–9
oxygen index (flammability) 68–9, 73

p-n junctions 208, 391
pad-mangle method 48, 49, 52
paraffin waxes 23–4, 36–9
Parkinson's disease 376–7
participatory design 13–15
partitions 183–4
passive electrical sensor fabrics 474, 477–9
passive insulation 25–6
passive non-transducer electrical fabrics 474, 476–7
passive smart fabrics 472–4
patch antennas 273–4, 404–5
patient charts 427, 429, 431
patterning of conductive elements 244–6

Index   499

Pauli exclusion principle 218
pentacene–based organic thin film transistors 260–2
percolation threshold 325, 333–5, 338
permeability *see* water vapour permeability
perpendicular routing 252
personal area network (PAN) 342, 369–70, 392, 477
personal assistants, general–purpose 418
perspiration rate 144, 145
  see also sweating
Phantom Fabric 247
phase change materials (PCMs) 2, 3–4, 11, 21–62, 390
  applications 29–30, 50–5
  function of textile structure with PCM 40–1
  future prospects in textiles and clothing 30–2
  historical background 21–2, 34, 39–40
  linear alkyl hydrocarbons 23, 36–9
  manufacture of textiles containing microPCMs 44–50
  mode of PCM performance in clothing 42–4
  outdoor clothing *see* outdoor clothing
  phase change technology 22–3, 34–6
  test methods 28–9
  textile treatment with PCM microcapsules 24–5
  thermal performance 24–8
  thermoregulating properties of fabrics with microPCMs 55–60
    heat transfer 56–7
    principle of measuring 57–60
phase change range 34–5, 41
Philips 2, 166
phlebology dressings 487
photochromic inks 201
photochromic lenses/spectacles 195
photochromic materials 193, 194–6, 471
  applications 195–6
  definitions of photochromism 194–5
photon-mode erasable optical recording 195
photonics 2
photons 207–8
photovoltaic devices *see* solar cells
physical coating methods 214
physical environment 11
physical properties
  PCMs 68–9, 72–3
  shape memory materials 117–18

physiological monitoring 375–7, 382–5, 400, 487
physiology 9–13
piezochromic materials 193, 194
piezoelectric films 386
piezoelectric materials 124, 471, 473
piezoresistive textiles 386
Pilot Ink Co. 199–200
planar inverted F antennas 273–4, 404
plasma polymerisation 229
plastic crystals 23, 36
plastic film-covered conductive fibres 320
plastic threaded chip carrier (PTCC) packages 255
platinum 349
polyacetylene (PA) 224–6, 246, 248
polyacetylene films 225–6
poly(3-alkylthiophene) 224
polyamide, silver-coated 348–9
polyaniline (PANI) 224, 226–7, 228, 230, 232–3, 234
  conductive polymer fibres/yarns 308–23
  conductive polymer textile sensors 325, 326–31
  PANI/HCSA 236
polybenzimidazole (PBI) fibres 212
polydimethylsiloxane (PDMS) 390
polyelectrochromism 202
polyetheretherketone (PEEK) fibres 212
polyethylene glycols (PEGs) 39–40
polyethylene/nylon 6 graft copolymer 110, 112–13
polyethylene terephthalate (PET)
  fibres and solar textiles 212
  PANI-coated PET conductive yarns 327–31
polyethylene yarns, PANI-coated 327–31
poly-3,4-ethylenedioxythiophene (PEDOT) 230–1
polyhydric alcohols 36
polyimide (PI) fibres 212
PolyLED 392
Polymer Technology Group 100
polymeric slit films 249, 264–5
polymers, shape memory *see* shape memory polymer films; shape memory polymers
poly(methylene-1,3-cyclopentane) polyethylene (PMCP) block copolymer 110, 114–15, 116
poly-N-vinylpyrrolidone 384
polynorbornene 110–11

poly(3-perfluorooctylthiophene) 230
polyphenylene 224, 246, 248
polyphenylene vinylene (PPV) 224, 230, 246, 248
polypropylene, photochromic 196
polypyrrole (PPy) 224, 227–9, 231, 234, 246, 248, 308
polypyrrole-coated textiles 283–307
  chemical properties 285
  conductivity 284–5
  conductivity changes 286–9
    extrinsic stability 287–9
    intrinsic stability 286–7
  mechanical properties 285–6
  nylon lycra 455–60
  results of stability enhancement 292–303
    annealing 291, 296–8
    large anion dopants 290, 298–9
    printing and chemical vapour deposition 293–6
    purification 296, 297, 298
    temperature and humidity ageing 299–303
  stabilisation of the polypyrrole 290–2
polytetrafluoroethylene (PTFE) membrane 471–2
polythiazyl 224
polythienyl vinylene 224
polythiophene (PT) 229–30, 246, 248, 325
polyurethane (PU)
  lamination of PU foam containing microPCMs onto a fabric 46, 47
  segmented 96, 110, 111–12
Popperian worldview 11–12
pore sealing 148
Porelle film 147
positional sensing 479–80
post-operative monitoring 427–8
post-operative recovery 379
power bus and data network 253–4
power handling to destruction 317–20
'power plastic' 209
power sources 2, 363, 393, 468
power transmission, wireless 392–3, 445–6
pragmatic dimension of research 12
prehospital emergency care 421–33
  cases and situations 422
  circumstances 422
  data security 427
  day surgery 427–8
  interpretation of monitored parameters 425
  mobile isolation 429–30
  optimal smart solution 430–1
    cables 431
    patient chart 431
    sensors 430–1
    teleconsultation 431
  patient charts 427, 429, 431
  protective covering 428–9
  telemedicine 425–6
  vital functions 422–3
    effects of transportation 426
    integrated monitoring 429
    monitoring 423–5
    selection of monitoring methods 425
pre-sciences 9–11
pressure mapping fabrics 487
pressure monitoring 386–7
  foot pressure 387, 388
  interface pressure 386–7
pressure sensor mat 387, 409–10
pressure sensors 408–10
  Detect fabric 478–9
  micro system technology 353
prevention of health problems 374, 377–9
printable electronics 236
printed circuit boards (PCBs) 240, 475
printing 293–6
processability
  ICPs 224
  incorporating SMMs into textiles 170–1
processor 1–2
programmable grid array 251–2
protective covering for emergency care 428–9
protective garments 51–3, 119
  workwear see WearCare project
protonation 310
proximity detection 478
pulsation 424
pulse oximetry 424–5
purification 296, 297, 298

quantum tunnelling composite (QTC) 387

Radio Frequency Identification (RFID) tags 400
raised wire connectors 255–6
recovered strain 97
rehabilitation 466–7
relative resistivity 336–8
relaxation time 108–9
reliability 372
remote controls, fabric 485
remote monitoring 3–4, 373–4

Repel film 147
repetitive strain injuries 379
research trends 3–4
residual strain 97
resistance
   conductive polymer textile sensors 328–31
      influence of temperature 329–31
   conductive yarns/fibres 363
   ECG measurement 441, 443
   electronic textile circuits
      evaluation 266–9
      washing and dry cleaning 270
   elongation sensors 411, 413
   material-locked joint 353
   measurement of respiration by Intellitex suit 437–40
   PANI-coated conductive yarns 313–15
      environmental effects 315–17
   wearable biofeedback systems
      control of initial resistance 455
      resistive force and hysteresis 458
resistance heating 480–2
resistance temperature devices (RTDs) 477
resistance welding 250
resistivity 217–18
   commercially available conductive fabrics 246, 247
   CPC textile sensors 332–3, 334–9
resistors 476, 477–8
resolution 344
Respibelt 384
respiration 423
   measurement by Intellitex suit 437–40
response time 457
responsivity 105–6
restrained recovery 85
reversible phase 108
ribbon cable connectors 255–6
robustness 372
   wearable biofeedback systems 459
'rubbery' effect 126, 129–30, 131
rubbery state 108–9, 168

S-glass fibres 212
safety helmets 53
sail, with solar cells 209
satellite solar panels 135
screen printing 49, 50
secondary prevention, medical 374, 377–9
SeeThru Conductive 247
Sefar Petex hybrid fabric 406–8
segmented polyurethane 96, 110, 111–12

self-stabilisation 292
semantic dimension of research 12
semiconductive polymers 227–30
semiconductor yarns/fabrics 482
semiconductors 207–8, 213, 246, 248
sensate liner 243
sensitivity of sensors 456
sensor integrated belt 441–5
sensor smart fabrics 473–4
sensors 1–2, 105–6
   based on conductive polymers 326–31
   CPC textile sensors 331–9
   development of a functioning wearable textile sensor 453–60
   electronic textiles 274–6, 408–14
   intelligent textiles as sensors for medical applications 382–8
   prehospital emergency care 430–1
   textile embedded 3–4
   wearable biofeedback systems 453–4
      sensitivity of sensors 456
Sensory Fabric 409
sensory systems 5–6
sewing machine 101
shape memory alloys (SMAs) 86–94, 100–1, 107, 124–42, 167
   applications 93–4, 134–8, 139
   definition and description 125–6
   engineering textile and clothing aesthetics 178–81
      fabric development 180
      potential exploitation and limitations 181
      programming SMAs 178–9
      utilising SMAs' aesthetic effects 181
      yarns 179, 180
   fabrication 92–3
   future trends 140
   history 86
   hysteresis 87–8
   properties 86–92, 126–32
      damping effect 126, 130–2, 137
      double memory effect 89–90, 126, 129, 130, 167, 179
      hysteresis 87–8
      mechanical properties 91–2
      'rubbery' effect 126, 129–30, 131
      single memory effect 89, 126, 128–9
      superelasticity 90–1, 126–7
      thermoelastic martensitic transformation 86, 88–9, 125–6
   shape memory effect 89–90
   types of alloys 132–4

502 Index

shape memory ceramics (SMCs) 94
shape memory coatings (SMCs) 168, 181–2
shape memory effect
   fabrics containing shape memory
      coatings 182
   SMAs 89–90
      fabrics containing 181
   SMPs 107–9
      fabrics containing 175–8
   SMPU films 153–4
shape memory gels 100
shape memory materials (SMMs) 2, 3–4, 85–103, 165–89, 392
   aesthetic interactive applications of shape changing smart textiles 182–4
   alloys *see* shape memory alloys (SMAs)
   ceramics 94
   engineering textile and clothing aesthetics 172–82
      shape memory coatings 181–2
      SMAs 178–81
      SMPs 172–8
   future prospects 100–1
   gels 100
   magnetic 94–5, 101
   mood changing textiles 184–6
   polymers *see* shape memory polymers (SMPs)
   principles 166–9
   technical requirements for shape changing textiles and clothing 169–72
      aesthetic degradation 171
      durability of shape changing effects 171
      processability 170–1
      requirements for clothing 171–2
      shape changing under stimuli 169–70
   typical end-uses 169
shape memory polymer films 143–64
   breathable fabrics 145–51
   water vapour permeability through 152–62
      glass transition temperature as transition point 152–6
      room temperature range soft segment crystal melting temperature as transition point 157–62
      soft segment crystal melting temperature as transition point 156–7
shape memory polymers (SMPs) 95–100, 104–23, 168
   applications 98–100, 118–20

   architecture of SMP network 109–10
   challenges and opportunities 120–1
   engineering textile and clothing aesthetics 172–8
      extrusion of SMP 172–3
      potential applications and limitations 178
      shape programming 173–5
      SMP effects in fabrics 175–8
      yarn and fabric formation 173, 174
   general properties 95–8
   potential use in smart textiles 115–18
      use of changes of physical properties 117–18
      use of functional properties 115–17
   principle 107–9
   suitable for textile applications 110–15
   thermomechanical cycle 96–7
shape memory polyurethane (SMPU) 110, 111–12
   water vapour permeability 152–62
shape programming
   SMAs 178–9
   SMPs 173–5
shell materials 38
Shieldex Supra 247
shirts
   Corpo Nove shirt 166, 169, 183
   Lifeshirt 274, 383, 384, 401, 435, 436
   wearable motherboard 383, 391, 435
Siemens 166, 203
signal integrity 242, 243, 269
signal pre-processing 405–8
signal rise times 267, 268
signal transmission lines 242–4, 266–9, 403–4
silicon rubber sensor 411, 412, 413
silicon rubber tube 352–4
silicon solar cells 208, 213
silk protein moisturisers 390
silver 220–1
silver-coated polyamide yarns 348–9
single memory effect 89, 126, 128–9
skin pH 385
skin temperature 42–3, 65–6
   exercise/rest tests for outdoor clothing 71–2, 78–9
Sky Station International 209
sleeve, sensing 275
slit-film technique 249, 264–5
SMART ambience 184–6
smart fibre 118–19
smart materials *see* intelligent textiles

smart phones 414–15
'smart shirt' wearable motherboard 383, 391, 435
smart sutures 107, 117–18, 119
Smartex 435
snap connectors 253, 255–6
Snowtex conductive fabrics 247
social environment 11
socket connectors 253
sodium dodecylbenzenesulphonate 299
soft segment crystal melting temperature
   room temperature range soft segment crystal melting temperature as transition point for WVP 157–62
   transition point for WVP 156–7
Softmail Fabric 247
Softswitch 387, 392, 409
solar cell arrays 210–11
solar cells 206–16
   background 206–7
   challenges 211
   conductive layers 213–14
   future trends 214–15
   how they work 207–8
   suitable textile constructions 211–13
   technological specifications 210–11
   textiles as substrates 209–10
solar tensile pavilion 209
SolarActive International 196
soldering 250, 255–6
solid-liquid PCMs 22–3, 35–6
solid-solid PCMs 23, 36
solution coating 233
solvatochromic materials 193, 194
spectacle frames 136
spider legs 352
spin-coating 214
sportswear 483–4
   SMMs 115, 120, 183
spray-coating 48, 49, 51, 214
stabilisation
   photochromic inks 201
   polypyrrole-coated textiles 290–2
      experimental results of stability enhancement 292–303
stainless steel electrodes 440–5
staples, medical 134
steady-state manikin tests 73
steel 220
   fibres 383–4
stents 135
stiffness
   bending stiffness 272

fabrics with PCM 68, 72, 73
stimuli 5–6, 471–2
   and response 105–6
   shape changing under 169–70
storage devices 468
storage modulus ratio 154–5
strain gauges 232, 437–40
   gauge factor 456
   linear dynamic range 457
   wearable biofeedback systems 453–60
strain sensitivity 283, 285, 293–303
strain sensors, flexible 244, 292–304
stretch recovery test 329
styrene-1,4-butadiene block copolymer 110, 113–14
sudden infant death syndrome (SIDS) 377–8, 436
sulphonated polyanilines 468
Super Textile Corp. 196
supercritical fluid, polymerisation in 234–5
superelasticity 90–1, 126–7
superstrate configuration 213
Supplex 146
surfactant 48
surgical protective garments 119
sutures, smart 107, 117–18, 119
sweating 42, 66, 79, 144–5
   perspiration rate 144, 145
   sweat accumulation in garments in exercise/rest tests 72, 79–80
Sympatex film 147
syntactic dimension of research 12
System-on-Textile technology 405–8

T-shirt, biomedical 435–6
tactile sensors 275, 276
teleconsultation 425–6, 431
telemedicine 363
   prehospital emergency care 425–6
telemonitoring 3–4, 373–4
see also monitoring
temperature
   body *see* body temperature
   changes in environmental temperature 64–5
   electrical properties of CPC textile sensors 336–7
   electrical properties of PANI–coated conductive yarns 315–16, 317
   and resistance of conductive polymer textile sensors 329–31
   skin *see* skin temperature
   thermal ageing test 299–303

temperature measurement fabrics 477–8
temperature profiles 27
temperature regulating factor (TRF) 56–7, 365
  measuring 57–60
temperature-sensitive inks 200–1
temperature-sensitive shape memory polymers 104–23
tensile fatigue 271–2
tensile strength 271–2
tension threshold of conduction 317
textile art 182
textile electronic bus systems 350
textile embedded sensors 3–4
Textile Testing and Innovations 31–2
textrodes 383–4, 440–5, 446
textronics 371, 393
thermal ageing 299–303
thermal barrier effect 41
thermal diffusion 150
thermal insulation 21, 25–9, 52, 64–5
  fabric insulation value 69, 72, 73
  thermal insulation value of clothing 70, 73
thermal spray process 93
thermal vibration (micro-Brownian motion) 98–9, 161–2
Thermasorb 24
thermistors 477
thermochromic inks 200–1
thermochromic materials 169, 193–4, 196–200
  applications 199–200
  definitions of thermochromism 197–9
  inorganic thermochromism 199
  liquid crystal thermochromism 198
  molecular rearrangement thermochromism 198–9
  organic compound thermochromism 197–9
thermoelastic martensitic transformation 86, 88–9, 125–6
thermogravimetric analysis (TGA) 29, 318–19, 320, 368–9
thermophysiological comfort 22
thermoplastic elastomer fibres 412–13
thermoplastic elastomeric (TPE) matrices 325
thermoplastic encapsulation techniques 363
thermoregulating effect 25–8, 40–1, 52, 65–6
  measurement of thermoregulating properties 55–60

thermoretractable tube-covered conductive yarns 320, 321
thickener 48
thickness
  fabric 68, 73
  film 151
thin film SMAs 93
thin film SMMs 101
thin film transistors (TFTs) 231, 256–65
  amorphous and nanocrystalline 257–8, 259, 260
  organic 258–62
  on threads or polymeric slit films 262–5
three-dimensional (3D) modelling 365–6, 367–8
three worlds, theory of 11–12
threshold point (filler concentration) 325, 333–5, 338
Thüringen Institute 363
time 315, 316
time domain reflectometry (TDR) 266
titanium 171
  see also nickel-titanium alloys
total heat transfer 144
transdermal drug delivery 390
transducers 105, 363
  sensor and effector smart fabrics 473–4
transformation temperature 107
transient environmental manikin tests 69–71, 74–8
transistors
  fibre-based 250–2, 262–4, 391
  thin film 231, 256–65
transmission lines 242–4, 266–9, 403–4
transparent conducting oxides (TCOs) 213, 214
transportation
  mobile isolation 429–30
  negative effect on vital signs 426
trapped air insulation 28
treatment, medical 374, 379–80
Triangle Research and Development Corporation (TRDC) 21–2, 40
TU Berlin 369
twinning 88–9
two-dimensional positional sensing fabric 479–80
two-way shape memory effect 89–90, 126, 129, 130, 167, 179

UCO Sportswear 30
ulcers, foot 387
ultrasound, medical 473

Index   505

unconstrained recovery 97
universal design ('design for all') 14
unobtrusiveness 372
upholstery 54, 485–6
urinary incontinence 379
usability testing 364–7
USB cables 271–2
users
    participatory design 13–15
    requirements 372–3, 381
    user-textile communication 392
UV light 316, 317
UV-sensitive SMP 168

vacuum consumable arc melting 92
vacuum drying 296, 297, 298
vacuum heating treatment 300, 301, 302, 303
vacuum induction melting 92
valence electrons 207, 218
ventilation 43
Verhaert 436
Versatech 147
vibration-controlling actuators 136
virtual prototypes 8
virtual working environment 8
visual analogue scales 15
vital functions 422–3
    body temperature 423, 425
    circulation 423
    consciousness 422
    integrated monitoring 429
    monitoring 423–5
    negative effects of transportation 426
    respiration 423
Vivometrics Lifeshirt 274, 383, 384, 401, 435, 436
voltage treatment 291–2
VTAM project 383, 435–6

wall hangings 183–4
Warema Renkhoff 209–10
washing
    effects on fabric-based circuits 269–70
    Intellitex suit 438–40
water 35
    extrinsic instability of polypyrrole-coated textiles 287–9
water vapour permeability (WVP) 115–17, 142, 149–50
    nonporous films 150–1
        film thickness and 151
        structural factors and 151

    through shape memory polyurethane 152–62
    glass transition temperature as transition point 152–6
    room temperature range soft segment crystal melting temperature as transition point 157–62
    soft segment crystal melting temperature as transition point 156–7
Wealthy Central Monitoring System 383, 384, 392, 435
wearable back manager 415–17
wearable biofeedback systems 450–70
    development 453–60
        compatibility with simple electronics 459–60
        control of initial resistance 455
        linear dynamic range 457
        resistive force and hysteresis 458
        response time 457
        robustness and environmental effects 459
        sensitivity 456
        textile substrate 455, 456
    functional electronics 460, 461
    future directions 467–9
        biomechanical applications 469
        electronic fibres 468
        electronics and interconnects 468–9
        integration of electronic materials into textiles 468
        novel power sources and storage devices 468
    Intelligent Knee Sleeve 379, 392, 450, 459, 462–7
    interconnections 460–1
    joint motion 452–3
    need for biofeedback technology 450–1
    other applications 467
    problems with current biofeedback devices 451
wearable computer 6, 486
wearable health assistants 399–420
    applications 415–18
        wearable back manager 415–17
        wearable heart manager 417–18
    approach 401–2
    context recognition technology 414
    electronic textile technology 402–14
        communication 403–5
        sensing 408–14

signal pre–processing 405–8
outlook 418
vision 399–400
wearable components 414–15
wearable heart manager 417–18
wearable mood change device 186
wearable motherboard 383, 391, 435
wearable technology (systems) 359, 369
WearCare project 359–68
   electronics 362–4
   methodology 360
   objectives 359–60
   textile materials 361–2
   usability testing 364–7
weave-masking technique 263–4
weave-patterned organic transistors 250–2, 262–4, 391
weaving 343–4
   solar textiles and woven fabrics 212
   see also woven electrical fabrics
webcam 186
weight, fabric 68, 72, 73
weight sensors 275–6
welding 93
wet/dry suit 119
wet spinning 24, 234
wheelchair users, outerwear for 14
white metal compounds 222–3
wicking 150

window blinds 53–4, 183–4
wire bonding 255
wireless communication 392–3, 445–6
wiring looms 477
wiring structures 406–8
workwear *see* WearCare project
wound dressings 487
woven electrical fabrics 242–4, 473, 474, 475–87
   active non–transducer fabrics 474, 482
   active sensor fabrics 474, 479–80
   applications 483–7
   disconnects 251–2
   effector fabrics 474, 480–2
   passive non-transducer fabrics 474, 476–7
   passive sensor fabrics 474, 477–9
woven stainless steel electrodes 441, 442, 443
wristwatch, communicative 244

yarns
   SMAs in 179, 180
   SMPs in 173, 174
   see also conductive polymer fibres/yarns
Young's modulus 333, 334

Zell Conductive Fabric 247
zinc 349
zinc-copper (brass) 86, 104, 125